Advanced Series in Agricultural Sciences 9

Josef Palti

Cultural Practices and Infectious Crop Diseases

With 43 Figures

Springer-Verlag
Berlin Heidelberg New York 1981

Dr. J. Palti
Agricultural Research Organization
The Volcani Center, P.O. Box 6
Bet-Dagan, Israel

ISBN-13: 978-3-642-68268-1 e-ISBN-13: 978-3-642-68266-7
DOI: 10.1007/978-3-642-68266-7

Library of Congress Cataloging in Publication Data. Palti, J. Cultural practices and infectious crop diseases. (Advanced series in agricultural sciences; 9) Bibliography: p. Includes indexes. 1. Plant diseases. 2. Micro-organisms, 3. Agriculture. 4. Agricultural ecology. I. Title. II. Title: Infectious crop diseases. II. Series. SB731.P27. 632'.3. 81-14533.AACR2

Softcover reprint of the hardcover 1st edition 1981

Typesetting, printing, and bookbinding: Brühlsche Universitätsdruckerei, Giessen
2131/3130-543210

To
my wife Nenette,
my children Noemi, Ronny and Daniel, and
my grandchildren Yuval, Nir, Ilan, Tamar, Shlomit,
Efrat and Eyal

Contents

Introduction 1

Part 1 Climate, Cropping and Crop Disease

1.1 Agro-Ecosystems, the Cultural Practices They Have Generated, and the General Impact of Such Practices on Crop Disease. . 4

1.1.1 Humid Agroclimates 4

1.1.1.1 The Humid Cool Temperate Agroclimates . . . 4

1.1.1.2 The Warm Humid Temperate Agroclimates . . 6

1.1.1.3 The Wet Tropics 6

1.1.2 Dry Agroclimates 7

1.1.2.1 The Cool Dry Temperate Agroclimates 7

1.1.2.2 The Warm Dry Temperate Agroclimates 8

1.1.2.3 The Semi-Arid Tropical Agroclimate 9

1.1.3 Agroclimates and Crop Disease – Outlook for the Future 9

1.2 Microclimate and Crop Climate 10

1.2.1 The Microclimate as Affected by Topography and Soil 11

1.2.1.1 Topography 11

1.2.1.2 Soil 13

1.2.2 Crop Factors Interacting with Microclimate, and the Resultant Crop Climate 14

1.2.2.1 Density of Plant Cover (Canopy) 14

1.2.2.2 Shade 15

1.2.3 When, Where and How Much Can Cultural Practices Influence the Crop Climate? 17

1.3 The Collective Approach to Disease Control: Epidemiological Considerations and the Role of Cultural Practices in Regional Management of Inoculum 18

1.3.1 The Cardinal Role of Inoculum and its Control for the Farming Community as a Whole 20

1.3.2 Various Crops Susceptible to the Same Pathogen – Which Crop is More Valuable to the Farming Community? 22

1.3.3 Crop Varieties Differing in Susceptibility – Dangers and Opportunities 23

1.3.4 Restricting Seasons, Locations and Growing Practices for Susceptible Crops 24
1.3.5 Minimizing Multiplication and Spread of Air- and Vector-Borne Inoculum 26

1.4 Soil, Soil Microbiota, and Soil-Borne Disease 27

1.4.1 Soil and the Growth of Underground Organs of the Crop 28
1.4.1.1 Texture and Depth 28
1.4.1.2 Soil Reaction 29
1.4.1.3 Soil Water 32

1.4.2 Soil Microbiota and Their Interaction with Soil-Borne Pathogens 34
1.4.2.1 Soil Microbiota Restricting Pathogen Development 35
1.4.2.2 Soil Microbiota Transmitting Plant Pathogens or Associated with Their Development 37
1.4.2.3 Plant Symbionts and Plant Pathogens 40

1.4.3 Suppressive or Resistant Soils 42

1.5 Stress, Strain and Predisposition 45

1.5.1 Types of Stress-Induced Strain 46
1.5.2 Temperature Strain 47
1.5.2.1 High-Temperature Strain 49
1.5.2.2 Low-Temperature Strain 50

1.5.3 Water Strain 51
1.5.3.1 Water Deficit Strain 52
1.5.3.2 Excess Water Strain 54

1.5.4 Other Strains 55
1.5.5 Combinations of Stresses and Strains 55
1.5.6 Charcoal Rot (*Macrophomina phaseolina*): Prototype of a Disease on Crops Under Stress 56
1.5.7 Summary of the Effects of Stress and Strain on Diseases Caused by Pathogens 57

1.6 Crop Age, Injury and Disease on Leaf and Fruit, with Special Reference to Disease in the Ageing Crop 58

1.6.1 Germination to Pre-Maturation of the Crop 58
1.6.1.1 Seedling and Early Growth Stage 58
1.6.1.2 Flowering or Full Vegetative Growth Stage . . 59

1.6.2 Maturation and Senescence 60
1.6.2.1 Physiological Changes 61
1.6.2.2 Changes in Crop Climate 61
1.6.2.3 Injury and Wound Parasites 62

1.6.2.4 The Time Element and the Mounting Inoculum 62
1.6.2.5 Disease in Ageing Tree Crops 63
1.6.2.6 The Ageing Crop as Source of Inoculum for its Neighbours 63
1.6.2.7 Disease Control in the Ageing Crop 64
1.6.2.8 *Botrytis cinerea* – a Pathogen of Soft and Senescent Tissues 66

1.7 Weeds and Crop Disease 66
1.7.1 Which Pathogens Spread and Survive Through Weeds? 67
1.7.2 Weeds Particularly Apt to Endanger Crop Health . . . 70
1.7.3 Effects of Cultural Practices on Weeds as Related to Disease Control 71

Part 2 Major Cultural Practices and Their Effect on Crop Disease

2.1 Cost/Benefit and Risk Assessment and the Complexity of Multiple Choice in Pest Control Decisions on the Farm . . . 73
2.1.1 Cost/Benefit Assessment 73
2.1.2 Risk Assessment 74
2.1.3 The Complexity of Multiple Choice in Pest Control Decisions 75

2.2 Sanitation 77
2.2.1 Aims and Limitations of Sanitation 77
2.2.2 Preventing the Introduction of Inoculum 79
2.2.2.1 Propagating Material 79
2.2.2.2 Irrigation and Drainage Water That Spreads Inoculum 83
2.2.2.3 Inoculum Introduced in Plant Debris, Compost and Manure 83
2.2.2.4 Inoculum Introduced and Spread by Equipment and Man 85
2.2.3 Elimination of Living Plants That Carry Pathogens . 87
2.2.3.1 Additional and Alternate Hosts 87
2.2.3.2 Groundkeepers and Volunteers 88
2.2.3.3 Roguing 88
2.2.3.4 Removal of Diseased but Living Parts from Trees 90
2.2.4 Destruction or Inactivation of Inoculum Deriving from Dry Plant Matter and Debris 90
2.2.4.1 Sanitation of Diseased Trees 91
2.2.4.2 Removal of Crop Debris or its Incorporation in the Soil 92

2.2.4.3 Burning or Flaming Crop Residue 93
2.2.4.4. Flooding Fields and Orchards 95
2.2.4.5 Disinfestation of Soil by Heat 96

2.3 Crop Sequence 98

2.3.1 Aims of Crop Sequence Management. 99
2.3.2 Economic Considerations 99
2.3.3 Choice of Crop Sequence in Relation to Other Farming Practices 101
2.3.4 The Benefits of Fallowing. 103
2.3.5 What Are the Chances of Improving Crop Health by Management of Crop Sequence? 105
2.3.5.1 The "Heritage" of the Preceding Crop 105
2.3.5.2 Survival of Pathogens Outside Crop Debris. . 107
2.3.5.3 Survival and Multiplication of Inoculum by Aid of Non-Hosts 109
2.3.5.4 Use of Varieties Resistant and Seasons Unfavourable to the Pathogens 110
2.3.6 Monoculture 111
2.3.6.1 Irreversible Disease Pattern. 112
2.3.6.2 Reversible Disease Pattern 113
2.3.7 Multiple Cropping 114
2.3.7.1 The Benefits of Multiple Cropping 114
2.3.7.2 Multiple Cropping of Field and Vegetable Crops 114
2.3.7.3 Intercropping in Orchards and Plantations . . 115
2.3.8 Decoy and Trap Crops 116
2.3.9 Crop Sequence and Crop Disease in the Future . . . 117

2.4 Soil Amendments and Mulches 118

2.4.1 Organic Amendments 119
2.4.1.1 Effects on Soil and Water Relations 119
2.4.1.2 Effects on Pathogens 119
2.4.1.3 Practical Application of Soil Amendments . . 123
2.4.2 Organic Mulches 123
2.4.2.1 Effects on Soil Temperature and Retention of Precipitation 123
2.4.2.2 Stubble Mulch and Cereal Debris in Relation to Cereal Diseases 124

2.5 Tillage 125

2.5.1 The Aims of Tillage 125
2.5.2 Tillage to Prepare Land for Sowing and Planting . . 126
2.5.2.1 Depth of Soil Favourable to Root Growth . . 126

2.5.2.2 Distribution of Inoculum in the Soil Profile and Over the Field 128
2.5.2.3 The Diverse Effects of Non-Tillage 128
2.5.2.4 Tillage Practices as Affecting Cereal Diseases in England 130
2.5.3 Topsoil Management. 131
2.5.3.1 Shaping the Topsoil and Earthing-Up (Hilling) 131
2.5.3.2 Elimination of Soil Crusts 132
2.5.4 Tillage as a Control Practice 132
2.5.4.1 Destruction of Weeds and Volunteer Plants Between Successive Crops 132
2.5.4.2 Weed Control While the Crop Grows 132
2.5.4.3 Reduction of Inoculum by Tillage. 132
2.6 Crop Nutrition 133
2.6.1 Balanced Nutrition and Nutrition Unbalanced for Economic and Crop Health Reasons 134
2.6.1.1 Size, Quality and Maturation of Yield 134
2.6.1.2 Crop Nutrition and the Pathogen 134
2.6.1.3 Manipulation of Nutrients 135
2.6.2 Fertilization with Nitrogen 135
2.6.2.1 General Effects of Nitrogen in Relation to Crop Disease 135
2.6.2.2 Nitrate Nitrogen and Ammonium Nitrogen. . 136
2.6.2.3 Timing of Nitrogen Fertilization 139
2.6.2.4 Toxic Effects of Nitrogen on Pathogens . . . 139
2.6.3 Fertilization by Phosphates, Potassium and Calcium . 140
2.6.3.1 Phosphates. 140
2.6.3.2 Potassium 140
2.6.3.3. Calcium 142
2.6.4 Minor and Trace Elements 142
2.7 Moisture Management in Non-Irrigated Crops 143
2.7.1 Practices Affecting the Soil Moisture Potential, and Their Effects on Crop Disease. 143
2.7.2 Optimization of the Use of Rainfall as Related to Crop Disease. 144
2.7.3 Management of Leaf Wetness in Rain-Fed Crops . . 145
2.8 Irrigation. 146
2.8.1 Effects of Irrigation on Soil and Foliosphere Climate. 146
2.8.2 Irrigation Effects on the Host Crop 148
2.8.2.1 Extension and Manipulation of Growing Seasons 148
2.8.2.2 General Effects of Irrigation on Crops and on Stress 149

2.8.3 Irrigation and the Pathogen 151
2.8.3.1 Survival and Management of Inoculum in Irrigated Crops 151
2.8.3.2 Dispersal of Inoculum in Irrigated Crops. . . 152
2.8.3.3 Attraction of Virus Vectors to Irrigated Crops 155
2.8.4 Techniques of Irrigation 155
2.8.4.1 Overhead Sprinkling May Promote Disease. . 155
2.8.4.2 Trickle Irrigation 158
2.8.5 Minimizing Disease by Irrigation Management . . . 159
2.8.5.1 Choice of Irrigation Technique 159
2.8.5.2 Timing of Water Supply 160
2.8.6 Irrigation and Crop Disease – Outlook 162

2.9 Rate of Sowing and Planting, and Density of Stand 162
2.9.1 Effects at Various Stages of Growth 163
2.9.2 Root and Shoot Contact in High-Density Crops . . . 165
2.9.3 Density and Disease in Tree Crops and Vines 165
2.9.4 Crowed Stands May Mean Lower Incidence of Certain Systemic Diseases 166

2.10 Sowing and Planting Dates and Manipulation of Flowering and Fruiting Periods 168
2.10.1 Date of Sowing and Incidence of Disease 168
2.10.2 Minimizing Virus Diseases by Varying Sowing Dates 170
2.10.3 Relative Growth Rates of Crop and Pathogen, and Levels of Crop Susceptibility, at Various Sowing Dates 171
2.10.4 Seasonal Massing of Air-Borne Inoculum 174
2.10.5 Management of Periods of Bud-Burst, Flowering and Fruiting to Minimize Disease 174
2.10.6 Manipulation of Periods when Crop Meets Pathogen 175

2.11 Harvesting Dates and Practices 175
2.11.1 Timing of the Harvest 176
2.11.2 The Danger of Inoculum Spreading at Harvest . . . 177
2.11.3 Harvesting Techniques as Related to Crop Injury and Disease. 178

2.12 Planning to Minimize Influx of Air- or Vector-Borne Inoculum to Neighbouring Crops. 180
2.12.1 What Makes a Neighbouring Crop a Dangerous Source of Inoculum? 180
2.12.2 Planning to Reduce Infection Risks from Neighbouring Crops 182

2.13 Pruning and Grafting 184
2.13.1 Pruning and Bark Ringing 184
2.13.2 Grafting 184

2.14 Effects of Physical Barriers on Crop Infection and of Optical Means on Virus Vector Control 185
2.14.1 Effects of Windbreaks and Tall Crops on Infection. . 185
2.14.2 Physical Protection of Crops from Infection by Soil-Borne Inoculum 186
2.14.3 Optical Means for Virus Vector Control 187

Part 3 Interactions Between Cultural Practices, Resistance Breeding, and Application of Chemicals: Integrated Control

3.1 Keep Inoculum Out – by Any Available Means. 191
3.2 Prevent Multiplication and Spread of Inoculum. 192
3.2.1 Use of Resistant Varieties to Delay Inoculum Build-up 193
3.2.2 Cultural Practices and the Timing of Pesticide Use. . . 193
3.3 The Proper Place for Cultural Practices in Integrated Disease Control 194
3.3.1 Economic and Human Factors 194
3.3.2 Pathogen, Crop, and Environmental Factors 195
3.4 Profit in Fungicide Applications, as Related to Cultural Factors 196
3.5 Cultural Practices and the Use of Herbicides and Physiologically Active Chemicals 200
3.5.1 Herbicides and Desiccants 200
3.5.2 Chemicals Applied to Affect Crop Physiology 202
3.6 Hop Wilt in England: Success of Integrated Control. 202
3.7 Some Thoughts on the Future of Integrated Disease Control and its Components in World Crop Production. 205
3.7.1 Bringing New Areas into Production. 205
3.7.2 Intensification of Production on Areas Already Cultivated 206
3.7.3 Widening the Geographic Range of Valuable Crops . . 206
3.7.4 Education of Farmers to Employ Cultural Practices for Disease Control 207

French, German and Spanish Translation of Some of the English Terms Used in this Book 208

References 210

Pathogen Index 231

Subject Index 239

Acknowledgements

It is my great pleasure to indicate here all those who have given me their generous help in the writing of this book.

The Agricultural Research Organization of the Ministry of Agriculture, Israel, has enabled me to write the book by granting me a nine-months' sabbatical leave for this purpose.

Professor Dr. J. Kranz, Tropeninstitut der Justus-Liebig Universität Giessen, conceived the idea, helped in planning, and unceasingly encouraged writing of the book by contributing ideas and by making constructive criticism.

Professor Dr. G. Schuhmann, President of the Biologische Bundesanstalt, Braunschweig, and Professor Dr. W. Gerlach, Director of this institution in Berlin, have placed all their facilities at my disposal, and given me the benefit of their expertise and friendship.

Professor Gerlach and his colleagues have also rendered me a most valuable service by checking the names of fungal and bacterial pathogens for accuracy.

Professor Dr. K. Domsch, Forschungsanstalt für Land- und Forstwirtschaft, Braunschweig, has kindly permitted me to peruse the voluminuos list of literature references he and his colleagues have collected in the preparation of the *Compendium of Soil Fungi* (Domsch et al. 1980), even before this had been published.

Professor Dr. I. Arnon, former Director of the Agricultural Research Organization, Israel, and a recognized authority on farming in arid zones, has allowed me to benefit from his advice.

Dr. F. Rapilly, Director of the Station de Pathologie Végétale, Centre National de Recherche Agronomique, Versailles, has been most responsive to all my requests during my stay at Versailles, and has arranged for my visits to research centres in France, at Dijon, Avignon, Antibes and Montpellier.

Professor Dr. G. Viennot-Bourgin, the doyen of French phytopathology, has been kind enough to let me consult him on various aspects of this book.

Mr. A. J. Johnston, Director of the Commonwealth Mycological Institute, Kew, was most helpful during my stay at the Institute.

I am indebted to Dr. J. M. Thresh, East Malling Research Station, Kent, U.K., for advice on matters relating to virus diseases, and for checking the names of viruses mentioned.

Many of my Israeli colleagues in research and advisory bodies have contributed ideas not only in the field of plant protection proper, but also in the many other disciplines on which this book touches, such as agrometeorology, tillage, and agricultural economics. Special thanks are due to Mr. S. Moeller and Mr. H. Shoham, both of the Extension Service of the Ministry of Agriculture, Tel Aviv, Israel, for detailed comment on the manuscript.

In order to identify all those colleagues in Israel, France, Britain, and Germany, who advised me, and many of whom have permitted me to quote from their experience, in the form of "personal communications", a list of their names and institutions is presented here. To all, I wish to express my sincere gratitude.

List of scientists and advisers who have been consulted and many of whom have agreed to have their findings quoted as "personal communications".

Abbreviations of institutions mentioned in this list:

ADAS	Agricultural Development and Advisory Service, Ministry of Agriculture, Fisheries and Food, Cambridge, U.K.
ARO	Agricultural Research Organisation, Bet Dagan, Israel
BBA	Biologische Bundesanstalt für Land- und Forstwirtschaft, Braunschweig and Berlin, German Federal Republic
CMI	Commonwealth Mycological Institute, Kew, Surrey, U.K.
CNRA	Centre National de la Recherche Agronomique, Versailles, France
EM	East Malling Research Station, Maidstone, Kent, U.K.
ES	Extension Service of the Ministry of Agriculture, Tel Aviv, Israel
FSA	Faculté des Sciences Agronomiques, Gembloux, Belgium
GERDAT	Groupement d'Etude et de Recherche d'Agronomie Tropicale, Montpellier, France
HU	Hebrew University of Jerusalem, Faculty of Agriculture Rehovot, Israel
IMS	Israel Meteorological Service, Bet Dagan, Israel
INRA	Institut National de la Recherche Agronomique, at the stations indicated: Antibes, Avignon, Dijon or Montpellier, France

Allard, C., CNRA, Versailles
Bell, T. S., ADAS, Cambridge
Bernaux, P., INRA, Montpellier
Beyries, A., INRA, Montpellier
Bouhot, D., INRA, Dijon
Brosh, S., ES, Tel Aviv
Butt, D. J., EM. East Malling
Cassini, R., CNRA, Versailles
Cayrol, J. C., INRA, Antibes
Dalmasso, A., INRA, Antibes
Davet, M., INRA, Montpellier
Delhotal, P., CNRA, Versailles
Faivre Amiot, A., CNRA, Versailles
Férault, A. C., CNRA, Versailles
Follin, J. C., GERDAT, Montpellier
Gerlach, W., BBA, Berlin
Gladders, P., ADAS, Cambridge
Golan, Y., ES, Tel Aviv
Grosclaude, C., INRA, Avignon
Grouet, D., CNRA, Versailles

Hadass, A., ARO, Bet Dagan
Harris, D., EM, East Malling
Hepner, J., HU, Rehovot
Katan, J., HU, Rehovot
Kroeber, H., BBA, Berlin
Lamarque, C., CNRA, Versailles
Lapierre, H., CNRA, Versailles
Laville, E., GERDAT, Montpellier
Lebrun, A., CNRA, Versailles
Leclant, F., INRA, Montpellier
Lecoque, H., INRA, Avignon
Lomas, J., IMS, Bet Dagan
Lot, H., INRA, Avignon
Louvet, J., INRA, Dijon
Messiaen, C.-M., INRA, Montpellier
Moeller, S., ES, Tel Aviv
Muller, R., GERDAT, Montpellier
Notteghem, M. J., GERDAT, Montpellier
Pappo, S., ARO, Bet Dagan
Pauvert, P., CNRA, Versailles
Ponchet, F., INRA, Antibes
Populer, C., Station de Phytopathologie de l'Etat, Gembloux
Rapilly, F., CNRA, Versailles
Ranan, D., Bank Hapoalim, Tel Aviv
Rieuf, P., INRA, Avignon
Ritter, M. P., INRA, Antibes
Rotem, J., ARO, Bet Dagan
Royle, D. J., Long Ashton, Bristol
Scotto la Massese, C., INRA, Antibes
Sewell, G. F. W., EM, East Malling
Shoham, H., ES, Tel Aviv
Signoret, P., INRA, Montpellier
Spharim, Y., ARO, Bet Dagan
Spire, D., CNRA, Versailles
Thresh, J. M., EM, East Malling
Tramier, R., INRA, Antibes
Vigoroux, A., INRA, Montpellier
Waller, J. M., CMI, Kew
Yarham, D. J., ADAS, Cambridge

Introduction

The development of a crop, and therefore its health, is always the result of interplay between biological and environmental factors, as influenced by human agency. In other words, crop health is a highly complex affair.

This book is concerned with only one group of agents affecting crop health, the pathogens, and not with animal pests or direct effects of physiological or weather factors. Even within this one group, however, the interaction of causal agents with environmental and biotic factors is highly complex.

No less complex is the effect of cultural practices on the crop and its health. There is probably no major practice that does not affect diverse facets of crop growth, which in turn affects crop/pathogen relationships. Thus tillage sequentially affects depth and rate of root development, hence nutrient uptake, hence general plant size and habit as well as crop climate and crop susceptibility. Irrigation affects all these parameters, and facilitates crop growth under diverse macroclimatic conditions, with all the ensuing implications for disease development.

In this book an attempt is made to superimpose one set of complexities, the cultural practices, on another such set, crop health. This may seem overambitious, not to say foolhardy, unless we remember that it has been done by farmers, consciously or unconsciously, ever since the beginnings of agriculture. We are here chiefly trying to rationalize traditional practices, review modern research on the development of further practices, and assess the place of the latter in integrated disease control.

With the exception of purely sanitary measures, few cultural practices have been developed expressly for the purpose of disease control, their principal aim being to mitigate climatic and nutrient constraints in order to obtain acceptable yields. The number and variety of cultural operations rises, up to a point, with the variety of constraints with which the crop has to contend. Where agriculture has developed in spite of various constraints, with the help of practices such as irrigation, soil amelioration, utilization of favourable soils, seasons or topographical "niches" etc., the scope for disease management (by modifying the practices to the detriment of the pathogen) is therefore much wider than in farming systems not greatly handicapped by constraints. We therefore begin the consideration of cultural practices in general, and the review of their effect on crop disease, with a brief look at the principal agroclimates of the world.

Since the scope of this book is wide, it is important to stress from the outset which subjects it does not attempt to cover:

a) Physiological and environmental disorders, except in as much as they affect host/pathogen relationships.
b) Resistance breeding, which is referred to only in Part 3 in the context of integrated control.

c) Post-harvest disease, excepting only reference to effects of some field or orchard operations (especially at harvest) on post-harvest problems.
d) Forest diseases.
e) Diseases of crops grown in glasshouses or under plastic tunnels, since management of crop and environment in glasshouses involves special techniques. The subject of glasshouse crops has recently been reviewed by Hanan et al. (1978).
f) Phytosanitary legislation, which is mentioned only in passing, as complementing cultural practices, especially where the communal approach to such practices is concerned.

The emphasis in this book is very much on what the grower can do, either by himself or in cooperation with the farming community. For this reason the main part of the book, Part 2, is divided according to individual farm operations. As mentioned above, these practices interact greatly, and considering them separately is to some extent unreal. Yet as growers approach problems in terms of operations, this is therefore the approach we have chosen.

In writing about cultural practices, it is often difficult not to state the obvious. We have not refrained from doing so where it helped to draw as complete a picture as possible of the influence of cultural practices on crop disease. It has also been attempted to point out the drawbacks of some important cultural practices in relation to plant disease. Since the purpose of farming is not prevention of disease, but raising crops (preferably with a minimum of disease), farmers will often – for economic, political, social or environmental reasons – adopt measures undesirable from the point of view of crop health, e.g., sowing out of season or too densely, "forcing" plants, irrigation by unsuitable methods etc. All those engaged in looking after the crop's health, however, must be aware of the extent to which these "undesirable" practices affect the crop disease situation, and must assess the risk involved with open eyes.

Agricultural practices cannot always be defined with full scientific precision, and their results are not always wholly "reproducible." The complexity of their effects, the difficulty of analyzing the often numerous interactions of factors producing them, the necessity of multiseasonal repetition of experiments – all these combine to make us base evaluation of results achieved by agricultural practices not only on experimental facts, but also on observations and impressions of scientists, advisers, and growers. Many such observations have not been published, in fact many of the most experienced and practical observers have published very little, and then frequently in journals not readily accessible.

All this applies very much to plant protection, as affected by agricultural practices. In recognition of this fact, much of the material for this book has been gathered by interviewing dozens of colleagues in research and advisory services in Israel, in the Federal Republic of Germany, in France, and in the United Kingdom. Identification of sources as "personal communications" is therefore frequent, and a list of such "communicators" is attached.

The literature survey for this book has been carried out mainly at (a) the Biologische Bundesanstalt, Berlin, where the computerized information service and library, under the direction of Professor Dr. W. Laux, has proved of special value in gathering data from eastern Europe; (b) the library of the Centre National de

Recherche Agronomique, directed by Mlle. Cagnac, where reports of French scientists from tropical countries were studied; (c) the library of the Commonwealth Institute of Mycology, Kew, Surrey, under the Head Librarian Miss S. Daniels, where a wealth of plant protection literature, not easily available elsewhere, was placed at my disposal; and (d) the libraries of the Agricultural Research Organization, Bet Dagan, Israel, headed by Mrs. N. Czerny.

Part 1 Climate, Cropping and Crop Disease

1.1 Agro-Ecosystems, the Cultural Practices They Have Generated, and the General Impact of Such Practices on Crop Disease

A number of overall classifications of agro-ecosystems or farming systems have been published. We shall, in the main, follow those proposed by Duckham and Masefield (1970) and by Thorne and Thorne (1979), supplemented by more specific references to particular zones, such as those by Arnon (1972) for dry climate farming. Neglecting the polar-subpolar, the cold-temperate boreal and the arid tropical zones, which are of little interest in the context of farming and plant disease, we shall here deal with the six zones listed below and shown in Fig. 1.1, reproduced from Thorne and Thorne (1979):

A. Humid Agroclimates
   1. The humid cool temperate agroclimate
   2. The humid warm and temperate agroclimates
   3. The wet tropics

B. Dry Agroclimates
   1. The cool dry (semi-arid and arid) temperate agroclimates
   2. The warm dry temperate agroclimates
   3. The seasonally dry tropics

1.1.1 Humid Agroclimates

1.1.1.1 The Humid Cool Temperature Agroclimates

In the zones characterized by these agroclimates, the growing season ranges from 4 to 8 months, with the 8-months' season restricted to regions with oceanic environment. During this season there is, in general, sufficient precipitation to support crop growth. Occasional gaps in rainfall may create moisture stress problems, but – by contrast – excessive concentration of rain in the warm season may also be a problem. The principal constraint for crop growth is, however, that of winter cold (Thorne and Thorne 1979).

As seen from Fig. 1.1, the humid cool agroclimates of this type comprise many of the countries with highly developed farming economies, especially in Europe, North America, and East Asia. Here the adaptation of crops and development of cultural practices to meet the constraints of the climate has made great progress.

The cultural practices developed in these zones largely aim at lessening risks of low temperature effects and at extending cropping seasons by selection of suitable

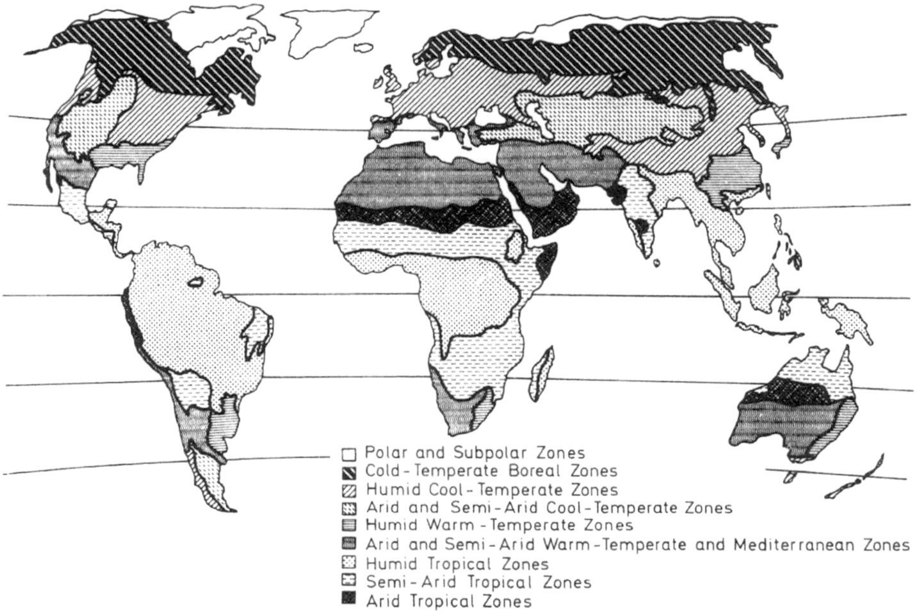

Fig. 1.1. Major climatic zones of the world. (Thorne and Thorne 1979)

soils, appropriate topography, and optimal sowing and harvesting seasons. From the point of view of crop disease control, such practices mitigating low temperature effects are not always welcome: the "cold incision" provided by cool winters effects a most important reduction of bacterial and fungal inoculum, and of virus vectors, to levels that permit build-up to epidemics during the subsequent growing season only after several cycles of renewed multiplication. Many plant pathogens tolerate winter cold much less well than many of their hosts, although major pathogens have, of course, developed mechanisms to survive the cold, e.g., the chlamydospores of some species of *Fusarium* or the sclerotia of species of *Verticillium*, *Sclerotinia*, and *Sclerotium*, and the survival in dormant host tissue of fungi such as the powdery mildew *(Podosphaera leucotricha)* of apples.

The practices evolved to minimize effects of winter cold on crops grown in the humid cool temperate climate must therefore be examined closely to avoid their providing "niches" for increased survival of inoculum from one main season to another*. No less important is the proper use of the normally disease-free "period of grace" with which crops are often blessed just after the period of winter cold. A concentrated effort made at this time to avoid or delay re-introduction of inoculum and to slow down its cycles of multiplication can have significant effects for the entire cropping season. This can be done by regional operations of sanitation and crop planning, to avoid early sowing of highly susceptible crops or varieties, by organized regional disease control by chemicals, or more simply by careful application to early crops of all cultural practices calculated to put the

* As stated in the Introduction, one of the important means of protecting crops from cold, viz. growing crops under cover, will not be considered here, as it involves a whole range of specialised practices

pathogen at a disadvantage: proper drainage, soil preparation, and sowing techniques to ensure rapid seedling emergence, crop rotation, and avoidance of proximity to potential sources of inoculum, such as the previous year's stubble. To block or delay by these means the build-up of inoculum in its early stages, and thus defer outbreaks of epidemics, is one of the fundamental tasks of plant protection in general, and of cultural practices in particular, in the farming systems of the humid cool temperate zone.

1.1.1.2 The Warm Humid Temperate Agroclimates

Zones with warm humid temperate agroclimates are found in all the continents, except Europe, and include intensively cultivated regions in the southern or southeastern parts of the USA, South America, South Africa, China, and Australia (Fig. 1.1). They are characterized by precipitation adequate for crop production for six or more months of the year, which usually permits a cool season and a warm season crop (Thorne and Thorne 1979).

The favourable climate makes it possible to grow many crops, or groups of crops, susceptible to the same disease, on a year-round basis. These not only include non-deciduous fruit crops, such as citrus and avocado and mango, but also cucurbits, tomatoes, and many other vegetable, forage and flower crops. To these must be added major industrial crops susceptible to the same soil-borne diseases, e.g., cotton and groundnuts, both affected by *Verticillium dahliae* and *Sclerotium rolfsii*. This constant availability of hosts, and the absence of a cold period to reduce inoculum levels periodically, present the farming community with a central problem of disease control: the management of inoculum and restriction of its build-up to epidemic levels. In some instances this can only be done by breeding resistant varieties, and in others chemical protection of the crops plays a major role. But wherever susceptible crops are grown in these climates, a basic consideration is the manipulation of microclimate and soil, combined with strict sanitation, i.e., adaptation of cultural practices to minimize disease.

Such sanitation and manipulation of the environment should include – in the warm humid temperate climate perhaps more than anywhere else – a regional approach to the elimination of sources of inoculum, to the planning of crops so as to avoid proximity of old and young plants liable to attack by the same disease, and to the control of weeds. The continuous wetness of this climate is also eminently favourable to the development of soil-borne pathogens, those mentioned above as well as dangerous parasites such as *Phytophthora cinnamomi* and *Phymatotrichum omnivorum*, and also of destructive nematodes.

Undesirable effects of the high rainfall, such as leaching of nutrients and soil erosion, can be mitigated by mulching and no-tillage practices. This, however, heightens the danger of pathogen transmission on stubble and other plant debris much more than in cooler and drier regions.

1.1.1.3 The Wet Tropics

These climates include zones on or near both sides of the equator in America, Africa, Asia, and Indonesia. Annual rainfall is in excess of 1,500 mm, distributed over most of the year, though a few months' dry season occurs in some regions in the outer latitudes of the tropics (Thorne and Thorne 1979).

In general, crops in the wet tropics can be grown throughout the year, and the potential for crop production is vast. But in much of this region, farming has remained stationary, at the subsistence level, over a long period. In the agro-ecosystems that have developed there, including shifting cultivation practiced by many millions of farmers, the emphasis is on reliable year-to-year production of modest-sized crops, and not on maximum, but annually fluctuating yields. This purpose is served by a diverse assortment of crops that have evolved over long periods and are adapted to endemic diseases, many of which are caused by pathogens not encountered in other climatic regions (Kranz 1966b). These endemic diseases occasionally break out into epidemics, whose prevention is the major concern of disease control in tropical subsistence farming. Putter (1979) has pointed out that this involves two principal lines of operation: (a) Removing sources of inoculum outside and within the field; such sanitation can be extremely thorough in subsistence farming, where the farmer tends his plants individually at short intervals and will readily remove leaves with disease symptoms as soon as these appear; (b) Reducing the infection rate, which under conditions of tropical subsistence farming is best done by selection of partially resistant plant material. Intercropping, increasingly advocated for subsistence farming, slows down spread of air- and vector-borne pathogens, and can be particularly useful in limiting spread of certain soil-borne, "simple-interest" diseases (Mukibi 1976).

Most of the research on epidemics in the wet tropics has been carried out on plantation crops, such as groundnuts, coffee, and tea. The posistion here differs radically from that in subsistence farming: varieties bred for their high yield are grown over large, continuous areas, diseases often occur in epidemics, and chemical control can be applied. The principal aim of cultural practices in disease control in such tropical crops is strict sanitation and the all-round manipulation of factors such as crop density, crop sequence, and proximity to other crops.

1.1.2 Dry Agroclimates

1.1.2.1 The Cool Dry Temperate Agroclimates

An important aspect of growing crops in all climates with dry seasons is moisture stress which, from the crop disease aspect, results in heightened susceptibility to facultative parasites (see Chap. 1.5).

The cool temperate climate, which comprises semi-arid as well as arid zones, is a typically continental climate found chiefly in parts of the USA, Canada, Siberia, and the Argentine (Fig. 1.1). This climate is characterized by long, cold winters, cold soils, short growing seasons, high summer insolations, high day-time temperatures in summer, and by potential evapotranspiration exceeding precipitation for much of the growing season (Thorne and Thorne 1979).

The principal concern of farmers in this zone is the conservation of water and optimal utilization of the short growing season, or – if possible – its extension. Farming in this climate provides the world with a major part of its winter cereals, and most of the cultural practices evolved there have to be evaluated, in the context of this book, in relation to their effect on diseases of wheat, barley, oats, and rye.

One important aspect of water management in these zones is the fact that a large part of winter precipitation consists of snow. Since strong winds are frequent in this season, much of this precipitation is lost as the snow is blown away, a loss that can in part be prevented by leaving cereal stubble to persist in the field, to hold the snow until it thaws. This practice, as well as all kinds of fallowing and residue mulching to absorb precipitation, are of overriding importance (Greb 1979). The fact that stubble and other residues also serve to transmit cereal diseases (*Fusarium* spp., *Pseudocercosporella herpotrichioides*) from season to season is a secondary consideration, but one that must not be overlooked.

Since rainfall is often erratic during the summer season, and evapotranspiration high, irrigation represents an obvious means of using the short season to best advantage, where the type of crop and availability of water makes this an economic proposition. Such irrigation poses some of the disease control problems which will be mentioned more fully in the section on warm dry temperate climates. However, in the cool dry climates the impact of irrigation on plant disease is less pronounced, because the season is too short for epidemic build-up of many pathogens, and lower temperatures put the latter at a disadvantage.

1.1.2.2 The Warm Dry Temperate Agroclimates

These climates may conveniently be divided into three:

a) *Arid* desert regions covering much of Africa above latitude 20°, and of southwestern or western parts of Asia (e.g., the Arabian peninsula), Australia, and North and South America. In these regions rainfall is minimal and serves only for production of some vegetation for grazing but not for cropping.

Crop production in such desert regions is possible only with irrigation. The air and soil climate of such irrigated fields is quite unfavourable for most pathogens, and only extremely drought-resistant fungi could threaten crops, for example powdery mildews and some rusts (Boughey 1947).

b) *Semi-arid* regions with a *rainy warm* season and a rainless cool season. These regions lie chiefly on the equatorial side of the large deserts, e.g., in a belt traversing Africa, south of the Sahara, or in parts of the Indian subcontinent. Most agriculture here is concentrated at higher elevations, in oases, or where irrigation can be practiced.

c) *Semi-arid* regions with *rainfall* in the *cool* season and a rainless, warm summer, the Mediterranean type of climate. A wide range of crops can here be produced in the rainy, cool winter-period and some important fruit crops can be grown without irrigation, especially the "biblical" olives, vine and pomegranates. The range of winter-grown crops has lately been expanded quite considerably by the practice of growing crops sheltered under plastic foil. In spring, crops can be grown, at least in part, on precipitation still available from winter, but summer and autumn-grown crops need irrigation.

In this climate, the "cold incision" that so much reduces inoculum in the winter of cool climates is replaced by a "dry incision" which fulfils the same function in summers – unless, of course, the crop is irrigated. On lands irrigated with some frequency, this advantage of once-a-year sanitizing by drought may be lost, although irrigation in a warm, dry air, even sprinkling, promotes diseases in general much less than summer rain.

The introduction of large-scale summer irrigation, on the one hand, and the annually increasing area of winter grops grown under plastic, on the other, have brought Mediterranean farming face to face with the problem of overseasoning inoculum control. Here, however, much more than in warm, humid zones, this can largely be handled by cultural practices. In the first place, irrigation and the growing of crops in tunnels are themselves practices that can be varied greatly to keep disease down, viz., by choosing the right type of irrigation and by proper aeration of covered crops. Moreover, choices of soil, topography, sowing density etc. can influence crop conditions decisively, in order to obviate early-season build-up of inoculum, especially where this is organized on a regional scale.

To counter the other principal threat of diseases to intensively grown Mediterranean crops, that of ever-mounting incidence of multi-host root-fungi and vector-borne viruses, resistance breeding is the best hope; but a number of cultural practices have also recently shown their value, e.g., mulching against infection by white-fly borne viruses (Cohen and Melamed-Madjar 1978).

1.1.2.3 The Semi-Arid Tropical Agroclimate

Semi-arid tropical regions border the wet tropics, usually to the north and south, in all continents with tropical climates (Fig. 1.1). These semi-arid regions are characterized by annual precipitation ranging from 500–1,500 mm, predominantly in the 2–7 months of the wet season. But even in these seasons rainfall is highly variable, and often very intense, causing serious problems of erosion (Thorne and Thorne 1979).

The disease control problems of crops grown in these zones, in principle, resemble those described for the warm dry temperate zones, but there are significant differences due chiefly to the higher temperatures and greater rain intensity. The variety of crops grown in the semi-arid tropics is more restricted, and tree crops, with few exceptions such as some palms and mango, are hard to grow. The large majority of plant pathogens find this climate most uncongenial, although smuts, some rusts, and soil fungi such as *Macrophomina phaseolina* and *Sclerotium rolfsii* are fairly common (Kranz 1966b). Little research has been carried out on the survival of pathogens in this climate and on their potential for causing losses.

1.1.3 Agroclimates and Crop Disease – Outlook for the Future

From the point of view of crop disease development, the agroclimates described above may be divided into three categories:

a) Agroclimates in which a definite incision of *cold* separates one growing season from the next, and inoculum of most pathogens is greatly reduced before the new season begins. This comprises the humid-cool and cool-dry agroclimates. Cultural practices play an important role in mitigating effects of cold that is not extreme. Thus selection of site and season for sowing or planting, overhead sprinkling or misting when chilling or slight frost threatens, and mulching to maintain soil temperature at a higher level, are helpful practices. But the most effective means of restricting or preventing loss from cold, the growth of crops under cover, is becoming ever more costly, and will therefore affect future cropping in cool climates only to a limited extent.

b) Agroclimates in which an incision of *drought* separates growing seasons, and thus again reduces seasonal inoculum. These agroclimates include the semi-arid temperate regions with rainfall in the warm or in the cool season, and the semi-arid tropical regions. The ever-widening use of irrigation in the dry season is in the process of nullifying the pathogen-curbing effects of the dry seasons wherever water is available. This is certain to have important effects on crop diseases in the coming years.

c) Agroclimates without either cold or drought incisions, i.e., the warm-humid and the wet tropical climates. These usually permit the growing of two crops per year, and the uninterrupted availability of crop hosts throughout the year must be taken into account. In fact, most of the crops grown by small farmers in these climates have reached some equilibrium with endemic pathogens. Plantation crops grown in these regions have to rely on some measure of resistance in the varieties grown, and on chemical control, and probably less on new developments in disease control by cultural practices.

1.2 Microclimate and Crop Climate

The microclimate is the outcome of macroclimate acting on limited environmental units, such as slopes or valleys, light or dark soil, or on units under different environmental management, e.g., tilled or irrigated versus non-tilled and non-irri-

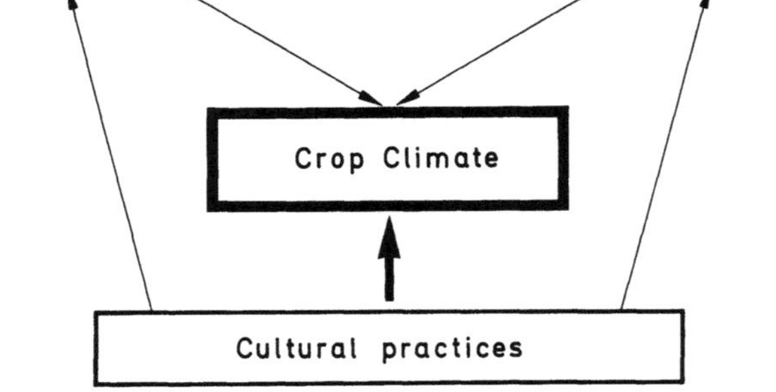

Fig. 1.2. Interrelations between macro-, micro-, and crop climate, and cultural practices

gated fields, or simply bare or cropped soil. We are here particularly concerned with climate interactions with the crop, and prefer the result of such interactions to be termed the crop climate.

If cultural practices are to be used effectively, their role in shaping components of the crop climate and their potential for changing it, must be clearly understood. As seen from Fig. 1.2 macroclimatic factors, such as rain, cloudiness or overall radiation exert a basic influence. There is little the farmer can do to change this, apart from choosing the most suitable sowing season. However, the microclimate can be managed to some extent by pre-sowing or pre-planting choices of topography and soil type, and expert management of the crop climate during crop growth is one of the fundamental aims of cultural practices.

1.2.1 The Microclimate as Affected by Topography and Soil

1.2.1.1 Topography

Most meteorological parameters are affected by the topography. Comparing values at the bottom of valleys and on neighbouring, open high ground, Seemann (Seemann et al. 1979) states that, on average, temperatures decrease by about 0.6 °C per 100 m ascent, with minimum temperatures decreasing less than maximum temperatures. Relative humidity is lower on the higher ground. But the most characteristic contrast lies in the daily amplitude of temperature; this, at least on days with inversions at night, is more restricted on higher than on lower ground.

Insolation. The amount of solar radiation received by the ground depends on its inclinations and compass orientation or direction of exposure. From records taken in Germany, Seemann (Seemann et al. 1979) presents data which are reproduced in part in Table 1.1. These indicate (a) the wide differences in insolation between the N, NE, and NW exposures on the one hand, and the S, SE, and SW exposures on the other; (b) the gradual rise of the amount of insolation on the S, SE, and SW exposures as the angle of the slope steepens from flat to 30° (except for the June records); and (c) the effect of the season: differences in insolation for various exposures and slopes are smallest in June, when all insolation values are highest. Effects of topography on shade will be discussed below, in connection with the effects of the plant canopy.

Wind is generated on slopes as a result of their heating up during the day and cooling down at night. The cooled air moves downwards to the bottom of the slope, to be replaced by slightly warmer air from the layer immediately above, which latter, on coming into contact with the soil, cools in its turn. Movement of the air is more rapid the steeper the slope, but it is generally slow. If it encounters obstacles, such as trees, it rises above them so that crops immediately beyond the obstacles are protected from the air flow. This is illustrated in Fig. 1.3a.

The effects of windbreaks, which can also be in the form of tall crops, will be described in Chapter 2.14.

Special wind conditions are sometimes observed in narrowing valleys, where a "nozzle" effect may produce strong currents. Research in Switzerland has, for example, shown the effect such winds may have on cereal disease.

Table 1.1. Total monthly insolation from direct sunlight in kcal/cm^2 in cloud-free weather for slopes of various inclinations and orientations for $\varphi = 50°$. (After Seemann in Seemann et al. 1979, p. 64)

Month	Inclination, slope	N	NE / NW	E / W	SE / SW	S	Difference between north and south slope
December	0° = Flat	2.0	2.0	2.0	2.0	2.0	0.0
	10° = Slope	0.7	1.2	2.0	3.0	3.5	2.8
	20°	0.0	0.6	2.0	4.0	5.0	5.0
	30°	–	0.1	2.0	4.8	6.0	6.0
March	0° = Flat	8.8	8.8	8.8	8.8	8.8	0.0
	10° = Slope	7.1	7.9	9.2	10.6	11.2	4.1
	20°	4.9	5.6	8.9	11.8	12.8	7.9
	30°	2.1	4.6	8.7	12.6	14.1	12.0
June	0° = Flat	18.6	18.6	18.6	18.6	18.6	0.0
	10° = Slope	17.8	18.1	18.7	19.4	19.5	1.7
	20°	16.5	17.0	18.4	19.7	19.6	3.1
	30°	14.2	15.0	17.5	18.8	18.8	4.6
September	0° = Flat	10.8	10.8	10.8	10.8	10.8	0.0
	10° = Slope	9.2	9.6	11.0	12.1	12.9	3.7
	20°	6.5	8.0	10.9	12.2	14.4	7.9
	30°	3.9	6.3	10.6	13.8	15.4	11.5

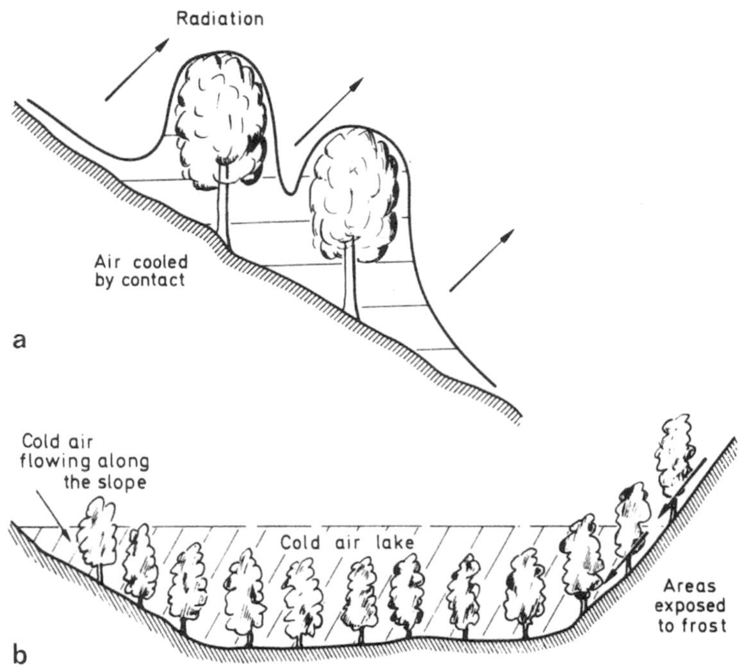

Fig. 1.3. a Flow of cold air around vertical objects (Primault in Seemann et al. 1979). **b** Cold air lake due to cold air flowing downhill, and the resultant danger of frost. (Primault in Seemann et al. 1979)

Dew. Wind conditions as well as slopes affect dew. The orientation of the slope has a marked influence on the temperature of the upper soil layer and the layer of air next to it. On E and SE slopes, in particular, the energy of the morning sun is partly used to evaporate dew, thus tending to shorten the dew period. On the other hand, the lower temperatures near the bottom of slopes and in valleys, together with frequent occurrence of low clouds there, increase the amount of dew precipitation and lengthen the hours of dew. Dew will be referred to in the chapters on moisture management and irrigation (Chaps. 2.7 and 2.8), as one of the basic climatic components affecting disease development.

Frost. In discussing relations between topography and frost, Seemann (Seemann et al. 1979) has stated that night cooling begins earlier on flat ground or in valleys, where true sunset may be earlier, than on slopes. On flat ground, the air moves little, but on sloping ground the cold air flows downhill as soon as it has gathered sufficient momentum, often beginning early in the evening. The well-known resultant dangers of frost to crops in low-lying areas are illustrated in Fig. 1.3 b, and are referred to in the section on low temperature strain (Chap. 1.5.2.2). One cultural practice that has lately become available in order to use slopes for growing crops susceptible to frost (bananas, avocado) in subtropical climates, is trickle irrigation (Palti and Shoham 1979). This will be referred to in the chapter on irrigation.

1.2.1.2 Soil

The soil climate results essentially from the interaction of three components: temperature, humidity, and aeration. Soil temperature is determined by sunlight absorbed by the soil surface and transmitted to underlying layers. At night, heat loss by radiation exceeds the influx, and the soil is cooled. The character of the soil determines its heat relationships: the amount of heat that can be absorbed (heat capacity) grows with an increase in humidity; thermal conductivity depends on porosity and humidity. Since air is much inferior as heat conductor to mineral particles or water, thermal conductivity rises with the ratio of the latter to the air content of the soil. The diurnal range of soil temperatures, i.e., their amplitude, is lower in overcast than in clear weather, and in light-coloured than in dark soils, and also depends on snow or plant cover (see below) (Chirkov in Seemann et al. 1979, p. 132).

A practical way in which farmers can influence soil effects on the crop climate is by mulching. Stubble mulching, the most widely used practice, has important effects on conserving moisture and avoiding erosion, and markedly affects temperature. Since stubble quite frequently carries considerable amounts of inoculum, its management is discussed in detail under the heading of soil amendments (Chap. 2.4).

Other methods of mulching are the application of dark materials, such as coal dust or tar-based sprays. These raise soil temperature and facilitate earlier sowing in spring. Mulching by plastic foil is gaining in importance; its influence on the soil climate depends on the colour of the foil, its close adherence to the soil surface, and the width of row over which it is spanned.

1.2.2 Crop Factors Interacting with Microclimate, and the Resultant Crop Climate

The influence of a crop on its environment has been summarized concisely by Yarwood (1959): "When young plants emerge from the soil, the climate near the ground is changed. Air movements are reduced, plant parts absorb sunlight and reduce the sunlight reaching the soil. Extremes of soil surface temperatures are reduced. Plants are conditioned in early life by a zone a few inches above the soil surface where the greatest extremes of temperature occur. The normal stratification of the lower atmosphere is further stratified by many plant surfaces."

We shall here consider some of these interrelated aspects chiefly from the angle of possible effects of cultural practices.

1.2.2.1 Density of Plant Cover (Canopy)

The density of the plant cover will here be considered only from the aspect of crop climate. Interaction of the density effects with disease proneness of the host and the development of disease will be discussed in the chapter on sowing and planting practices and the density of stand (Chap. 2.9).

The practices that determine density of plant cover are the rate of sowing and planting, the application of water and nutrients, the shaping of bushes and trees by pruning and training, and the frequency of harvesting or mowing crops that yield successive waves of growth, such as forage crops, mint, and flower crops. Each of these practices is treated in some detail in the second part of this book.

The effect of a given density of plant cover on the crop climate is dependent on a variety of factors. These include direction of the crop row in relation to the prevailing wind, distance of foliage from the ground, and the topographical and soil factors mentioned above, prominent among them the rate at which the top soil dries after application of water. An overriding influence may be exerted by the macroclimate, e.g., the frequency and intensity of rainfall and high winds. In fact, in unirrigated crops, crop density probably affects disease development to a marked extent only where macroclimatic factors are not extreme. Irrigation obviously influences the plant cover, as does also the size of the field or plantation, since in small fields border effects are of much greater importance.

What are the climatic parameters chiefly affected by density of plant cover? The amplitude of fluctuations of temperature and relative humidity values, for diurnal or longer periods, are reduced under plant canopies once these have achieved a certain density. This is true for conditions in the soil as well as in the foliosphere. Moisture in the surface soil is retained for longer periods, making for increased humidity in the foliosphere. Reduction of wind impact has the same effect.

The effects on dew are manifold and somewhat contradictory. Reduced amplitude of temperature and RH tends to produce less dewfall, but relatively high RH over moist soil may increase dewfall, and the reduction in wind may mean that dew persists for longer periods. In addition, the plant canopy affects the distribution of dew, as much of the foliage is shielded from above.

Effects on the distribution of rain are most pronounced in tree crops with dense crowns. As illustrated in Fig. 1.4, reproduced from Primault (Seemann et al. 1979,

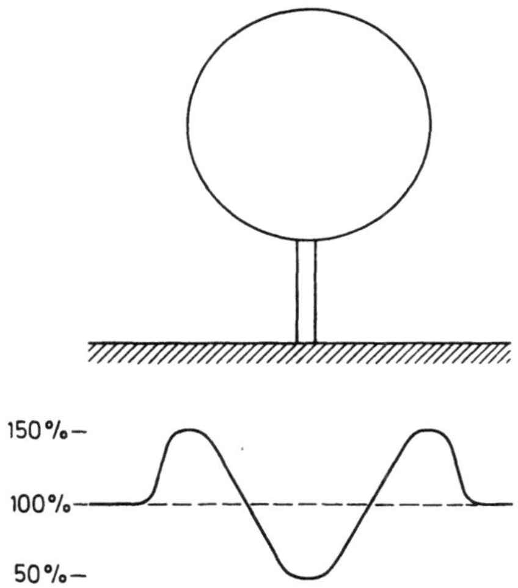

Fig. 1.4. Distribution of precipitation under a tree with a dense canopy. (Primault in Seemann et al. 1979)

p. 156), the rain draining off the periphery of the crown causes a highly uneven distribution of water. Some of the water never reaches the ground, but evaporates from the leaves. The water that gets to the ground may, however, hit the soil in streamlets that could cause considerable splashing of soil onto the lower parts of trees; this greatly favours soil-borne pathogens such as *Phytophthora citrophthora*, causing brown rot and gummosis of citrus trees and fruits. In any case, the distribution of rain under dense tree canopies differs widely from that of irrigation water applied by any method other than over-the-tree sprinkling.

In tropical plantation crops, density of plant cover may have a twofold effect. In the rainy season, when rain runs down limbs and trunks, wet soil and foliage will take longer to dry under dense cover, and prolonged periods will favour many diseases, such as the coffee berry disease *(Colletotrichum coffeanum)*. Conversely, in seasons poor in rain but rich in dew, dense plant cover will shield lower organs from dew formation, and will thus reduce the proportion of shoot growth in danger of attack by pathogens requiring films of water for their development (Waller, pers. comm.).

1.2.2.2 Shade

Three types of shade may be distinguished, differing greatly in the duration and degree of shading they afford.

1. Shade Given by Slopes. This, with sunrise, is at first general, over the whole crop, then over a steadily diminishing part of the crop, and in any case limited in its duration. For the farmer, such shade is a matter of pre-sowing or pre-planting choice: if he chooses a slope with northern exposure he may expect relatively long hours of shade.

2. Shade Given by Vertical Objects Bordering the Field, e.g., rows of trees or hedges, buildings, or tall neighbouring crops, or by tall crops sown at intervals in the crop (maize, sunflower). These will in general shade only the periphery or part of the field for limited daily periods. Such shade can be managed by pre-sowing choices, such as sowing or not sowing near trees or houses, and by timing the sowing of taller shade crops so that their maximum development coincides with the period at which the main crop is most in need of the shade.

3. Shade Given by Horizontal Objects, and in particular by canopies of crops or shade plants planted closely enough to overshadow the whole crop. In these cases shade is permanent; it can be managed by pre-planting choices of shade trees and, more importantly, by all the practices mentioned above as affecting crop density, including the rate of sowing, watering, nutrition, pruning etc.

The morning shade provided by slopes and vertical objects suffices to increase attacks of potato late blight *(Phytophthora infestans)* and many downy mildews, and has been reported by Lukens and Mullamy (1972) to increase incidence of *Drechslera maydis* in maize. Palti and Netzer (1963) have described a case in which a row of cypress trees of varying height bordered the southern side of a tomato field in the Jordan Valley; when the field was attacked by *P. infestans*, the row of cypresses was "photographed" in the tomato field, with those rows opposite taller trees (and shaded for longer hours in the morning) affected much more intensively than rows opposite lower trees. On the other hand, Kranz and Aust (1979) failed to find a significant effect of an artificial "tree" on leaf wetness in a barley field in Germany. However, the prolongation of the shade period by two hours per day, coupled with some effect on spore deposition, sufficed to raise incidence of barley mildew (*Erysiphe graminis* f.sp. *hordei*) above that in the surrounding field.

Shading by trees, which according to their density may give the effects of vertical or horizontal objects, is of particular importance in tropical crops. Reviewing the benefits of shade trees in such crops, Wellman (1977) includes not only direct effects, such as reduction of light filtering down to lower storeys of leaves, closure of stomata for longer periods, cooling of foliage and reduced dew deposits, but also numerous indirect effects: protection from winds, provision of organic matter by leaves shed by the shade trees to make a mulch, as well as extraction of nutrients from deeper soil layers and "recycling" them (as mulch), absorption of excess water in low sites, and prevention of erosion.

However, shading effects in tropical crops are not all beneficial. As seen from Table 1.2, shading of coffee and cocoa plantations helps to control some diseases, but promotes some other major diseases, especially rust and coffee berry disease in coffee and pod rot in cocoa. Shading due to horizontal objects also shelters the crop from rain. Populer (1972) holds that this may contribute to the increased incidence of *Oidium* on rubber under shade in Zaire, since rain washes off the conidia.

Achieving the right balance of shade, with due regard to the amount of shade required by the crop itself for optimal growth, is therefore an intricate matter that demands a great deal of insight and experience; it may on occasion constitute the focal point of cultural practices in tropical crops.

Little experimental work is available on the effect of shade from horizontal objects on crop disease in temperate countries. In their above-mentioned study, Kranz and Aust (1979) found that a horizontal screen suspended over a barley field reduced wetting of foliage by dew and brief showers, thus promoting spore dispersal of *E. graminis* f. sp. *hordei* and incidence of the mildew.

Table 1.2. Effects of shade on development of diseases in some tropical crops

Crop	Disease	Pathogen	Country	Effect of shade	References
Coffee	Rust	*Hemileia vastatrix*	Brazil Kenya	Shade favours diseases	Schieber (1975)
Coffee	Rust	*Hemileia vastatrix*	Nicaragua	Shade favours diseases	Llano (1973)
Coffee	Coffee-berry disease	*Colletotrichum coffeanum*	Africa	Shade favours diseases	Steiner (1973)
Coffee	Leaf spot Pink disease Wilt	*Cercospora coffeicola* *Corticium salmonicolor* *Fusarium oxysporum* f. sp. *coffeae*	Central America	Shade reduces incidence of disease	Wellman (1972)
Cocoa	Canker	*Ceratocystis fimbriata*	World-wide	Shade favours disease, if aeration impeded	Schieber, in Kranz et al. (1977)
Cocoa	Leaf blight	*Colletotrichum gloeosporioides*	Ghana	Shade favours diseases	Dakwa and Danquah (1978)
Cocoa	Pod rot	*Phytophthora palmivora*	West Africa	Shade favours diseases, but cocoa needs some shade	Muller (1974)
Cocoa	Mistletoe	*Tapinanthus bangwensis*	Ghana	Some shade needed by the crop, lack of shade favours mistletoe	Ahenkorah et al. (1974)
Cocoa	Die-back	?	Uganda	Shade good for cocoa, as strong evaporation aids die-back	Trocmé (1972)
Oilpalms	Seedling blast	*Pythium splendens* *Rhizoctonia lamellifera*	Nigeria	Shading helps control blast	Aderungboye (1973)
Tea	Blister blight	*Exobasidium vexans*	India	Shade favours disease	Rangaswami (1975)
Rubber	Powdery mildew	*Oidium* sp.	Zaire	Shade increases disease by deflecting rain, which washes conidia off unshaded leaves	Populer (1972)
Citrus	Wither-tip	*Colletotrichum limetticola*	Central America	Shade reduces incidence of disease	Wellman (1972)

1.2.3 When, Where and How Much Can Cultural Practices Influence the Crop Climate?

It may be as well to state first when and where most cultural practices can least be expected to influence the crop climate: in climates with few seasonal weather changes and all-year-round rain, in regions with uniform soil conditions and essentially flat land, and on crops grown under conditions approaching their natural habitat or the conditions for which they have been bred. However, in such climates and under such conditions, sanitation is of great importance, as diseases, once inoculum is blown in or otherwise gains a foothold, may then spread most rapidly in the uniform crop.

As the diversity of cropping conditions increases, so does the scope for cultural practices to be of value for disease control: use of slopes, taking advantage of different types of soil, adding water, sowing at different seasons etc. Cultural practices can most markedly influence crop climate, for better crop growth and health in:

a) annual crops grown out of season in climates with contrasting seasons (hot and cold, rainy and dry); steps can then be taken to mitigate extreme weather conditions, e.g., minimizing heat and cold stress by choosing appropriate sowing dates and topography, and managing moisture by mulching, tillage and irrigation;

b) perennial crops grown under conditions differing greatly from their natural habitat, e.g., crops originating under tropical jungle conditions, but grown where natural tree cover is not available (cocoa), or temperate-zone crops grown in warmer climates in which their dormancy is not uniformly broken (deciduous fruit); for these crops, conditions can be improved by measures such as choice of topography and moisture management, as well as provision of as much shade as is needed, and manipulation of flowering and fruiting periods.

The extent to which cultural practices can change the crop climate depends primarily on the macroclimate and the adaptation of crops to it. Some practices can induce really basic changes, foremost among them irrigation to provide water, culture under cover to provide warmth, and shading where radiation is excessive. Such practices are often the sine qua non of crop growth, and unless they can be applied, the crop can simply not be grown in the season in question.

Almost all other practices, however, effect only limited changes in the crop climate, nudging it in the direction favourable to the crop: sowing later in spring ensures somewhat higher soil temperatures for better germination of some crops; proper fertilization enables the crop to make somewhat better use of the soil water available. Many practices thus tend to alleviate environmental and nutrient stresses, but cannot prevent them from affecting the crop under conditions highly unfavourable to it. Such practices are therefore useful under conditions suboptimal to crop or disease development, but rarely under marginal conditions.

1.3 The Collective Approach to Disease Control: Epidemiological Considerations and the Role of Cultural Practices in Regional Management of Inoculum

The effective fight against crop diseases is increasingly becoming a communal concern, at least in advanced farming economies. This is due to developments in chemical control, advances in biological control of pests, better understanding of the role of inoculum multiplication in regional outbreaks of disease, and considerations of "enlightened self-interest" of a better-educated farming community. Such collective action may be taken by common consent of the region's farmers, or by legislative action.

Norgaard (1976) has pointed out that collective pest management strategies are advantageous due to economics of scale in information gathering, processing, and decision making, especially where monitoring devices and computer modelling of pest management is concerned.

The principal communal aspects relating to chemical control will only briefly be mentioned here:

a) pollution hazards, especially where fields or orchards are close to residential areas, roads, waterways, or fishponds;
b) prevention of the development of resistance in insects, mites and fungi, to widely used pesticides; this is an aspect of great importance, to which much more consideration should be given in future;
c) avoidance of damage to beneficial fauna and flora, especially in integrated pest control management.

The role of cultural practices in the protection of crops from disease includes preventing the introduction of inoculum into the region, and the elimination of additional and alternate hosts for the pathogens that can attack important crops. These aspects, which comprise on-farm as well as communal sanitation, will be dealt with in the chapter on sanitation (Chap. 2.2).

Apart from such direct, one might say "physical," attack on the pathogen, there are ways of restricting establishment and spread of air-borne inoculum by cultural practices, that can be carried out effectively only by communal effort. These comprise restrictions in the seasons, locations and growing practices for

Table 1.3. Principal effects of cultural practices on factors relating to size of inoculum and rate of infection

	Amount of primary inoculum y_0	Length of suscepti-bility phase	Length of latent period p	Period of infectious-ness i	Apparent rate of infection r	Rate of inoculum build-up Rc
Sanitation (incl. healthy propagating material)	+	+				+
Crop sequence	+	+				+
Tillage	+	+				+
Crop nutrition		+	+	+	+	+
Irrigation	+	+		+	+	+
Choice of sowing season	+	+	+	+	+	+
Sowing practices (depth, density)	?	+	+[a]	+	+	+
Harvesting practices	+					
Proximity to infection source	+	+				+

[a] If temperature and/or humidity are markedly affected

susceptible crops, as well as operations aimed at retarding the multiplication of inoculum and its spread within the region.

Vanderplank (1975) has defined the component factors relating size of inoculum and rate of infection as amount of primary inoculum (y_0), and length of latent period (p), period of infectiousness (i), apparent rate of infection (r), and rate of inoculum build-up (Rc). To this may be added a factor denoted by Aust et al. (1980) as length of the host's susceptibility phase. An attempt has been made in Table 1.3 to indicate which of these factors is markedly affected by the various cultural practices to be discussed here.

As shown in the table, all practices except those concerned with crop nutrition and perhaps sowing techniques, can affect the amount of primary inoculum and of its subsequent rate of build-up. These two practices, however, affect all other aspects of infection; choice of the sowing season may affect every one of the processes of infection and multiplication of inoculum.

The length of period over which the host remains susceptible is affected by all factors acting on size and build-up of inoculum, since massive amounts of inoculum can frequently overcome age-related host resistance to disease (Aust et al. 1980).

Understanding these effects of cultural practices, especially in their relation to wind- and vector-borne inoculum, leads to an appreciation of the importance of inoculum control and of the role the farming community as a whole can play to achieve it.

1.3.1 The Cardinal Role of Inoculum and its Control for the Farming Community as a Whole

There will be no disease in the region, unless viable inoculum (propagules capable of infecting) meets susceptible crop under conditions favouring infection. This includes effects of regional macroclimatic and of in-field crop climate conditions. That much is commonplace. But, as mentioned above, what is not always realized, is how much the farming community as a whole can do to prevent that meeting. This evidently involves an integrated approach, with due regard to the use of resistant varieties, crop-free periods, chemical means of protection and eradication, and the cultural practices that will be discussed here (see Chap. 3.1).

Approaches to inoculum control differ in various agroclimates and agrosystems. In subsistence farming in the tropics, the emphasis will be on constant removal of infected plant organs, mainly by hand, from the year-round growth of crops (Putter 1980). This is a tradition-bound operation by the individual farmer rather than by the community.

In temperate climates, cold winters reduce amounts of inoculum sharply. However, it is generally assumed that during the growing season sufficient inoculum is present to start an epidemic of the more virulent diseases, such as potato late blight *(Phytophthora infestans)* and apple scab *(Venturia inaequalis)*, as soon as conditions are favourable. Hence no mention is usually made of the inoculum factor in many of the disease prediction schemes working with fair success in temperate regions, and only recently are efforts being made to improve prediction of late blight by assessing the amount of available inoculum (Croxall and Smith 1976).

In semi-arid climates, winter cold is replaced by summer heat and drought as the factor effecting sharp seasonal reduction of inoculum in unirrigated crops, for which reason highly drought-susceptible bacteria rarely affect such crops. Many

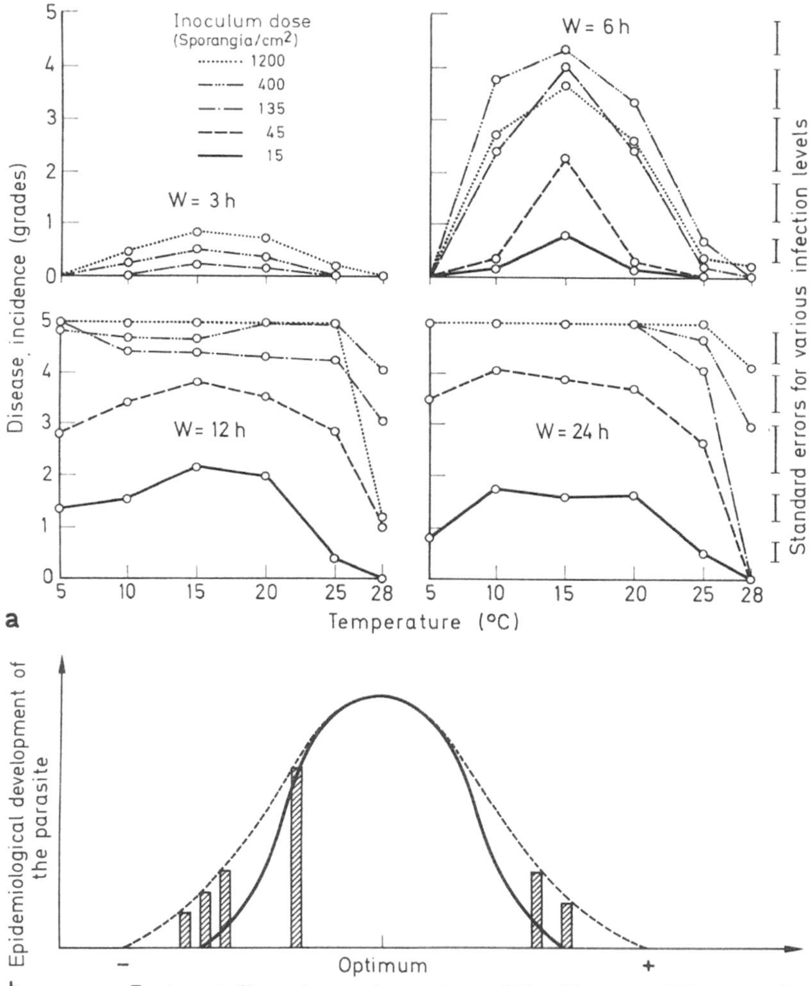

Fig. 1.5. a The danger inherent in massive inoculum build-up: compensation phenomena in the development of late blight *(Phytophthora infestans)* on potatoes. Large amounts of inoculum (135–1,200 sporangia/cm^2) induce an appreciable amount of infection even where wetness periods are very short (W = 3 h), and large-scale infection where wetness periods are somewhat longer (W = 6 h) (Aust et al. 1980). **b** Schematic illustration of the extension of pathogen development periods by compensation phenomena: Where a given factor (e.g., spore load, wetness period) has reached what is normally considered its minimum level, below which the pathogen would not develop, it may yet support infection (or sporulation etc.) when aided by the compensatory effect of another factor present at near optimal level. At the other end of the scale, a factor present at a level somewhat beyond the accepted maximum (e.g., temperature), may yet permit infection to carry on as other factors are present at optimal levels. (Aust et al. 1980)

pathogens, however, manage to survive summer in dormant state, are blown in or brought in from neighbouring climates, or persist on crops grown throughout most of the year under irrigation or under cover. Such pathogens often meet conditions of humidity and temperature that lead to explosive outbreaks of disease, as soon as the size of inoculum has assumed major proportions. Foremost among these diseases are the downy mildews and blights caused by allied fungi (Palti and Rotem 1981), where the control of inoculum size assumes an importance far exceeding that to be found elsewhere. Rotem (1978) and Aust et al. (1980) have shown that presence of large amounts of inoculum can compensate pathogens for the environmental limitations they encounter in subtropical climates. Thus, relatively brief periods of leaf wetness, insufficient to complete infection where inoculum is limited, will suffice for infection in the presence of massive inoculum, e.g., of *Phytophthora infestans* on potatoes or *Pseudoperonospora cubensis* on cucumbers (Fig. 1.5 a). This implies that, unless inoculum is controlled early in the season, the disease will prevail with greater severity during the main season even under definitely sub-optimal conditions. It also means that, once inoculum has snow-balled to massive amounts, the disease will be harder to control by chemicals and will carry over into succeeding months with temperature and moisture conditions marginal for pathogen development (Fig. 1.5 b). It follows that careful attention has to be paid in planning fields so that proximity of younger fields to older fields is avoided, as the latter may serve as sources of inoculum, and that equally careful consideration is given to spray schedules to prevent inoculum build-up.

The phenomenon of compensation by massive inoculum for other factors limiting epidemic development of disease is, of course, neither restricted to semi-arid countries nor to downy mildews and their allies, nor to environmental factors. Aust et al. (1980) have shown that in central Europe the barley mildew (*Erysiphe graminis* var. *hordei*) will overcome adult resistance of barley leaves to infection, if inoculum is present in sufficient amounts. Roelfs et al. (1972) have demonstrated that heavy inoculum pressure can overcome certain degrees of resistance of wheat varieties to *Puccinia graminis* f. sp. *tritici* and *P. recondita* f. sp. *tritici*. Golan (pers. comm.) has observed that pears, not normally subject to attack by powdery mildew (*Podosphaera leucotricha*) in Israel, are attacked when in proximity to the heavily mildewed apple variety Ana". With regard to clubroot of Brassicae, caused by *Plasmodiophora brassicae*, Colhoun (1961) determined that high light intensity will facilitate attack of the disease even with low spore loads, which would be insufficient for such attack at low light intensity.

1.3.2 Various Crops Susceptible to the Same Pathogen – Which Crop is More Valuable to the Farming Community?

Many important pathogens, some obligatory (especially viruses) and many more facultative, can attack a variety of crops, so that one crop acts as source of inoculum to another. If a minor crop grown on a limited area can cause widespread infection of a major crop covering large areas, the question then arises, whether the minor crop is worth growing. It may be valuable to individual growers, but looked at from the point of view of the farming community as a whole, can growing of the minor crop be justified? In many cases, the answer must obviously be in the negative. Thus in the Pacific North-West of the United States it has been realized that peaches, not a major crop in the region, are important overwintering hosts of the *Myzus persicae* aphid. This aphid in spring picks up the Western Yellows and Beet Yellows viruses from any available source, especially from weeds growing in ditches, and spreads them early in the season to the widely grown sugarbeet crop. Thus the peaches and the beets are evidently incompatible (Henneberry and George 1969). Cucumber Mosaic Virus, mostly originating from cucurbits, has been found to affect the valuable banana crop in Israel (Nitzany, unpubl.) as well as in the Antilles (Laville, pers. comm.); the obvious measure to take is exclusion of cucurbits from major centres of banana growing. In the tropics, in regions in which both cocoa and coffee can be grown, and both these crops are susceptible to *Ceratocystis fimbriata*, one crop may far exceed the other in value. In some parts of Central America, conditions are only marginal for coffee, but good for cocoa: in such cases the suggestion has been made to reduce the general level of inoculum in the region by eliminating the coffee (Schieber 1969). Similarly, in the Sudan, the cotton leaf curl virus spreads from the minor kenaf *(Hibiscus cannabinus)* crop to cotton; the kenaf should clearly be eliminated (Follin, pers. comm.).

1.3.3 Crop Varieties Differing in Susceptibility – Dangers and Opportunities

Varieties in many major crops differ so greatly in their susceptibility to certain species or strains of pathogens that their role in communal approaches to disease control merits special consideration.

It is most important that herbaceous crops sown early, or early varieties of plantation crops, should not be highly susceptible, as this may promote disease in all later crops. This has already been mentioned in relation to *Phytophthora infestans* blight on potatoes and tomatoes, and is equally true for all herbaceous crops in which there are varieties resistant to downy mildew fungi (e.g., maize, onions, melons, soybeans), rusts, or viruses (potatoes). Use of resistant hybrids is of special importance for maize, a crop with a short growth period grown often in successive sowings. Fortunately, hybrids are available that are resistant to major maize blights and various downy mildews and rusts (Nyvall 1979). This subject is referred to in Chap. 3.2.1.

In fruit crops, early varieties – valuable as they are for obtaining higher revenue for early-season yields – can have disastrous effects on later varieties in the whole region. A case in point is the "Ana" apple variety, produced in Israel by crossing a local variety with "Delicious;" this apple sprouts early as it does not require winter cold to terminate its dormancy, but is extremely susceptible to powdery mildew *(Podosphaera leucotricha)* and scab *(Venturia inaequalis)*. In the vicinity of this variety, other, later apple varieties are scabbed and mildewed early and profusely, and even fairly resistant varieties contract powdery mildew (Golan, pers. comm.;

Pappo, pers. comm.). In England the apple variety Worcester Pearmain, resistant to mildew, is preferred to the susceptible Jonathan as pollinator in apple orchards (Butt, pers. comm.).

Where strains of pathogenic species are concerned, differences in varietal susceptibility may be used by communal or even country-wide action to lessen danger of disease.

In the United States, the suggestion has been made by Browning et al. (1969) that the annual progress of oat rust *(Puccinia coronata)* from Texas to the central and northern states could be checked by sowing only rust varieties with rust-resisting genes in a wide belt across this "Puccinia path."

In Europe, the suggestion has similarly been made, that farmers over entire wheat-growing regions should cease for 2–3 years growing wheat susceptible to certain strains of stripe rust *(Puccinia striiformis)*, in order to avoid dangerous build-up of these strains (Lemaire, pers. comm.).

Containment of a disease by the use of resistant varieties has been proposed by Heale et al. (1980), in relation to *Verticillium* wilt of lucerne in England. This approach seems valid for restricting and progressively eliminating soil-borne diseases of various crops.

1.3.4 Restricting Seasons, Locations, and Growing Practices for Susceptible Crops

In agro-climates with cold or dry seasons, in which inoculum is sharply reduced (see Chap. 1.1), the pathogen at the end of such seasons requires availability of susceptible tissue, as well as a certain period of time, to pass several cycles of multiplication before an epidemic begins to flourish. Only after inoculum has reached a certain quantitative threshold will an outbreak of disease occur. This has been demonstrated for *Phytophthora infestans* on potatoes by Guntz (1959) in France and by Croxall and Smith (1976) in England. In the case of air-borne pathogens, the period required to reach the threshold can often be as short as 2–3 weeks. What can be done to delay establishment of primary inoculum and its subsequent multiplication?

With regard to inoculum blown in from outside the region, a fundamental principle of communal disease prevention is the limitation of periods in which susceptible crops are available. This may be done by ceasing to grow an early-season crop which could serve as the "breeding ground" for diseases for more valuable later crops, or for later sowings of the same crop over much wider areas. Examples from temperate climates are prohibition of sowing winter barley, where summer barley is more important, to avoid build-up of powdery mildew *(Erysiphe graminis)*; or elimination of early potatoes which could generate massive inoculum of late blight to attack later crops early in their growth.

Nurseries are a likely source of air- or vector-borne inoculum to infect crops over entire regions, since seedlings in nurseries are sown early in the season, and densely. This applies to many fungal diseases, such as *Phytophthora* blights on tomatoes and eggplants, and downy mildews on tobacco, Brassicae, lettuce, beet, and onions. It is equally important in the case of virus diseases, e.g., cassava mosaic (Waller, pers. comm.). All such nurseries should be strictly excluded from regions in which the crops are subsequently grown.

The aim of the communal approach to inoculum control is thus clearly to avoid overlapping sowings of crops susceptible to the same disease and to programme crops so as to obtain *host-free periods*. This concept has been advanced in California in relation to virus diseases of sugarbeet (Kontaxis 1973) and of celery (Milbrath and Ryan 1938).

Special problems are presented by biennial seed crops, as these may contract a variety of diseases during their long growth span and serve as major sources of inoculum for younger crops. Seed crops of sugarbeet are a dangerous source of the downy mildew caused by *Peronospora farinosa* (Byford and Hull 1967) and of viruses. The immediate vicinity of the crop is, of course, in greatest danger, which diminishes fast at a distance of about 100 m to the infected seed crop (Broadbent 1969). Seed crops of onion have been shown convincingly by Rondomanski (1966) in Poland to be the primary source of downy mildew attacks *(Peronospora destructor)*. Seed crops of Brassicae and carrots serve to spread viruses (Broadbent 1969). The decision to grow seed crops therefore has in many cases to be taken by the farming community as a whole, and may involve deciding against a market garden crop of the same species (see also Chap. 2.12).

No less important to the community is the removal of infected crops immediately after harvest, or in some cases early harvesting to obviate further multiplication of inoculum (see Chap. 2.11). Thus the region-wide removal of the crop by compulsory harvesting and ploughing of diseased sugarcane fields, before specified dates, has been required to prevent distribution and build-up of leaf scald *(Xanthomonas albilineans)* or red rot *(Glomerella tucumanensis)* in Queensland, Australia (Hughes 1976). Similarly, in Western Africa, the spread of the cotton virescence mycoplasma by leaf-hoppers can be checked effectively only if cotton residues are promptly destroyed after harvest (Follin, pers. comm.).

The question whether certain crops should be grown by ratooning (multiannual culture) should also be subject to communal decision. Ratoon crops commonly serve for the perpetuation and multiplication of pests as well as diseases, which may then affect the annual sowings of the same crop. Examples are sugarcane, in which the above-mentioned diseases and the smut caused by *Ustilago scitaminea* multiply (Waller 1967), and cotton in which black arm *(Xanthomonas malvacearum)* and virus diseases build up after the first year's growth.

Another way of restricting availability of susceptible crops is the exclusion, especially early in the season, of growing sites in which crops are particularly likely to be attacked by disease, or of growing practices conducive to early outbreaks. As stated in the chapter on topographic effects on crop climate (Chap. 1.2), low-lying fields are often disease-prone, especially to diseases such as downy mildews, while neighbouring slopes are not, if they are exposed to drying winds and radiation. This has been shown to apply to a variety of diseases, including the downy mildews of peas *(Peronospora viciae)* (Wilson 1971) and of tobacco *(P. tabacina)* as reported by Rotem (1978). The farming community should insist that crops in danger of being attacked by such diseases must not be grown in such disease-prone localities before a specified date.

Cultural practices, such as techniques of irrigation and density of sowing, may decisively affect the chances of a crop to "catch" and multiply inoculum blown-in from outside the region (see Chap. 2.7 on moisture management and Chap. 2.9 on sowing practices). Overhead irrigation is well-known to favour numerous diseases, including *Cercospora* and other leaf-spots, downy mildews, soil-borne *Phytophthora* diseases, and many others. Nurseries of vegetables, flowers and field crops, sown densely and kept moist, are most susceptible to such pathogens. Farmers have to appreciate that irrigation techniques other than overhead sprinkling are preferable to exclude disease from some early crops, and that nurseries of certain crops, e.g., tobacco, should better not be grown in the region in which the crop is to be raised. If appearance of disease can be delayed by such adaptation of cultural practices, they are well worth a communal effort or, if need be, legislative ruling.

1.3.5 Minimizing Multiplication and Spread of Air- and Vector-Borne Inoculum

Development of inoculum of pathogens known to exist in the region, or expected to arrive there in spite of all efforts at their exclusion, can be minimized by certain pre-sowing and pre-planting practices, as well as by in-growth management of environmental and nutrient factors. As far as air-and-vector-borne pathogens are concerned, these measures can be effective only if carried out on a regional scale (see also Chap. 2.12).

The first pre-requisite is, of course, to secure absolutely disease-free propagating material (see Chap. 2.2.2.1). How successful a communal effort can be in this respect has been shown in France, where the destructive onion yellow dwarf virus has been eliminated from the principal centre of garlic culture by strict supervision of the health of propagating sets (Lot, pers. comm.).

The decisions to be taken *prior to* sowing or planting include many of those mentioned in the previous section, chiefly choice of site and soil, attention to proximity of potential sources of infection, including the destruction of volunteers, wind protection, and direction of rows, pre-sowing manure and fertilizer application, installment of a suitable system of irrigation and – last but not least – careful choice of sowing or planting dates and of varieties. Where crops could either be sown or planted, the decision which practice to adopt may also be the decision of the community. Since crops can usually be planted later than sown, the collective decision to plant may make it possible to gain a host-free period early in the season.

Decisions to take *at the time* of sowing or planting relate to crop density, sowing for subsequent thinning or for final stands, depth of sowing or planting, and irrigation for crop germination or for rapid rooting of transplants.

In-growth management involves fertilizer top-dressings, tillage, judicious application of water (rate and frequency) and constant sanitation. In fruit crops, removal of infected flowers and fruits and cutting out of cancers and infected spurs must be carried out repeatedly; these may liberate spores into the air, as in the case of infection of deciduous fruit by *Sclerotinia fructigena* (Anon 1973c). All these measures must be integrated, where required, with chemical control.

Harvesting practices relevant to inoculum control are:

a) proper timing of one-time harvest or of successive pickings;
b) frequency and clipping height in cutting crops that regenerate growth (lucerne, setaria, lawn, etc.);
c) haulm destruction and other practices to minimize impact of inoculum on the harvested produce, and to prevent its build-up in the soil.

After harvest, sanitation is the primary requisite; isolated clumps of rejected produce left in the field may constitute cardinal sources of infection for the following season's crop over a whole region. This has long been known in relation to clumps of potato tubers infected with *Phytophthora infestans*. It is equally true with regard to virus diseases of mangolds, even when clumps are covered with straw and soil, and such diseases spread from the clumps to all beet crops in the region. Another case in which sanitation can be effective only when practiced on a communal scale is the destruction of volunteers, e.g., where barley is affected by yellow dwarf virus (Bayon and Ayrault 1977).

Special importance, at all stages of the crop's growth, attaches to the control of weeds that transmit pathogens. Communal efforts are particularly important in this respect, as they have to include non-agricultural land, rights-of-way, ditches and housing or industrial sites. A striking example of a weed that transmits a multitude of diseases is Johnson grass (*Sorghum halepense*): it serves as the source of numerous major diseases of maize and sorghum, including four smuts, several leaf spots, and most importantly the dwarf mosaic and chlorotic dwarf viruses of maize (see also Chap. 1.7).

The above list of sanitation measures best handled on a communal basis could be extended at will, but these examples may suffice to show the fundamental importance of this approach.

The reason for enumerating all these practices in this chapter on the communal approach to disease control is that *everything* influencing multiplication of inoculum should be the concern of the community. Those practices particularly relevant to communal management of air-borne inoculum are avoidance of proximity to sources of infection, often older crops, the choice of appropriate sowing seasons and methods of irrigation, and meticulous attention to harvesting practices. The latter may often decide the level of inoculum at which the next season's crop begins its growth.

1.4 Soil, Soil Microbiota, and Soil-Borne Disease

The properties of soil most relevant to the subject of cultural practices and crop disease are:

a) the soil as medium for root growth and for the development of underground organs such as potato tubers and groundnut pods;
b) the soil as habitat for microflora and microfauna;
c) the soil in relation to nutrients and water, including its ion exchange and water holding capacities and pH level.

The last of these points will chiefly be dealt with in the chapters on crop nutrition and water management. A brief review of aspects of soil climate relevant to crop disease development is included in Chap. 1.2.1.2.

1.4.1 Soil and the Growth of Underground Organs of the Crop

The suitability of a soil for unimpeded root growth, one of the essentials in the development of a healthy crop, and for formation of underground yield components, is determined by its texture, depth, composition, and reaction. All of these are to some extent amenable to change by cultural practices.

1.4.1.1 Texture and Depth

The texture of soil depends on the composition and fineness of its mineral components, as well as on the amount and composition of organic matter it contains. Texture and depth, often together with slope, determine the water-holding capacity and temperature relations of the soil. As mentioned in greater detail in the chapter on organic soil amendments, increase in the organic matter content will improve the water-holding capacity of sandy soils, but will ameliorate heavy clay soils by facilitating percolation of water to greater depth. The water-holding properties of a soil, in their turn, affect its temperature relations.

Stover (1959) has rightly stated that the texture of soils is associated with so many complex factors, that its effect on the growth and survival of fungi is hard to define. Soils of heavy texture, especially when containing appreciable amounts of organic matter, are often richer in their bacterial and actinomycete flora than soils of light texture.

Where the prevailing temperature range is favourable to root growth, and water is available, development of the crop's underground organs depends largely on availability of an adequate supply of oxygen and the absence of physical impediments, such as crusts and plough-soles. Tillage and enrichment of soil with organic matter are the accepted means of ensuring oxygen supply and doing away with impediments. As far as crusts are concerned, the recent reports on successful use of synthetic polymers as soil conditioners raise hopes that an additional means of securing good tilth may soon be available, at least for the more valuable crops (De Boodt 1979).

If texture is favourable, the deeper the soil, the more opportunity it affords crops with deep-going roots to realize their growth potential, provided the deeper soil layers are also favourably structured. It must, of course, be borne in mind that, with increasing depth, soils change in their atmosphere (less oxygen, more carbon dioxide), their organic matter and moisture content, and often also in their reaction. Such conditions will support root growth of many crops, but populations of root-infecting fungi decrease rapidly with depth, and are rarely found below the 50–60 cm level. The relation between soil texture and water potential is mentioned in Chap. 1.4.1.3.

The depth of soil opened up by subsoiling (40 cm or deeper), enabling cotton plants to deepen their root systems, has in fact been found to alleviate the cotton stunt disease complex associated with the nematodes *Hoplolaimus columbus* and *Meloidogyne incognita* (Bird et al. 1974).

1.4.1.2 Soil Reaction

Most crops can grow within pH limits wider than those conducive to development of their major root pathogens. This affords the farmer a means of combating disease by manipulation of the soil reaction.

In general, actinomycetes such as *Streptomyces scabies*, the cause of potato scab, grow best at soil reactions close to neutral, while fungi are more prevalent in acid than in neutral soils. Black (1968) holds that the reason for this is not so much that fungi grow better under acid conditions as that they are more adaptable to hydrogen-ion activity than bacteria or actinomycetes, and hence suffer less from competition or antagonism under acid conditions.

Table 1.4a lists the preferred pH values for development of some pathogens. These do not necessarily represent values at which the pathogen grows best in cul-

Table 1.4a. Soil reaction effects on some major soil-borne fungal and bacterial pathogens

Crop	Disease	Pathogen	Relation of pathogen to pH level	References
Wheat	Stalk rot	*Fusarium culmorum*	Most frequent in neutral or slightly alkaline soils	Domsch et al. (1980)
		F. equiseti	Most frequent at pH approximating 7.0	Domsch et al. (1980)
Cotton	Root rot	*Phymatotrichum omnivorum*	Prevalent at pH 8.2	Domsch et al. (1980)
	Wilt	*Fusarium oxysporum* f. sp. *vasinfectum*	Prefers low to moderate pH levels	Smiley (1975)
Potato	Scab	*Streptomyces scabies*	Prefers high pH levels	Smiley (1975)
	Rot	*Pseudomonas solanacearum*	Develops above pH 5.0	Stevens and Stevens (1952)
Tomato	Wilt	*Fusarium oxysporum* f. sp. *lycopersici*	Prefers low to moderate pH levels	Smiley (1975)
Lucerne	Wilt	*Verticillium albo-atrum*	Optimal pH: 5.5–6.5	Heale and Isaac (1963)
Pea	Root rot	*Aphanomyces euteiches*	Favourable pH: 5.4–7.5	Papavizas (1966)
Tobacco	Root rot	*Thielaviopsis basicola*	Optimal pH: 5.7–5.9	Stevens and Stevens (1952)
Various	Collar rot	*Sclerotium rolfsii*	Prefers low pH	Smiley (1975)
Brassicae	Club-root	*Plasmodiophora brassicae*	Optimal pH: 5.0–6.0	Garrett (1970)
Peach	Bacterial canker	*Pseudomonas syringae*	No infection at pH 6.6–7.2, infection at pH 5.6–6.1	Weaver and Wehunt (1975)
Avocado, mango, coffee, cocao, banana etc	Black rot	*Rosellinia bunodes*, *R. pepo*	Optimal pH 5.2	Gibson (in Kranz et al., 1977)

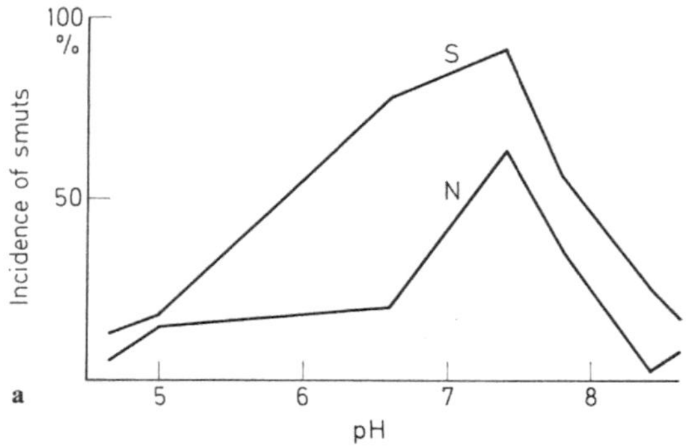

a

Fig. 1.6a. Effect of soil reaction of covered smut *(Ustilago kolleri)* on two varieties of oats, *Avena sativa* var. *victor (curve S)* and *A. nuda* var. *inermis (curve N)* (Gaeumann 1951)

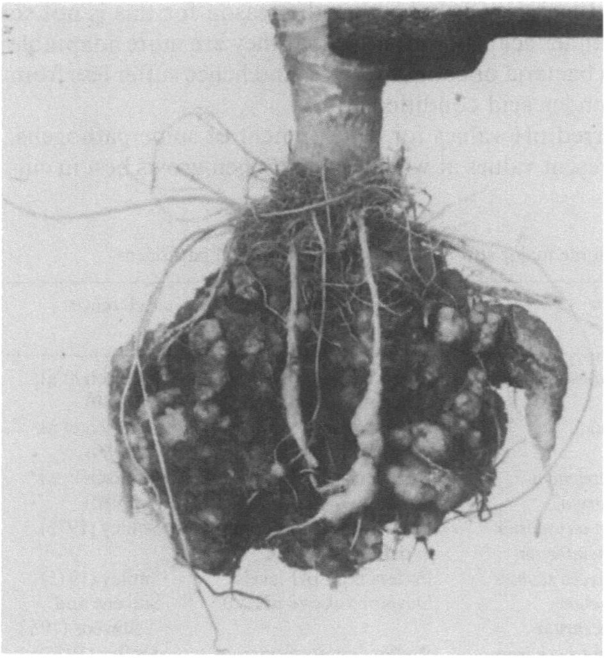

b

Fig. 1.6b. Cauliflower plant affected by club root caused by *Plasmodiophora brassicae*, a pathogen greatly favoured by acidic soil reaction. (By courtesy of the Division of Plant Pathology, Agricultural Research Organization, Bet Dagan, Israel)

ture, but levels at which it gains an advantage over competitors or antagonists in the soil flora. It is therefore not surprising that some root fungi have different pH optima for their development in different soils. *Fusarium oxysporum* f. sp. *cubense* was found to develop optimally at pH 5.0 in acidified sandy loam, but at up to pH 7.0 in sandy loam (Stover 1956).

Moreover, in addition to the effects of soil reaction on the pathogen, its effects on host growth and susceptibility must be taken into account. The curves depicted in Fig. 1.6a are probably the result of pH effects on both host and pathogen, i.e.,

Table 1.4b. Relations of nematodes to pH levels in soils

Nematode species	Relation to pH level	References
Tylenchorhynchus dubius *Pratylenchus crenatus*	More frequent in acid than in neutral or alkaline soils	Norton (1979)
Pratylenchus alleni *Hoplolaimus galeatus*	pH 6 preferred to pH 4 or pH 8	Norton (1979)
Pratylenchus penetrans	Optimum pH 5.2, less growth at 6.4, much less at 4.4 and 7.3	Norton (1979)
Heterodera schachtii	Occurs in both neutral and acid soils	Norton (1979)
Heterodera avenae	Prefers pH 7.5 to pH 5.0	Duggan (1963)

growth of oats and the smut *Ustilago kolleri* (Gaeumann 1951). The difference between the two oat varieties is also of interest: although incidence of smut in both reaches a peak at about pH 7.5, there is a marked difference in disease at pH levels of 5.5 to 6.7.

Yearly alteration of the soil reaction as a means of controlling *Pseudomonas solanacearum* on potatoes has been recommended in Florida, USA. The soil is treated with sulphur in June to give a pH of about 3.8 and remains below pH 4.0 until November. This reduces pathogen population adequately. In November the soil is then treated with limestone to bring the pH to a level above 5.0 for satisfactory production of a potato crop (Stevens and Stevens 1952).

As regards nematodes, the data presented in Table 1.4b, most of them gleaned from Norton (1979), indicate that most of the species listed grow best at pH values of 5.0–6.0. The exception is *Heterodera avenae* which prefers higher values.

A word of warning is indicated with regard to the application of lime to soils to raise their pH value. Such applications may block the uptake of elements such as iron, manganese and boron by crops, and cause some nutrient stress.

Adjustment of soil reaction in favour of the crop can be effected by materials especially applied for this purpose: lime or other materials rich in calcium to raise the reaction in acid soils, and sulphur to lower that in alkaline soils. Frequently, however, the pH level can be shifted in the desired direction by selection of appropriate forms of fertilizer. Thus, where nitrogen is to be applied, choice of anhydrous ammonia or of ammonium chloride or sulphate will tend to lower pH, that of calcium nitrate will raise it.

A case in which management of pH is important, but poses a dilemma, has been described by Jones and Woltz (1972). On tomatoes, increases in pH are conducive to higher incidence of *Verticillium albo-atrum*, lower pH to that of *Fusarium oxysporum* f. sp. *lycopersici*. At more alkaline values the *Fusarium* suffered from lime-induced deficiency of micronutrients supplied in part by superphosphate. The solution suggested to evade this dilemma is to apply superphosphate sparingly so that micronutrient supply remains low, which inhibits the *Fusarium*, but neither is the pH level raised to favour the *Verticillium*. Indeed, a case of tight-rope walking to keep tomatoes healthy.

1.4.1.3 Soil Water

From the point of view of crop growth, soil water is the most important aspect of the soil environment. It directly affects development of crops, and exerts an indirect influence through its actions on microflora and microfauna in general, growth of soil-borne pathogens, and susceptibility of plants to diseases of roots and shoots.

In discussing soil water, the most useful concept is that of the soil's water potential. This has been investigated and described by Cook and Papendick (1970, 1972) in a number of publications from which much in the following paragraphs has been gleaned. They describe soil water potentials as follows: "Consider an organism in contact with pure, free water and assume that the amount of work that must be done by that organism to obtain a given quantity of water under this condition is zero. Now consider the same organism having to obtain the same quantity of water from unsaturated soil. Such water is bound by certain forces associated with the presence of soil. The difference in work that must be done on the soil water relative to pure, free water in order for the organism to obtain water is the soil water potential. By definition, the soil water potential decreases with decreased soil water content. Total soil water potential may be divided into components based on origin of forces acting on the water. These are (1) *osmotic potential* due to solutes in the soil water; (2) *matric* or *capillary potential* which includes both adsorption and capillary effects due to the solid phase; (3) *gravitational potential* caused by elevation differences from the reference; and (4) *pressure potential* caused by external atmospheric pressure on the soil water. The components are additive, the sum being equal to the total water potential."

Water potential is commonly expressed in units of pressure, i.e., bars or atmospheres (1 atm = 1.013 bar). In a system containing liquid water in equilibrium with its vapour, measurements of the relative humidity can be used to estimate the water potential. If the relative humidity limiting growth of an organism is known, the water potential at which its growth may cease can be inferred, as shown in Fig. 1.7. Its relations to the water content of the soil is a major aspect of soil texture. Coarse-grained soils (sands) retain less water at a given potential than do finer-textured soils; moreover, the flow of unsaturated water is greater in soil with fine pores than in soil with coarse pores. Thus, the outwards gradient of water potential from the surface of roots or seeds absorbing water would be less in heavier soils. This means that in sandy soil the immediate vicinity of an absorbing plant surface may be drier, and therefore less conducive to infection by organisms such as *Pythium* spp., than that of clay soils (Cook and Papendick 1970).

The effects of nitrogen fertilization on water potential of soil will be referred to in Chap. 2.6.2.

The lower limit of available water for higher plants is often considered to be approximately −15 bar. At what percentage of soil moisture this water potential is reached depends, inter alia, on whether the soil is wetted up, e.g., by irrigation of dry soil, or dried down, e.g., by drainage of wet soil.

Cook and Papendick (1972) have compiled lists of soil fungi favoured by dry and by wet soil, respectively (Table 1.5). Those favoured by dry soil occur as a rule over a wider range of soil types, including medium and light soils, than the fungi

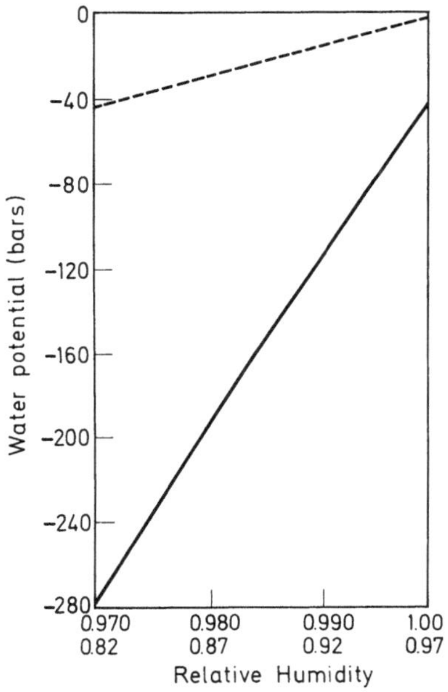

Fig. 1.7. Water potential-equilibrium relative humidity relationship at 25 °C. The *dashed curve* shows water potential for relative humidities between 1.0 and 0.970. A reduced scale is used for the *heavy curve*, which shows water potential for relative humidities between 0.97 and 0.82. (Cook and Papendick 1970)

Table 1.5. Diseases favoured by dry and by wet soil, respectively. (After Cook and Papendick 1972)

Crop	Disease	Pathogen
A. Diseases favoured by DRY soil		
Cereals	Seedling blight	*Fusarium graminearum*
		F. culmorum
Wheat	Root and foot rot	*F. culmorum*
Clover	Seedling blight	*F. roseum*
Sweet potato	Stem rot	*F. solani* f. sp. *batatas*
Pea	Root and stem rot	*F. solani* f. sp. *pisi*
Sweet potato	Pox	*Streptomyces ipomeae*
Potato	Common scab	*S. scabies*
Sorghum	Charcoal rot	*Macrophomina phaseolina*
Cotton	Charcoal rot	*M. phaseolina*
Wheat	Seed decay	*Penicillium* and *Aspergillus* spp.
B. Diseases favoured by WET soil		
Cotton	Root rot	*Phymatotrichum omnivorum*
Wheat	Take-all	*Gaeumannomyces graminis*
Wheat	Cephalosporium stripe	*Cephalosporium gramineum*
Wheat	Bare-patch	*Rhizoctonia solani*
Tobacco	Black root rot	*Thielaviopsis basicola*
Various	Armillaria root rot	*Armillariella mellea*
Various	Southern blight	*Sclerotium rolfsii*
Various	Cottony rot	*Sclerotinia sclerotiorum*

favoured by wet, mostly heavy soils; this applies, however, only to non-irrigated crops.

The water potential of the soil has a profound effect on the microbial composition and activity of the soil, and on the relations between soil pathogens and their antagonists and competitors. Bacteria are relatively inactive at water potentials approaching the wilting point of plants, while some important fungal pathogens are still quite virulent at that point, especially many species of *Fusarium*. This ability to develop at water potentials at which bacteria are inactive and hosts weakened explains to a large part the higher incidence of many *Fusarium* wilts and rots on fairly dry, or alternately wet and dry, soil than on wet soil. The same applies to diseases, mainly charcoal rots, caused by *Macrophomina phaseoli (Sclerotium bataticola)* on many hosts in dry soils (Shokes et al. 1977), as described more fully in Chap. 1.5.6.

Apart from its effect on growth and nutrition of host and soil microbiota, including pathogens, soil water plays an important role as a medium of locomotion of many motile stages of plant pathogens, e.g., larvae of nematodes and zoospores of *Pythium* and *Phytophthora* fungi. Thus a low soil water potential may preclude the spread of infection by such pathogens until the soil is re-wetted.

Disease control by management of soil water will be discussed in Chap. 2.7.1.

1.4.2 Soil Microbiota and Their Interaction with Soil-Borne Pathogens

A rich flora and fauna of soil microbiota lessens the danger of epidemic outbreaks of diseases caused by soil-borne pathogens. As Wilhelm (1965) has states: "Biological complexity of the soil assures associative, competitive, and antagonistic relations which limit population explosions and thus bring about balance. The more numerous the kinds of organism, and the greater their number, or perhaps the shorter their generation time, the more stationary the balance."

The pathogenic components of the microflora that here concern us most are bacteria and fungi and some viruses transmitted by organisms in the soil; the microfauna components of interest are chiefly the nematodes.

Garrett (1970) has divided root-infecting fungi into specialized and unspecialized groups. The un-specialized fungi are not host-specific and attack mostly juvenile and ageing tissues; many, if not all, such fungi can probably exist in the soil as saprophytes in the absence of living host tissues, often with a high degree of competitive saprophytic ability. They are thus intermittent parasites on the roots of many plant species, with an alternative mode of existence as saprophytes. The specialized fungi, on the other hand, are generally host-specific, have lost most of the properties that are necessary for life as a competitive saprophyte but are incompatible with the establishment of a successful host-parasite relationship; they also develop more slowly than fungi of the first group. Distribution of the specialized fungi is governed by that of their hosts. It must, however, be stressed that "specialization" is not a precise concept, and that the single-host specialization of many strains of *Fusarium oxysporum* differs widely from specialization of *Verticillium albo-atrum*.

The microbiota of the soil will here be considered under three headings:

a) the soil flora and fauna that restricts pathogen development as it competes with, is antibiotic or lytic to, or predatory of, soil-borne pathogens;
b) the organisms that transmit pathogens in the soil, or are associated with their development;
c) beneficial micro-organisms, e.g., effects of mycorrhizae and of nitrogen-fixing bacteria on soil-borne plant pathogens.

1.4.2.1 Soil Microbiota Restricting Pathogen Development

The role of competitive and antibiotic microbiota in soil in disease management has been treated comprehensively by Baker and Cook (1974) and by Cook (1977). The means of taking advantage of the soil microbiota for this purpose are chiefly tillage, debris management, soil amendments, application of organic or mineral nutrients, crop sequence, and the application or withholding of water. All these are fundamental cultural practices which will be treated in the second part of this book.

The aims of managing the soil microbiota for disease control have been so well defined by Cook (1977), that we would like to quote his definition in full:

"a) *Reduction of the Inoculum Density* of the ppathogen, including its starvation by crop rotation, treatments that predispose sclerotia to microbial attack, e.g., by intermittent wetting and drying of soil or by flooding, promotion of the lysis of fungal organs by organic amendments and by using fungi against nematodes.

b) *Replacement of a Pathogen in Plant Refuse.* Many pathogens *(Pseudocercosporella herpotrichioides, Armillariella mellea)* use plant residues as refuge and as food base. By manipulating the environment, e.g., soil reaction or water potential, to the advantage of antagonists or competitors, the pathogen may be deprived of its base.

c) *Suppression of Germination and Growth of Pathogens*, possibly by self-inhibitors produced by the pathogen in a certain biotic or abiotic environment (e.g., temperature, pH, water potential).

d) *Protection of an Infection Court*, the slowing-down or prevention of pathogen infection by microbial production of antibiotics or other substances that will suppress germination of the pathogen or inhibit pathogenesis. This may take place where a non-pathogen or weak pathogen takes early possession of the infection court, as mycorrhyzae do, to exclude the virulent pathogen.

e) *Stimulation of a Resistance Response in a Potential Host*, which can be effected by inoculation with another organism, often closely related to the pathogen."

To what extent can these aims be attained with the knowledge now at our disposal? The answer is: to no large extent, and it would indeed be surprising if it was otherwise. The problems are so complex, relevant research has practically been limited to only some of the most advanced countries in temperate regions, and the volume of such research has, for obvious reasons, been much more restricted than, for instance, that on pesticides.

It must be realized that after some decades of relevant research, all the information now at hand concerning management of crop disease by way of soil microbiota refers to specific conditions and specific host/pathogen combinations. It does not, and perhaps cannot, lend itself to generalizations. As will also become apparent in the chapter on soil amendments, large-scale field and laboratory research in diverse agroclimates, very much including the tropics and subtropics, is needed to advance this hopeful approach to disease control to the stage of wide, practical use.

This is not intended to detract in any way from the value of achievements in research on antagonism and biological control. The status of this research has recently been reviewed by Cook (1979). Some of the work that has reached the stage of field application will be mentioned here.

Treatment of Propagating Material. In France, a considerable amount of research has been carried out on the use of avirulent strains of *Gaeumannomyces graminis* to control take-all. Inoculated into seed, such a strain has been shown to protect wheat from aggressive strains of the fungus and, in some varieties, to advance development of the reproductive stage (Lemaire et al. 1979). Since seed inoculation is a fairly simple process, this approach seems practical and deserves research in other countries. Another case in which seed treatment looks hopeful is that of applications of oospore suspensions of *Pythium oligandrum* to beet seed to reduce incidence of damping-off (Vesely 1979).

Antagonists and Competitors in the Soil. Practical success in using a particular strain (no. 84) of *Agrobacterium radiobacter* to control crown gall *(A. tumefaciens)* has been reported from many countries. Several companies are now distributing commercial preparations, mostly as finely ground peat, for application to bare-rooted plants, graft unions, and cuttings of apple, pear, stone-fruit, raspberry, rose, chrysanthemum, and some other ornamentals. Seed treatment of apple and peach by dipping in water suspensions of strain 84 has also given good results (Moore 1979). The control achieved seems mainly due to the production by strain 84 of a substance termed bacteriocin 84. However, recent reports from Greece attribute a breakdown in control of crown-gall by strain 84 to the appearance of new, recombined forms of *A. tumefaciens* which combine genes for virulence with genes for bacteriocin production (Panagopoulos et al. 1979).

Seed inoculation with nitrogen-fixing bacteria, and its effect on plant diseases, are mentioned in the next chapter.

Again referring to take-all of wheat and barley, it has been found in England that the non-pathogenic parasite *Phialophora radicicola*, which develops on grasses, will restrict growth of *G. graminis* on succeeding cereal crops. Thus, using a grass ley as a break crop in cereal cultivation provides an effective means of reducing incidence of take-all (Yarham 1979b) (Shipton 1977, 1979).

Following extensive work under controlled conditions on the antagonistic effect of *Trichoderma harzianum* on *Sclerotium rolfsii* and *Rhizoctonia solani*, Chet et al. (1979) in Israel found that application of a wheat-bran culture of the antagonist in the field to soil naturally infested with *S. rolfsii* reduced incidence of the pathogen on groundnuts. This confirmed earlier work by Backmann and Rodriguez-Kabana (1975), who applied granules inoculated with *T. harzianum* to groundnut fields in Alabama, USA. Effects of mycorrhizae on crop diseases are mentioned in Chap. 1.4.2.3.

1.4.2.2 Soil Microbiota Transmitting Plant Pathogens or Associated with Their Development

Many of the organisms that move actively through the soil are suspect as carrying pathogens. Thus bacteria are transported in the guts of earthworms, root aphids have been shown to be capable of transmitting viruses; the maggot of *Hylemya cilicrura*, the widespread seedcorn maggot, can disseminate soft rot bacteria such as *Erwinia carotovora* (Kevan 1965), and *Verticillium albo-atrum* can be spread by the bulb-mite *Rhizoglyphus echinopus* (Price 1976). Transmission by soil fauna of phytopathogenic fungi, bacteria and viruses and association of such fauna with disease development has, however, been found significant chiefly in relation to a group which itself contains many pathogens, the nematodes.

Nematodes. Figure 1.8 indicates the genera of nematodes known to transmit, or otherwise associate with, other plant pathogens, including some shoot-infecting nematodes *(Ditylenchus, Aphelenchoides, Anguina)*. Transmission in the soil is most important in the case of the ectoparasites *Xiphinema*, *Longidorus*, and *Trichodorus*. The viruses transmitted by this group are listed in Table 1.6 and include incitants of important diseases of grape-vine and small fruit. Not all these viruses are transmitted solely by nematodes: some can also be seed-borne, especially those carried by species of *Longidorus* (Taylor and Cadman 1969).

Table 1.6. Viruses transmitted by nematodes[a]

Crop	Vector	Virus
Grapevine	*Xiphinema index*	Arabis mosaic
	X. americanum	Tomato ringspot
Raspberry	*X. diversicaudatum*	Arabis mosaic, Strawberry latent ringspot
	Longidorus elongatus, L. macrosoma	Raspberry ringspot
Red currant	*L. elongatus*	Raspberry ringspot
Black currant	*Xiphinema* sp.	Arabis mosaic
Peach	*X. americanum*	Tomato ringspot
Strawberry	*X. diversicaudatum*	Arabis mosaic, Strawberry latent ringspot
Beet, celery, lettuce, potato	*Longidorus* spp.	Various strains of tomato ringspot
Potato	*Trichodorus christiei*	Tobacco rattle
Tobacco	*Trichodorus* spp.	Tobacco rattle
Pea, lucerne	*Trichodorus* spp.	Early browning
Rhubarb	*Xiphinema diversicaudatum, X. coxi*	Arabis mosaic
Anemone	*X. americanum*	Tobacco ringspot
Gladiolus	*X. americanum*	Tobacco ringspot
	Trichodorus similis	Tobacco rattle

[a] Data from Taylor and Cadman (1969)

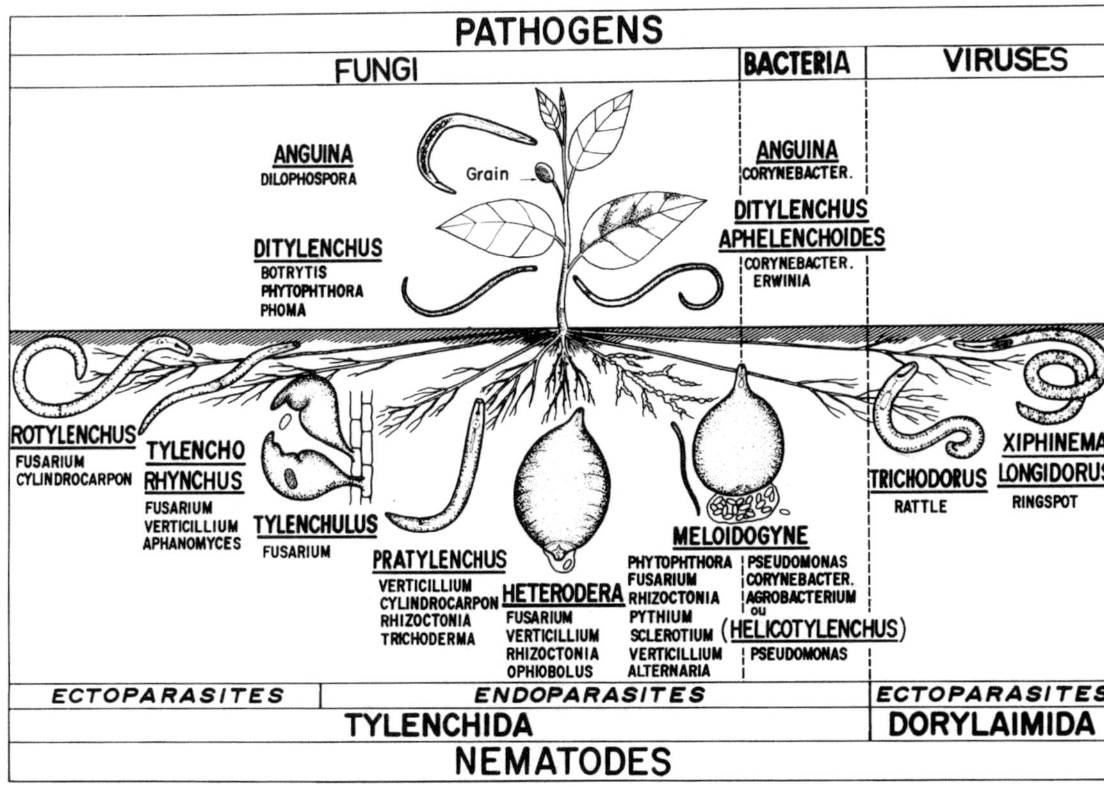

Fig. 1.8. The principial associations between genera of nematodes and fungal, bacterial and viral phytopathogens. (After Dalmasso et al. in Louvet 1972)

Adults and larvae of the nematodes listed in the table transmit viruses with equal efficiency. The viruses do not multiply in the nematodes, but are retained by them for weeks or months, much longer than they survive in vitro (Thresh 1974). This enables the virus to persist until roots of a later crop are available.

The best-known cases of other types of association between nematodes and soil fungi are the relations between species of *Meloidogyne* and *Fusarium oxysporum*. In a number of important crops, in which some varieties are resistant to this fungus, the resistance is broken in soils infested with *Meloidogyne*, probably because penetration of the rootlets by the nematode facilitates entry of the *Fusarium* (Louvet 1972). These and other associations between nematodes and soil fungi are set out in Table 1.7. Many of the other associations between nematodes and fungi or bacteria indicated in Fig. 1.8 are, as far as we know, only of minor importance.

Fungi Transmitting Viruses. A number of soil fungi of the lowly groups of Chytridiales and Plasmodiophorales transmit virus diseases (Table 1.8). None of these fungi possesses a true mycelium. *Spongospora subterranea* is an important pathogen of potatoes in its own right. All the fungi listed transmit viruses preferably on heavy soil with high water content, but the potato mop top and wheat spindle streak diseases are also not infrequent on sandy, well-drained soils (Lecoq 1979). None of these diseases is transmitted by seed; all have narrow host ranges, except tobacco necrosis, and are difficult to transmit mechanically. Only the potato X virus has a distribution wider than that of its vector fungus.

Table 1.7. Some associations between nematode infestations and enhanced development of fungal or bacterial diseases

Crop	Disease	Nematode	Fungus or bacterium	Reference
Cotton	Wilt	*Meloidogyne* spp.	*Fusarium oxysporum* f. sp. *vasinfectum*	Louvet (1972)
Tomato	Wilt	*M. javanica*	*F. oxysporum* f. sp. *lycopersici*	Louvet (1972)
Tobacco	Black shank	*M. incognita*	*Phytophthora parasitica* var. *nicotianae*	Louvet (1972)
Potato	Wilt	*Pratylenchus thornei*	*Verticillium dahliae*	Krikun and Orion (1979)
Tomato	Wilt	*Trichodorus christiei*	*V. albo-atrum*	Conroy and Green (1974)
Tomato, okra	Root rot	*Meloidogyne incognita*	*Rhizoctonia solani*	Golden and Van Gundy (1975)
Tobacco	Root rot	*M. incognita*	*Pythium ultimum*	Melindez and Powell (1970)
Soybean	Root rot	*M. incognita* *Heterodera glycines*	*Pythium* spp. *Rhizoctonia* spp.	Schenck and Kinloch (1974)
Citrus	Root rot	*Tylenchulus semipenetrans*	*Fusarium solani*	Van Gundy and Tsao (1963)
Plum	Bacterial canker	*Criconemoides xenoplax*	*Pseudomonas syringae*	Mojitahedi et al. (1975)

Table 1.8. Virus diseases transmitted by soil-borne fungi

Vector	Virus disease	Distribution	References
Chytridiales			
Olpidium brassicae	Lettuce Big Vein	Australia, Asia, Europe, North America	Teakle (1969)
	Tobacco Necrosis (also affects beans, potatoes and tulips)	Japan, Australia, Europe, North America	Teakle (1969)
	Tobacco Stunt	Japan	Teakle (1969)
Plasmodiophorales			
Polymyxa graminis	Wheat Spindle Streak Mosaic	North America	Slykhuis (1976)
	Wheat Soil-borne Mosaic	Italy, Japan, North America	Slykhuis (1976)
	Barley Yellow Dwarf Mosaic	Japan	Slykhuis (1976)
	Oat Mosaic	USA, England	Slykhuis (1976)
Polymyxa betae	Beet Necrotic Yellow Vein (causing rhizomania)	Europe	Vuittenez (1980)
Spongospora subterranea	Potato Mop Top	Europe, South America	Teakle (1969)

1.4.2.3 Plant Symbionts and Plant Pathogens

The plant symbionts of practical importance in cropping are mycorrhizal fungi and nitrogen-fixing bacteria (*Rhizobium* spp.).

Mycorrhizae. Both mycorrhizal symbionts and root pathogens such as species of *Phytophthora*, *Pythium*, *Rhizoctonia*, and *Fusarium*, are intimately concerned with succulent, fine feeder roots. Where an ectomycorrhizal mantle develops on such roots before they are infected by pathogens, these latter are rarely able to penetrate the roots. Moreover, the mycorrhiza changes the microbial rhizosphere population, apparently to the disadvantage of the above pathogens. Most studies on ectomycorrhiza/pathogen relationships have been carried out on forest trees, and are beyond the scope of this book. Among fruit tree symbionts, the *Scleroderma bovista* ectomycorrhiza on pecan *(Carya illinoensis)* has been found to inhibit five *Phytophthora* and four *Bythium* species associated with feeder root necrosis (Marx 1973).

The relation between endomycorrhiza and plant disease has been reviewed by Schoenbeck (1979). These fungi form haustorium-like structures, called arbuscules, in host cells and hyphal swellings (vesicles) in or between host cells, hence the term vesicular-arbuscular (VA) mycorrhiza. Schoenbeck has listed pathogens, fungi as well as nematodes, infections or populations of which are reduced by endomycorrhizae (Table 1.9). On the other hand, he states that resistance of crops to some diseases is somewhat weakened by VA mycorrhiza, e.g., resistance of tomatoes to tomato mosaic virus and potato virus X, that of beans to rust *(Uromyces phaseoli)* and anthracnose *(Colletotrichum lindemuthianum)* and of cucumber to powdery mildew *(Erysiphe cichoracearum)*.

The general effect of mycorrhiza on crop disease is undoubtedly beneficial. The direct protectant action of ectomycorrhiza is probably less important than the gen-

Table 1.9. Effects of endomycorrhizae on soil-borne pathogens of some crops. (Adapted from Schoenbeck 1979)

Crop	Pathogen	Effect of mycorrhiza
Soybean	*Pythium ultimum*	None
	Phytophthora megasperma	Fewer plants killed
	Meloidogyne incognita	Fewer galls, increased yield
Cotton	*Thielaviopsis basicola*	Less stunting
	Meloidogyne incognita	Reduced stunting
	Pratylenchus brachyurus	Fewer nematodes
Tobacco	*Olpidium brassicae*	Reduction of infection
	Thielaviopsis basicola	Less stunting, chlamydospore production inhibited
	Meloidogyne incognita	Fewer nematodes
	Heterodera solanacearum	Fewer nematodes, lower dry weight
Tomato	*Fusarium oxysporum*	Less stunting, reduction of infection
	Meloidogyne incognita	Fewer nematodes
Cucumber	*Fusarium oxysporum*	Less stunting, reduction of infection
	Meloidogyne incognita	Fewer nematodes, increased fresh weight
Strawberry	*Cylindrocarpon destructans*	Less stunting, reduced infection
Citrus	*Phytophthora parasitica*	Reduction of damage

eral effect of most mycorrhizae on crop growth, and thus indirectly on proneness to disease: mycorrhizal symbionts may avoid or alleviate stress by relieving nutrient deficiency, increasing "rooting density" by supplementing roots with an extensive and rapidly formed absorptive apparatus (which is, moreover not susceptible to root pathogens), by compensating for root loss by disease, by lengthening root life, and by altering host metabolism to raise resistance to root pathogens (Bowen 1978; Schoenbeck 1979). The contributions mycorrhizae can make to crop health have not so far been sufficiently studied; artificial infection of seeds may present a practical way of introducing them into crops, much as nitrogen-fixing bacteria are applied.

Nitrogen-Fixing Bacteria. A limited number of studies has been carried out in recent years, mostly in the USA, to determine effects of species of *Rhizobium* producing root nodules on legumes, on soil-borne pathogens of these crops. In view of the increasing use of rhizobia specially adapted to specific crops, and their success in raising yields, this research deserves attention.

On soybeans, Mew and Howard (1969) found effects of *Rhizobium japonicum* on development of root diseases caused by *Fusarium oxysporum* to depend on the soil's pH level. At pH 7.0 and 7.6, but not at pH 5.2, ample rhizobial nodulation greatly reduced, or even eliminated, the root rot. Tu (1978) showed that simultaneous application of *R. japonicum* and *Phytophthora megasperma* to soybean plants growing in pots protected the soybeans from the root rot. On lucerne, Tu (1980) has likewise shown that rhizobia prevented development of root rots due to *F. oxysporum* and *P. megasperma;* by thus keeping the roots healthy, the rhizobia protected the lucerne from winterkill. It must, however, be pointed out that, once legume roots become infected by root rots, the latter are no longer affected by rhizobia and interfere with rhizobial activity.

Rhizobia interact with nematodes in various ways. On soybeans, *Meloidogyne hapla* and *Pratylenchus penetrans* increase the number of bacterial nodules, but reduce their size, while *Heterodera glycines* decreases nodule numbers and weights and thus reduces N-fixation. On the other hand, on white clover, *Meloidogyne javanica* and *H. trifolii* reduce nodule number and longevity, but this does not affect the rate of N-fixation (Bowen 1978).

An interesting relationship between *R. japonicum* and nematodes of the genus *Cephalobus* has been determined by Cayrol (pers. comm.) in France. The non-stylet, saprophytic nematode lives exclusively on *Rhizobium*, and disgorges it during the passage through the soil. This serves to distribute the bacteria more uniformly and to deeper layers than they usually reach.

1.4.3 Suppressive or Resistant Soils

Soil suppressiveness, or the resistance offered by certain soils to the survival, saprophytic growth and pathogenic activity of phytopathogenic fungi, is a phenomenon which recent research has done much to clarify, but which is still only partly understood. It is associated with soils of certain types, almost invariably finely textured soils, but how far such soils exert their effect through soil microbiota antagonistic to, or competitive with, pathogens is not always clear.

A list of pathogens reported to have been suppressed in certain soils, appears in Table 1.10.

The greatest body of information available on soil suppressiveness relates to the cause of so many wilts, *Fusarium oxysporum*. At least eleven formae speciales of this fungus have been found impeded in some soils, and soils impeding one forma specialis generally impede others as well.

Because the phenomenon is complex and seems worthy of more attention than it has received in the past, we shall here describe in some detail the results of relevant research carried out in France by Louvet and his associates (Louvet et al. 1975; Rouxel et al. 1979; Alabouvette et al. 1979). In this work, an alluvial soil with montmorillonite clay, deposited in parts of the Rhone Valley, has been found highly resistant to *Fusarium oxysporum*, and elucidation of the nature of this resistance has proceeded in various directions.

That microbial action is at the base of such resistance has been established by the fact that

a) biocidal treatments (disinfection by steam or methylbromide) destroyed resistance; and
b) the resistance principle could be transferred to soils of similar composition even if they had previously been conducive to *Fusarium* wilts, by adding to the conductive soil about 10% of resistant soil, and repeating this dilution at intervals again and again, until the proportion of originally resistant to conducive soil was as low as 1:10,000.

The nature of the microbial action involved was found to base on the prevalence of non-pathogenic strains of *F. oxysporum* and *F. solani*. Analysis of the principal components of the fungal flora of the soils, inoculation of previously steamed

Table 1.10. Some pathogenic fungi prevented by soil suppressiveness from attacking crops

Crop	Pathogen	References
Vegetables		
Cucumber	*Fusarium oxysporum* f. sp. *cucumerinum*	Alabouvette et al. (1979)
Melon	*F. oxysporum* f. sp. *melonis*	Alabouvette et al. (1979); Smith and Snyder (1972)
Cucurbits	*Phomopsis sclerotioides*	Moody and Guidrat (1977)
Tomato	*F. oxysporum* f. sp. *lycopersici*	Alabouvette et al. (1979)
Potato	*Streptomyces scabies*	Menzies (1959)
Pea	*F. oxysporum* f. *pisi*	Baker and Cook (1974)
Bean	*F. oxysporum* f. sp. *phaseoli*	Burke (1965b); Furuya et al. (1979)
Radish	*F. oxysporum* f. sp. *raphani*	Alabouvette et al. (1979)
Sweet potato	*F. oxysporum* f. sp. *batatas*	Smith and Snyder (1972)
Lettuce	*Olpidium brassicae*	Westerlund et al. (1978)
Various vegetables	*Pythium ultimum*	Hancock (1979)
	Pythium spp.	Bouhot and Perrin (1980)
	Pseudomonas solanacearum	Béreau and Messiaen (1975)
Field Crops		
Wheat	*Fusarium culmorum*	Baker and Cook (1974)
	Rhizoctonia solani	Baker and Cook (1974)
	Gaeumannomyces graminis	Shipton et al. (1973)
Lentils	*Fusarium avenaceum*	Lin and Cook (1979)
Flax	*F. oxysporum* f. sp. *lini*	Tu et al. (1975)
Peppermint	*Verticillium albo-atrum*	Nelson (1950)
Fruit and Flower Crops		
Avocado	*Phytophthora cinnamomi*	Broadbent and Baker (1975)
Banana	*F. oxysporum* f. sp. *cubense*	Reinking and Manns (1933)
Carnation	*F. oxysporum* f. sp. *dianthi*	McCain et al. (1980)
Cyclamen	*F. oxysporum* f. sp. *cyclaminis*	Grouet (pers. comm.)

soil with each major component of the flora, and subsequent planting of wilt-susceptible melons, gave the results shown in Fig. 1.9a. Large percentages of plants remained healthy only in soils inoculated with *F. oxysporum* or with a mixture in which this species was included, and *F. solani* also had some effect. The role *F. solani* plays here has been further clarified in similar inoculation tests, the results of which appear in Fig. 1.9b. They show that either of the two species could effect resistance, but best results were obtained when the two were applied together.

Is this development of disease resistance in the soil linked with the presence of plant roots, and why is it limited to specific soils? It appears that originally suppressive soil never loses its suppressiveness even in the absence of plants. However, conducive soil made suppressive by addition of suppressive soil has been shown to lose its resistance within about a year, unless plants are grown in it. The physical characteristics that make certain clays, especially montmorillonite clays, into media with a potential for development of resistance, are not yet clear.

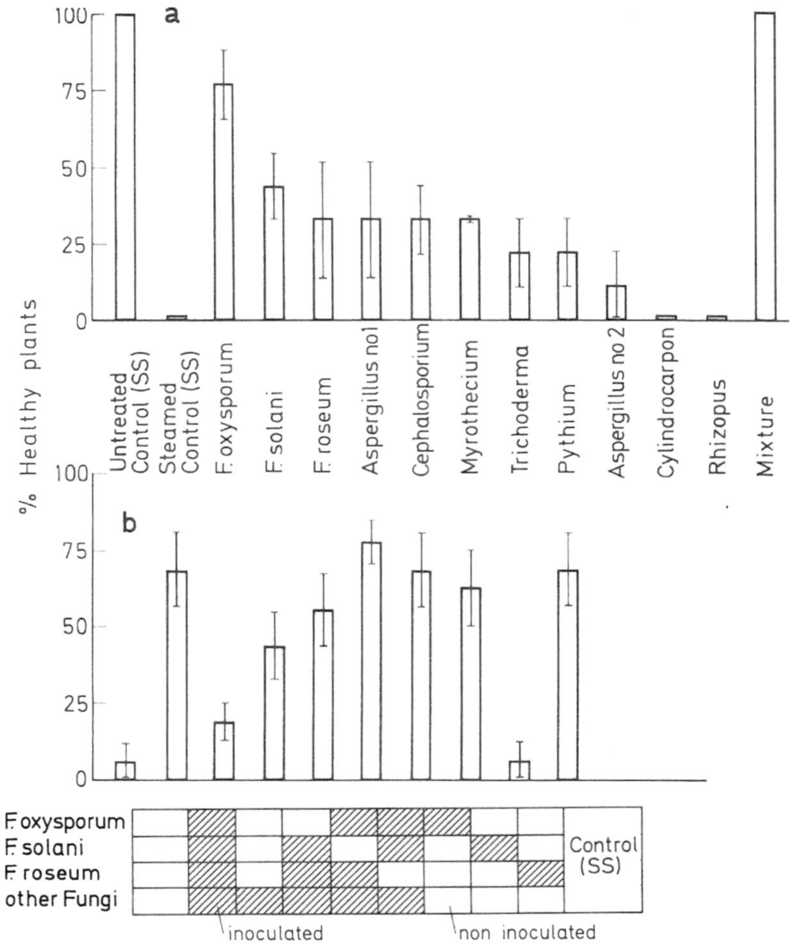

Fig. 1.9. a Role of the principal fungi occurring naturally in suppressive soil, in suppressing infection by *Fusarium oxysporum* f. sp. *melonis* on susceptible muskmelons grown in steamed soil (Alabouvette et al. 1979). **b** Role of three species of *Fusarium* occurring naturally in suppressive soil, in suppressing infection by *F. oxysporum* f. sp. *melonis* on muskmelons grown as above, when the species of *Fusarium* were applied singly or in mixture. (*ss* = suppressive soil) (Alabouvette et al. 1979)

The work described above represents a rare attempt to use the soil itself as a means of control of root pathogens. It shows that the receptivity or resistance of soil to pathogens apt to attack the crops to be grown should be studied before the crops and their sequences are chosen, and certainly before disinfectants are applied.

Agricultural soils should be carefully surveyed to determine whether they are suppressive and if they are, this characteristic should be husbanded as an important asset (Toussoun 1975). Cultural factors relating to soil suppressiveness need to be studied in detail. The practical conclusions that can be drawn at this stage are applicable largely to greenhouse crops: given the right type of soil, but with suppressiveness eliminated by disinfection, the grower of a susceptible crop, e.g., tomatoes attacked by *F. oxysporum* f. sp. *lycopersici*, can expect this soil to re-develop a high degree of resistance to *Fusarium* wilt diseases of other crops; he should then be able to grow other crops susceptible to this type of wilt, e.g., melons or carnations, free from such wilt.

Suppression of another *Fusarium* disease has recently been reported by Lin and Cook (1979) from the United States: they found that root rot of lentils caused by *F. avenaceum* is suppressed by the natural flora of certain soils. Here again, suppressiveness could be eliminated by soil disinfestation and could be restored by adding 1% of non-treated, suppressive soil to the previously disinfested soil. However, in this soil – as distinct from the French soils – suppression was less apparent as inoculum increased.

Bouhot and Perrin (1980), studying the behaviour of *Pythium* spp. in 15 different soils in France, have been able to distinguish between three groups: (a) conducive soils, in which inoculum potential increased with time; (b) tolerant soils, in which inoculum remains at a low level, and (c) suppressive soils, in which inoculum level decreases in time by one of two mechanisms: either *Pythium* did not develop at all, or the *Pythium* that developed was non-pathogenic. Bouhot (1979) has worked out techniques for estimating not only the inoculum density in the soils but pathogenic potential of fungal populations in the soil. Introduction of such techniques into routine assessment of soil properties before valuable crops are sown could help greatly in avoiding loss from fungi with so wide a host range.

The causes for the reported suppression of species of *Verticillium* and *Phytophthora* have not been studied so intensively. In some cases, there is so far no evidence for the involvement of suppressing organisms. Suppression of *Pseudomonas solanacearum* in the French West Indies on clay soils of the montmorillonite type has been ascribed solely to the physical properties of these soils (Béreau and Messiaen 1975).

The special case of suppression, up to a point, of the take-all pathogen of cereals (*Gaeumannomyces graminis*) after some years of monoculture, will be discussed in the chapter on crop sequence (Chap. 2.3.6). One cardinal difference between the suppression of this, as compared with *Fusarium* disease, is that resistance to *G. graminis* can be induced by monoculture in diverse types of soil, and is not limited to soils with specific physical properties.

1.5 Stress, Strain and Predisposition

Stress is a factor tending to limit the growth of organisms, or of certain phases of growth, by imposing a strain on the developmental processes of the organism. The extent to which strain is generated depends on the strength of the stress factor and the length of time it operates, as well as on the capability of the organism to resist this factor, i.e., strain is the result of interplay between two forces. Strains often render organisms more highly susceptible to the attack of pathogens, or predispose them to such attack, but pathogens themselves may also experience strain.

Viewing stress and strain from another angle, it can be said that unless the stress exerted by a parasite results in a strain on the crop, the action is parasitic, but not pathogenic. This means that if and as long as a plant can tolerate the action of the parasite without ill-effect, there is little reason to interfere in their relationship. Mussell and Malone (1979) have discussed stress and disease tolerance, as exemplified by various hybrids and varieties of maize, oats and sugarcane in relation to their respective *Helminthosporium* blights. They make a strong case for breeding tolerant, not immune varieties of crops as this would not push the pathogen into a corner from which it can (and will) escape only by generation of new, aggressive races.

This, however, is somewhat outside the scope of this book. We shall here consider stress, and the strain it causes, as acting in three ways:

a) As cause of conditions directly unfavourable to crop growth and, in general predisposing it to the attack of certain pathogens; this is the most important effect of strain and the one chiefly dealt with in this chapter.
b) As producing conditions unfavourable to full growth of pathogens, e.g., limiting colonization of host tissues and/or sporulation, and often promoting formation of resting structures. This applies largely to nutrient deficiencies and the strain of crop senescence.
c) As the result of initial pathogen attack, which aggravates the environmental stress on the plant (Duniway 1973). This is likely to happen in particular as the plant begins to age and the strain of senescence comes into play; disease development may then snowball until the pathogen, if obligatory, runs out of tissue to colonize and consequently faces nutrient stress of its own making.

The subject of plant response to environmental stresses has been treated comprehensively by Levitt (1972). Schoeneweiss (1975, 1978) has discussed environmental, as well as nutritional and defoliation stress, in relation to predisposition to plant diseases. He has pointed out that the most pronounced effect of stress, from the point of view of plant disease, is that it predisposes plants to attack by facultative, especially weak and non-aggressive, parasites and thus widens the spectrum of potential aggressors. Such parasites frequently occur on crops in latent form, but cause disease only when the host suffers from stress-induced strain.

The most obvious effects of environmental stress on crops are those causing direct damage, such as injury by frost, sunscald, high winds, flooding, etc. The ideal way to obviate such damage is by breeding varieties resistant to excessively high or low levels of temperature, water supply and radiation. Cultural practices may also contribute greatly to avoidance or mitigation of such damage, for instance by the choice of favourable topography and sowing dates. However, both the physiological disorders caused by environmental and nutrient stresses, and the breeding of resistant varieties, are outside the scope of this book, which is limited to the discussion of pathogens and the diseases they cause.

1.5.1 Types of Stress-Induced Strain

Levitt (1972) has distinguished between elastic and plastic strain induced by stress factors, the criterion being the reversibility of their effects.

a) *Elastic Strain* – a physical or chemical change reversible by removal of the stress. Such strain does not cause immediate permanent injury, and the damage it causes is proportional to its intensity and duration. If intense enough for a sufficiently long period, elastic strain may turn into plastic strain (see below). In relation to disease, elastic strain is a relative concept; it will facilitate disease development if the host suffers more than the parasite or if parasite and strain effects combine to interfere with host growth, e.g., by aggravating moisture deficiency.

The task of cultural practices in relation to elastic strain on crops is to alleviate its intensity and shorten its duration by a wide range of operations, usually applied during crop growth, especially for management of moisture and nutrition.

b) *Plastic Strain* – an irreversible chemical or physical change, often resulting in the death of plant organs or whole plants. Plastic strain affects the potential of damage from pathogens chiefly in two ways: (1) if most or all tissues are affected and the plant as a whole cannot recuperate, by furnishing points of entry for invasion of moribund or dead tissue by facultative parasites, which may then persist on plant debris or in the soil; (2) if only individual organs (such as flowers) or only a certain proportion of various organs are affected, and the plant as a whole can recuperate, by producing new, often unseasonal and highly susceptible growth to replace the organs injured or killed.

Cultural practices can mainly do two things to lessen damage from plastic strain: Firstly, reduce the likelihood of stresses coming to act on the crop, most commonly by pre-planting measures such as choice of topography and soil types, or provision of windbreaks, but also by in-growth mitigation of strains, e.g., sprinkling to reduce danger of frost. The second approach to lessening losses from plastic strain is to promote desirable re-growth to make up for tissues lost, e.g., irrigation or nitrogen fertilization, where these are appropriate measures to stimulate growth.

1.5.2 Temperature Strain

The stresses of both high and low temperature impose strains chiefly where the vigour of growth of host and pathogen diverge greatly at given temperatures. Leach (1947) has shown that the two crops, spinach, and watermelon, may be predisposed to two fungi, *Pythium ultimum* and *Rhizoctonia solani*, under contrasting conditions of temperature. Spinach, a low-temperature crop, escaped pre-emergence damping-off at 4 °C, when it grew relatively better than the *Pythium* and much better than the *Rhizoctonia*; at 12°–20 °C *Pythium* grew better than spinach but soil temperatures had to rise to above 20 °C before *Rhizoctonia* could affect spinach severely. On the other hand, the high temperature watermelon crop grew best and showed little damage from either fungus at soil temperatures above 20 °C, and was affected severely by both pathogens only at lower temperatures. As pointed out by Baker and Cook (1974), infections may be most severe, not necessarily at temperatures optimal to the pathogens but at temperatures relatively less unfavourable for the pathogen than for the host. This is clearly shown in Fig. 1.10 a, b.

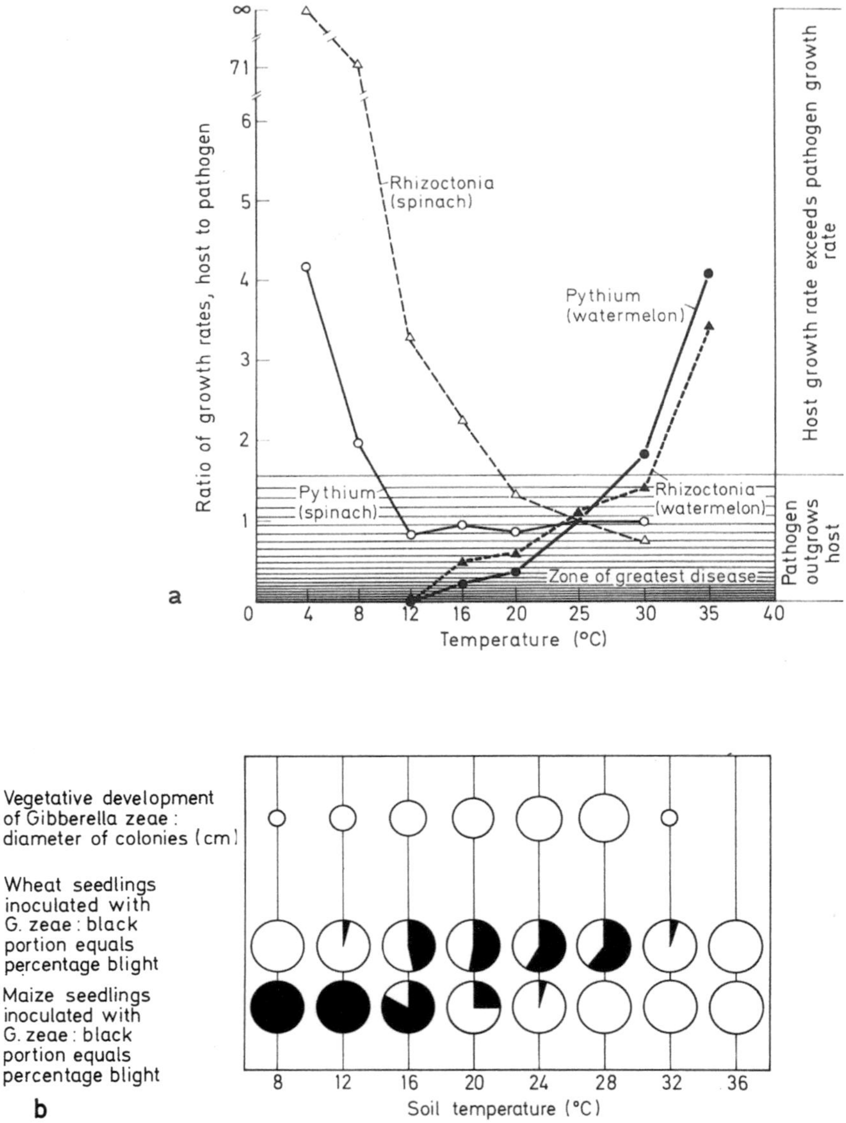

Fig. 1.10. a Ratios of emergence rates of spinach and watermelon seedlings to growth rates of *Pythium ultimum* and *Rhizoctonia solani* in relation to temperature (After Leach 1974 as cited by Baker and Cook 1974). **b** A graphic summary of the development of *Gibberella zeae (G. saubinettis)* on agar at various temperatures and of seedling blights of maize and wheat growing in infested soil at a range of constant soil temperatures. The *black portion of the circle* in the *two lowest rows* shows the average percentage of diseased seedlings. Note that in the case of wheat the optimum for disease was close to that for fungus growth, while in the case of maize (corn) the optimum was some 16° to 20 °C lower. (After Dickson 1923)

1.5.2.1 High-Temperature Strain

Most strains induced by high temperatures are of the *elastic* type. A good example of such strain is the predisposition of potato tubers to attack by *Macrophomina phaseolina*, as described in Chap. 1.5.6.

Other examples of temperature-induced susceptibility of crops towards soil-borne fungi are included in Table 1.11. As shown in laboratory experiments, however, crops may also be predisposed by heat to obligate foliar pathogens, e.g., *Erysiphe cichoracearum* affecting *Phaseolus vulgaris* (Yarwood 1956), and to viruses such as TMV and other viruses on bean (Yarwood 1956; Wu and Wildman 1963); but high temperature strain is also known to mask certain viruses.

A remarkable effect of heat strain is that it may render plants liable to attack not only by weak pathogens, but even by agents not normally capable of causing them disease. Thus *Phytophthora cactorum* and *Drechslera sativa* have been shown to affect a non-host, soybeans, if the latter was predisposed by heat (Chamberlain and Gerdemann 1966).

Table 1.11. Predisposition of crops to soil-borne diseases owing to temperature strain, where levels of temperature give pathogen growth a relative advantage over crop growth

Crop	Disease	Pathogen	Soil temp. (°C) at which crop is predisposed	References
Wheat	Root rot	*Fusarium culmorum*	24–28	Ross (1965)
	Snow mould	*Gerlachia nivalis*, *Typhula idahoensis*	0 or below	Bruehl et al. (1966)
	Crown and foot rot	*Fusarium graminearum*	18–28	Dickson (1923)
Barley	Stripe	*Drechslera graminea*	10	Ross (1965)
Maize	Crown and foot rot	*Fusarium graminearum*	8–16	Dickson (1923)
	Seedling stunt	*Pythium* sp.	11	Hoppe (1949)
Groundnuts	Seedling blight, Heat canker	*Diplodina gossypina*		Nyvall (1979)
Spinach	Damping off	*Pythium ultimum*	12–20	Leach (1947)
	Damping off	*Rhizoctonia solani*	above 20	Leach (1947)
Watermelon	Damping off	*Pythium ultimum*	12–20	Leach (1947)
	Damping off	*Rhizoctonia solani*	below 20	Leach (1947)
Potato	Charcoal rot	*Macrophomina phaseolina*	30–35	Littauer (1944)
	Wart	*Synchytrium endobioticum*	13–15	Ross (1965)
Tobacco	Root rot	*Thielaviopsis basicola*	17–23	Walker (1969)
Soybean	Root rot	*Phytophthora megasperma* var. *sojae*	heated to 44 °C for 1 hour	Chamberlain and Gerdemann (1966)
Avocado	Root rot	*Phytophthora cinnamomi*	15–27	Zentmyer and Bald (1977)
Apple	Root rot	*Sclerotium rolfsii*	30–35	Avizohar-Hershenzon and Palti (1962)
Peach	Root rot	*Armillariella mellea*	15–25	Bliss (1946)
Rose	Root rot	*Armillariella mellea*	10–28	Bliss (1946)

To avoid or minimize elastic strains, due to heat, predisposing crops to soil-borne pathogens, two cultural practices are widely adopted, irrigation and mulching. Sprinkling the crop until a few days before harvest gives good control of *Macrophomina phaseolina* on potatoes (Littauer 1944). Mulching has achieved remarkable success in protecting apple roots from the attack of *Sclerotium rolfsii* in Israel (Lavee, quoted by Avizohar-Hershenzon and Palti 1962). Although the fungus was found to be superficially present on roots in mulched as well as unmulched plots, the reduction in soil temperature effected by the mulch prevented penetration of the roots.

A positive effect of high temperatures on relationships between cucumber and the root-knot nematode *Meloidogyne hapla* has been reported by Cardin (1979). At soil temperatures of 24 C and above, cucumber resistance was expressed in an increase in the proportion of male nematodes. This was so pronounced that the author suggests growing cucumbers at high soil temperatures in glasshouse rotation with crops highly susceptible to *M. hapla*, such as parsley.

Plastic strain due to high temperature generally assumes the form of sun-scald of the bark of trees and scorching of leaves, flowers, and fruits, and cannot be separated from radiation effects. In addition to the direct loss of tissue, such scalding may have even more severe indirect effects by providing points of invasion for tree canker pathogens (e.g., *Schizophyllum*). The scorched parts of the plant may provide food bases for subsequent invasion of leaf and fruit pathogens, such as *Alternaria* spp. on tomato, or agents of collar rot, such as *Sclerotium rolfsii*.

The most common practices to prevent plastic strain due to high temperature and radiation are planting (a) more closely, (b) in rows with minimum exposure to sunlight (e.g., rows running north–south rather than east–west) and (c) on slopes shaded for part of the day.

1.5.2.2 Low-Temperature Strain

As mentioned above, temperature strain is often largely the result of host growth and reactions being less adapted to given temperature levels than pathogen growth. This very much applies to low-temperature effects on crops. Thus, at relatively low temperatures, *Verticillium* in the Ivory Coast will attack cotton varieties somewhat resistant at higher temperatures (Follin, pers. comm.). In France, cool temperatures have been found to predispose peaches to *Pseudomonas mors-prunorum* f. sp. *persicae* (Vigouroux 1979) and maize has been predisposed to *Fusarium culmorum* by 4–5 nights with minima below 7 °C.

As regards effects of temperature near or below the freezing point, Levitt (1972) distinguishes between low-temperature strains causing *chilling* effects, which occur above the freezing point and are often reversible (elastic strains), and those causing *freezing* effects, commonly irreversible (plastic strains). Freezing strain mostly results from a rapid fall in temperature and the extent of injury caused, at least by temperatures only just below the freezing point, is often related to the rate of thawing rather than the lowest temperature obtained. Table 1.12 lists a number of pathogens that take advantage of frost injury to attack crops.

Chilling and freezing strain of low temperatures to subtropical crops, such as avocado and citrus, is largely preventable by the choice of frost-free locations,

Table 1.12. Fungal and bacterial pathogens that take advantage of frost injury to attack crops

Crop	Disease	Pathogen	References
Fruit Crops			
Peach	Nodal canker	*Leucostoma cincta, L. persooni (Cytospora leucostoma)*	Dhavantari (1978) Grosclaude and Dupouy (1963)
	Blight	*Pseudomonas mors-prunorum* f. sp. *persicae*	Vigouroux (1979)
Pear	Blossom blight	*Pseudomonas syringae*	Panagopoulos and Crosse (1964)
Apple, pear	Canker	*Nectria* spp.	Gaeumann (1951)
Grapefruit	Storage rot	*Alternaria citri*	Schiffmann-Nadel et al. (1975)
Citrus	Twig blight, die-back	*Botryodiplodia theobromae*	
Field Crops			
Pea	Blight, foot rot	*Mycosphaerella pinodes*	Allard (pers. comm.)
Wheat, triticale	Black chaff	*Xanthomonas translucens*	Waller (pers. comm.)

especially slopes. Sprinkling such plantations as temperatures approach the freezing point may make a difference of 2°–4 °C and thus save crops, if continued until well after thawing has set in. This is apt to reduce the incidence of diseases gaining entrance through moribund tissue, e.g., *Botryodiplodia theobromae* on citrus. However, sprinkling may actually be harmful if temperatures fall to levels where water sprinkled onto the tree freezes, as the load of ice may then cause limbs to break off.

Sprinkling or misting is likewise advocated in temperate climates to delay bud break of deciduous fruit and thus lessen danger of damage from late spring frost (Mittelstaedt 1979). Where such frosts occur, they greatly facilitate entry of many of the pathogens mentioned in Table 1.12.

As mentioned above, cultural practices designed to stimulate regrowth can contribute a great deal to recouping losses from occasional frost, where growing conditions not long after the frost enable the crop to recover. Thus, potatoes and root crops may make extensive foliage re-growth in frost-damaged fields. Such unseasonal, tender growth is, however, often exceptionally susceptible to diseases such as *Phytophthora infestans* and downy mildews, and this must be taken into account by farmers stimulating re-foliation by application of nitrogen or water.

Some pathogens are known to aggravate temperature strain: *Fusarium avenaceum* depresses winter-hardiness of barley varieties (Smith and Olien 1978).

1.5.3 Water Strain

The strain caused by deficient or excessive water supply is certainly among the most common and most important factors that may predispose crops to disease. This is true for almost all agroclimates. In particular, strains deriving from lack of moisture are, of course, prominent in semi-arid countries on unirrigated crops. As attempts widen to grow irrigated crops in these countries, another type of water strain – more subtle but of real significance – will come to the fore: where irrigation is applied at long intervals because the available water is made to "stretch" over wide areas, or where water is expensive, or simply where farmers are inexperienced in water use, periodical strain will be suffered by many crops. This subject will be dealt with at greater length in the chapter on irrigation (Chap. 2.8).

Water stress as a predisposing factor in plant disease has been discussed comprehensively in a recent contribution by Schoeneweiss (1978) to a general review of water deficits and plant growth. He has pointed out that it is often difficult to separate moisture effects from effects of temperature and nutrition, as both the latter affect the rate of water absorption and loss.

1.5.3.1 Water Deficit Strain

The effects of prolonged drought, extending over whole seasons, on unirrigated crops will not be discussed here. It largely affects forests, some other tree crops and pasture, and effects of such drought are rarely amenable to mitigation by cultural practices.

We shall deal with strains imposed on pathogen and crop by less extreme seasonal variation in rainfall and by short-term water deficits.

Predisposition to disease as consequence of water deficit evidently applies to a large number of host/pathogen combinations. It is particularly dangerous where a crop weakened by a relatively moderate deficit is liable to attack by pathogens capable of infection at this or even lower deficits. An extreme case is that of *Fusarium culmorum* which attacks wheat at very low soil moistures and is in fact known to American farmers as causing "dryland foot rot" (Cook 1973). *Macrophomina phaseolina* is another pathogen capable of attacking its many hosts at soil moistures lower than those conducive to good crop development (Shokes et al. 1977), as described in Chap. 1.5.6.

The fact that, in regions with intermittent rain, *Pyricularia oryzae* is often observed to attack rain-fed rice more severely than paddy rice also indicates that the pathogen profits from stress induced in the rice by temporary lack of water.

As seen from Table 1.13, many additional destructive root pathogens can take advantage of the predisposition of their hosts to disease owing to water-deficit strain. Although some of the diseases they cause are most widespread on crops growing in wet soil, the infection potential of the pathogens may be realised to the full under dry conditions, when host resistance is reduced. This is also illustrated in Fig. 1.11.

Obligate parasites are not in general favoured by weakening of their hosts and are, on the contrary, more aggressive to vigorously growing crops (see Chap. 2.6 on crop nutrition and Chap. 2.7 on moisture management). There are, however, exceptions: tobacco is predisposed to blue mould *(Peronospora tabacina)* by low soil moisture (Rotem et al. 1968), and in some cases the combined effect of water deficits and of obligate parasites may advance development of disease. Ayres (1977) found that in water-stressed peas the gradient of osmotic potential between roots and leaves is lower than normal; leaf infection by powdery mildew *(Erysiphe*

Table 1.13. Predisposition of crops to diseases owing to water deficit strain

Crop	Disease	Pathogen	References
Field and Vegetable Crops			
Wheat	Stalk rot	*Fusarium culmorum*	Cook and Papendick (1972)
Maize	Stalk rot	*Fusarium moniliforme* and others	Sumner (1968)
Sorghum	Charcoal rot	*Macrophomina phaseolina*	Edmunds (1964); Solel (1967)
Rice	Brown spot	*Cochliobolus miyabeanus*	Herrera and Seidel (1978)
	Blast	*Pyricularia oryzae*	Ou (1972)
Bluegrass	Dollar spot	*Sclerotinia homoeocarpa*	Couch and Bloom (1960)
(*Poa pratensis*)	Leaf spot and root rot	*Drechslera sativa*	Endo (1972)
Safflower	Root rot	*Phytophthora cryptogaea*	Duniway (1977)
Cotton	Root rot	*Macrophomina phaseolina*	Ghaffar and Erwin (1969)
Urena (*U. lobata*)	Fibre blight	*Physalospora urenae*	Tramier, unpublished
Tobacco	Downy mildew	*Peronospora tabacina*	Rotem et al. (1968)
Tomato	Wilt	*Fusarium oxysporum* f. sp. *lycopersici*	Gaeumann (1951); Foster and Walker (1947)
Fruit and Plantation Crops			
Prune	Canker	*Cytospora leucostoma*	Bertrand et al. (1976)
Filbert	Blight	*Xanthomonas coryli*	Moore et al. (1974)
Walnut	Root rot	*Armillariella mellea*	Grente (in Viennot-Bourgin 1967)
Mango	Die-back	*Botryodiplodia theobromae*	Laville (pers. comm.)
Coffee	*Fusarium* bark disease	*Gibberella stilboides*	Waller (pers. comm.)

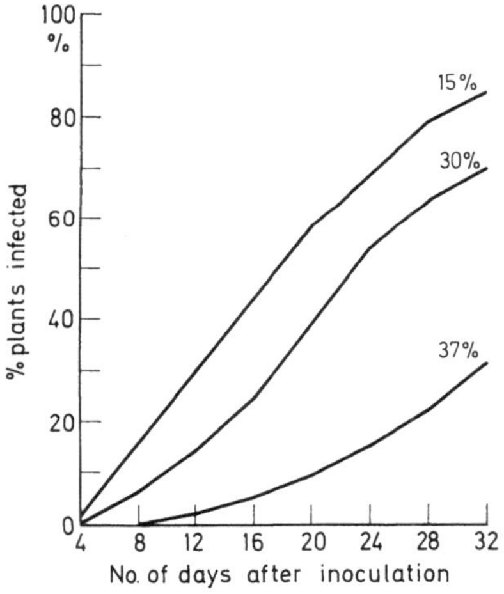

Fig. 1.11. Predisposition of tomato plants to *Fusarium oxysporum* f. sp. *lycopersici* by water deficit strain. Before inoculation plants were kept at 15% (dry), 30% (optimal), or 37% (wet) soil water content, respectively; after inoculation they were kept at 30% soil water content. (Gaeumann 1951)

pisi) reduces the potential still further and thus exacerbates the effects of even moderate water stress. However, if the stress is pronounced, development of the mildew decreases.

The most prominent beneficiaries of water-deficit strain in crops are undoubtedly the facultative parasites. Schoeneweiss (1978) has cited numerous examples of diseases caused by such pathogens and favoured by lack of water, and a list of such diseases is presented in Table 1.13.

The growth stage at which water strain develops is of great importance. In tree crops, special attention has to be paid to avoidance of water deficiency in their early stages of growth, when they may be most susceptible to attack by pathogens that do not readily attack mature trees. In wheat, water strain at the jointing stage increases susceptibility to take-all *(Gaeumannomyces graminis)* more than at the flowering or dough stages (Cook and Papendick 1972).

Current research on relief of water-deficit strain by application of chemicals holds promise also for the reduction of pathogen-related diseases associated with such strain. Plaut and Halevy (1966) found that certain growth retardants improve the generation of shoots and general recovery of wheat exposed to conditions of wilting. This may be of practical interest in protecting crops that repay the outlay, from the direct effect of the disease-induced aggravation of water-deficit strain.

The importance of "ecofallowing" in conserving soil moisture and thus reducing water strain on sorghum crops has been emphasized by Doupnik and Boosalis (1980) in relation to stalk rot *(Fusarium moniliforme)*. Where moisture stress is due to soil compaction, however, subsoiling may lessen pathogen attack. Under such conditions attacks of the nematode *Hoplolaimus columbus* on cotton was reduced by subsoiling (Norton 1978).

A controlling effect of low water stress on infestation of tomatoes by the root-knot nematode *Meloidogyne hapla* has been noted by Couch and Bloom (1960). At a soil moisture level near the permanent wilting percentage, eggs of the nematode hatched but larvae were unable to migrate in the soil. However, infestation by root-knot nematodes accentuates water stress of crops such as melons, tobacco and eggplant (Scotto de Massese, pers. comm.).

Other pathogens that aggravate water deficiency effects on their hosts include *Phytophthora cinnamomi*. According to Sterne et al. (1978), this fungus induces such a major change in the water transport system of avocado trees that typical control mechanisms, such as closure of stomata, could no longer correct the water deficit.

1.5.3.2 Excess Water Strain

The effects of excess water strain on predisposition of crops to disease have been studied much less extensively than the effects of water deficits. Reviewing what is known on this subject, Schoeneweiss (1975) in effect concludes that excess water, due to flooding or lack of drainage, has chiefly indirect effects by weakening the host in general: feeder roots may necrotize, oxygen deficiency is induced and may inhibit water and nutrient uptake, and this in its turn may result in chlorosis, wilting and subsequent predisposition, especially to attack of facultative parasites.

Under conditions of excess water strain even varieties considered resistant may become vulnerable, as has been shown in relation to root rot of bean caused by *Fusarium solani* f. sp. *phaseoli* (Miller and Burke 1975).

Another effect of excess water available to the crop is the development of water-soaked tissue and of leaves with thin cuticles. Such organs are prone to attack by many bacteria, and by fungi developing preferably on soft tissue, such as downy mildews.

1.5.4 Other Strains

Among the strains experienced by crops, great importance attaches to nutrient strains. In their extreme form, these cause deficiency diseases, which are not within the scope of this book. A certain lack of nutrients, however, and perhaps more often imbalance between nutrients, can markedly influence disease development, often through effects on host susceptibility. Avoidance of nutrient strain is one of the major objectives of cultural practices. This subject will therefore be treated at length in a special chapter devoted to crop nutrition (Chap. 2.6).

A special kind of strain can be imposed on winter crops by snow cover. Bruehl et al. (1966) state that in winter wheat, prevention of photosynthesis by deep snow while respiration is low but continues, leads to eventual exhaustion of carbohydrates that tends to predispose the crop to snow moulds *(Gerlachia nivalis, Typhula idahoensis)*.

The stress which Schoeneweiss (1975) has termed "defoliation stress" may be caused by man or by disease. Crops that are mowed or grazed repeatedly may be subject to this type of exhaustion strain. Siddiqui et al. (1968) have reported that increased frequency in cutting clover *(Trifolium pratense)* markedly heightened its susceptibility to species of *Fusarium*, especially *F. oxysporum*. Diseases that cause premature shedding of leaves, such as plum rust *(Tranzschelia discolor)* or powdery mildew of chilli *(Leveillula taurica)*, can certainly cause the same strain, but we are not aware of studies linking this strain to further development of these or other diseases.

One strain no living organism can escape, which may perhaps also be called an exhaustion stress, is that of senescence. The implications of the ageing of individual organs or of whole plants with regard to disease development are of fundamental concern in inoculum management and in the planning of crops. This aspect will be discussed in the chapter on crop age (Chap. 1.6).

The relation of plant stress to virus diseases is twofold. Greater number of winged aphids are produced on plants under physiological stress, i.e., the spread of virus diseases in plant populations may be promoted; on the other hand, virus concentration is less pronounced in plants developing under stress conditions (Swenson 1968).

It should be noted that cultural practices themselves, if improperly applied, can induce strains, mostly of the elastic type, that affect incidence of disease. Examples are the strains caused by sowing too deeply or too densely, irrigating at excessively long intervals, and unbalanced nutrition.

1.5.5 Combinations of Stresses and Strains

Stresses frequently interact to cause increased strain. High soil moisture, for example, interferes with nutrient uptake. This combined action of stress factors is an important contributing factor to plant disease. A good example is the stalk-rot of maize caused in the Mid-West of the United States by *Gibberella zeae*, *Fusarium moniliforme*, and *Diplodia zeae*. Boosalis et al. (1969) indicate that this rot is generally associated with plant stresses due to the practices of intensive farming: high soil moisture, very high plant populations, imbalanced nutrition, and injuries due to weather, pest or disease.

The most common combined stress acting on crops is undoubtedly that of extreme environmental factors acting on plants that also suffer the strains of ageing (see below).

1.5.6 Charcoal Rot *(Macrophomina phaseolina)*: Prototype of a Disease on Crops Under Stress

The charcoal rot, caused by *Macrophomina phaseolina* on numerous herbaceous and woody crops, will serve as an example for host predisposition by strain caused by individual stress factors and by their combined action. The fungus is a soil- and seed-borne facultative parasite. Its mycelium is very sensitive to soil-borne bacteria and actinomycetes; its sclerotia are sensitive to fungistasis, which contributes to its low competitive saprophytic ability.

Dhingra and Sinclair (1977, 1978) have compiled an exhaustive annotated bibliography of *M. phaseolina* for the years 1905–1975, and have summarized the biology and pathology of the fungus.

Crop Age and Maturation. Association of charcoal rot development with flowering and maturation has been demonstrated for many crops, such as grain sorghum (Edmunds and Voigt 1966), sunflowers, opium poppies, and soybeans. Under environmental conditions highly conducive to charcoal rot, its development has been inhibited on sorghum when pollination was precluded, and on soybeans formation of sclerotia has been prevented by removal of flowers. Groundnut pods are invaded only when mature. There are indications that in many crops, such as jute, clove, lucerne, potato, and cotton, plants become more susceptible as they age (Dhingra and Sinclair 1978).

Temperature. M. phaseolina is a high-temperature fungus, with optimal growth in culture at 30°–37 °C. Sclerotia, but not mycelium, even survive soil temperatures above 55 °C. The fungus therefore possesses a relative advantage at high temperatures over crop plants which develop better at lower temperatures, and that includes most of its major hosts. Even at moisture levels adequate for crop growth, groundnut kernels and maize stalks are much more susceptible at higher than at lower temperatures and this is most pronounced in the case of potato tubers (Littauer 1944). Tubers are affected where soil temperatures rise above 30°–35 °C. This is a strictly localized effect, limited to that part of the tuber actually exposed to such temperatures, i.e., usually those tubers or parts of tubers nearest to the soil surface (see Fig. 1.12).

Humidity. The fact that charcoal rot on all its hosts is aggravated, and on many hosts is conditioned, by low soil humidity is explained by (a) increased susceptibility of hosts suffering from the strain of water deficiency, and (b) reduced activity of microorganisms competing with, or antagonistic to, *M. phaseolina* in dry soil.

Combined Age-Temperature-Humidity Strain. There is no doubt that the most destructive attacks of *M. phaseolina* occur when crops are subjected to a combination

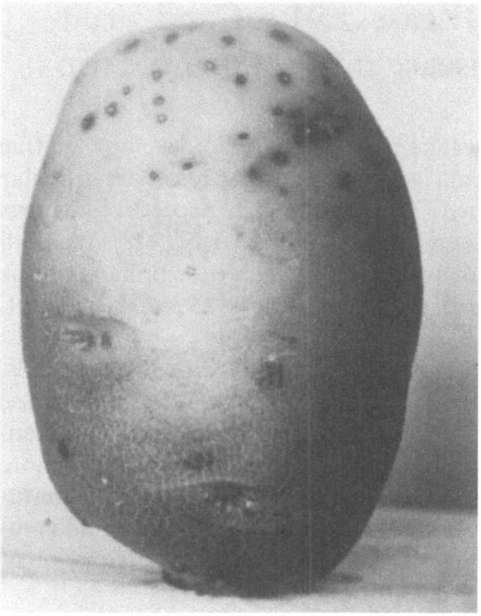

Fig. 1.12. Potato tuber affected by charcoal rot caused by *Macrophomina phaseolina*. The white spots in the centre of black rings are lenticels surrounded by masses of black sclerotia. The tuber is chiefly affected at its top, where it was nearest the soil surface and exposed to highest temperatures. (Littauer 1944)

of strains. Lack of water, especially when coupled with high soil temperatures, hastens maturity of most crops. At the same time plants may suffer from non-availability of nutrients, among which potassium seems to be of particular importance. The all-round debilitating effect of all these factors, coupled with the limited activity of the saprophytic soil flora in hot and dry soil, exemplify the effects of stress conditions on infectious crop disease. It has been suggested that in hot climates, mixing cotton with shading crops may reduce incidence of charcoal rot on cotton (Rangaswami 1975).

1.5.7 Summary of the Effects of Stress and Strain on Diseases Caused by Pathogens

The subject recurring in all the above sections on stress and strain is the extent to which they further the attack of facultative pathogens. Effects on obligate pathogens have been noted more rarely, e.g., water-deficit strain predisposing their hosts to tobacco blue mould or pea powdery mildew (Table 1.13). The great majority of diseases clearly advanced by crop strain are, however, the various groups of fungi that are facultative pathogens.

Most of these fungi fall into two groups:

a) soil-borne fungi, including most of the major causes of seedling diseases and of rots of roots and underground yield (potato tubers, groundnut pods etc.);
b) most of the fungi that preferably attack the senescent parts of the shoot.

These are the two groups we have to bear in mind, before all others, in assessing the dangers of strain for pathogen attack on crops.

1.6 Crop Age, Injury, and Disease on Leaf and Fruit, with Special Reference to Disease in the Ageing Crop

The changes in proneness to disease that take place as the crop develops are due to a complex of physiological, environmental and time-related causes, in particular where disease of leaf and fruit is concerned. The physiological aspects of changes in susceptibility with time have recently been reviewed by Populer (1978). We shall here attempt to approach the subject from a practical angle, with emphasis on the environmental forces and inoculum control, which can, at least to a certain extent, be influenced by farm operations.

Two phases in crop growth, or in the annual cycle of growth of perennials, stand out as presenting the greatest danger of pathogen attack. One is the phase of germination and seedling emergence, or of early sprouting of buds on perennials; the growth then made is delicate and has not yet acquired some of the more protective tissues and layers that are formed later. The second phase is that of maturation and senescence when natural degradation of ageing tissues sets in and combines with other factors to heighten the disease proneness of plant organs, or of the plant as a whole.

1.6.1 Germination to Pre-Maturation of the Crop

1.6.1.1 Seedling and Early Growth Stage

The diseases threatening the seedling, before and after its emergence, and the young plant, are numerous and often amenable to control by cultural practices. Separate chapters will be devoted to this critical stage of crop growth, in the second part of this book (Chaps. 2.8.5.2 and 2.9.1). Here we shall follow the young crop, after initial establishment of annuals or the post-sprouting stage of perennials, to the stage before the yield matures or that at which "immature" yield (flowers, lucerne foliage etc.) is harvested.

Vigorous growth, i.e., addition of new tissues at a relatively rapid rate, should be characteristic of the early phase of growth. Turgor is strong. Rows or tree crowns have not yet closed, so that aeration is good, but this also means exposure to wind injury and excessive radiation (scald). Hail injury may, of course, occur at this as at any other phase, and is apt to cause most damage to unyielding, turgid tissues. With the important exception of some fruit trees, the crop at this stage bears mainly foliage, but no or very few flowers, and certainly no fruit.

At this early stage of growth, danger to the crop is chiefly presented by two groups of pathogens: (a) Pathogens, mostly soil-borne, which can attack the tender stems of seedlings before or after emergence from the soil, but are unable to do so once the stem lignifies; these include all the damping-off organisms, and fungi such as *Sclerotium rolfsii* on cotton and *Leptosphaeria maculans*, causing blackleg or rape (Delhoval, pers. comm.). (b) Obligate pathogens, such as downy mildews, rusts and viruses, as well as powdery mildews of some perennial crops, such as rubber (Populer 1972). These affect young tissues rapidly and profusely. In open fields, however, unless weather conditions are near-optimal for the pathogens, their attack at this stage is checked to some extent by the aeration and radiation to which the young plant is subject before foliage grows dense, and the consequent limitations of wetness.

The means of restricting disease at this stage are largely pre-sowing or pre-planting choices of site and sowing season. Apart from sensible irrigation regimes, however, there is little the farmer can now do to protect his crop by in-growth cultural practices, and susceptible crops will most likely have to be protected by chemicals.

1.6.1.2 Flowering or Full Vegetative Growth Stage

The next stage in the crop's growth is characterized by ever greater density of foliage, flowering in all its phases, and incipient growth of fruit. Dense foliage reduces aeration, and somewhat lessens the danger of direct injury from radiation, wind and hail, but flowers and young fruit may be particularly susceptible to pathogens, and some organs, especially leaves, are now present at various levels of age and susceptibility.

This has been brought out very clearly by Populer (1978) in his work in Belgium with late blight *(Phytophthora infestans)* on some potato varieties. As shown in Fig. 1.13, the leaves present on individual potato plants at a given time, when subjected to inoculation in the field 88 days after planting, showed a wide range of susceptibility: the top leaves (leaf position 19–20 and above) and the lowest persisting leaves (leaf position 14 and below) gave normal reactions, while leaves of intermediate age (positions 15–19) gave hypersensitive reactions.

At this stage of the growth of young crops, some of the more rapidly developing pathogens have had sufficient time to complete one or several cycles of multiplication: *Phytophthora infestans* on potatoes and *Drechslera maydis* on maize can complete a cycle in as few as 4 days (Cowling 1978). Thus, inoculum loads are rising

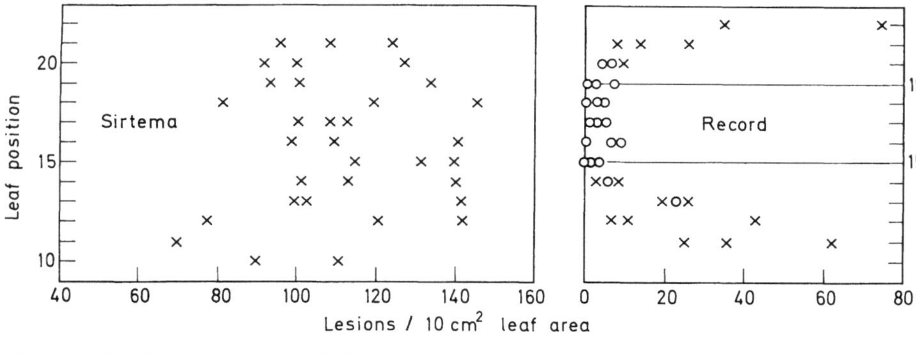

Fig. 1.13. The wide range of susceptibility to *Phytophthora infestans* of potato leaves of the Sirtema and Record varieties, as related to the position of the leaf (counted upward) on the stalk. The number of normal (x) and hypersensitive (o) lesions are given for each of the leaves of three stalks. On the variety Record, leaf growth was still active on positions above 19; it had finished during the week ending on inoculation day for positions 19 to 15, and for more than a week on lower positions. Planting date 18th April, spray-inoculation on 11th July 1972. (Populer 1978)

and can compensate pathogens for somewhat suboptimal environmental conditions and for greater resistance of varieties or age-induced resistance of the older leaves of susceptible varieties (see Chap. 1.3).

The stage of flowering is of special interest. Diseases such as loose smut of wheat *(Ustilago tritici)*, which can attack only at this stage, require thought before sowing: susceptible younger crops must not be sown near (especially to the leeward side) of crops likely to constitute sources of smut inoculum. *Sclerotinia laxa*, too, infects cherries and other species of *Prunus* at the time of flowering. Here again, since the radius to which spores of the fungus spread is strictly limited, care must be taken not to plant susceptible crops in close proximity, especially crops that flower at different periods. The only other measure applicable is strict sanitation, by burning or burying infected organs.

After flowering, flowers of many crops shrivel and shed. Their presence on or near foliage, fruit or stem provides a food-base for the attack of facultative parasites, foremost among them *Botrytis cinerea*, but also fungi such as *Sclerotium rolfsii*.

At or just after the period of flowering, diseases often appear that are caused by pathogens requiring longer time spans to establish themselves in the crop, or that cannot readily infect young tissues, e.g., powdery mildews on many herbaceous crops and *Cercospora*, *Septoria* and other leaf spots. Minor injuries caused by insects, wind (leaves rubbing against each other) and other agencies now begin to play a role by furnishing means of entry to wound parasites.

This may be the stage at which economically valuable, but physiologically immature, yields are harvested: cut flowers, foliage of forage and spice crops harvested in successive mowings, or the flowers or leaves of spices or tea.

The measures farmers can take to minimize risks of disease at this stage of growth relate chiefly to moisture management in the foliosphere, and will be described in the chapters on this subject (Chaps. 2.7.3 and 2.8.5).

1.6.2 Maturation and Senescence

Plants mature as they produce viable reproductive units, and eventually the plant organs, or the plant as a whole, pass into the stage of senescence. At this stage many life processes, such as respiration and photosynthesis, slow down, turgor weakens, and the composition of plant tissues undergoes various changes. At the stage of senescence, catabolic deterioration of tissues sets in, and microbiota are the agents of this process. These include many fungi and bacteria, among them numerous facultative parasites, which thus participate in one of the essential cycles of nature. Such parasites, therefore, can hardly fail to become much more prominent in number and in their action as the plant senesces (Kranz 1966a). We shall consider the role of facultative parasites on plant debris in the soil in a later chapter. Here we are concerned with the changes that occur in the crop as it approaches the end of its life span, the stage at which many of the most important food and fibre crops are harvested. In multiannual crops, the rate at which disease mounts with age often decides how long the crop is worth keeping: this applies, for example, to the rise of ratoon stunting and smuts in sugarcane, and of *Verticillium* wilt or anthracnose in lucerne.

Farkas (1978) has reviewed the relation between senescence and plant disease, with full details of the knowledge available on the physiological changes induced by disease in ageing the plant prematurely. He deals only briefly with the aspect that is of greatest importance to us here: the disease proneness of the ageing crop. As mentioned before, all changes in disease proneness result from the interplay of numerous factors. This is particularly true in the final stages of growth. Because of their immediate relevance to size and quality of yield, these factors will be reviewed in some detail.

1.6.2.1 Physiological Changes

As the plant ages, changes take place in the cells (often the cell wall), in entire tissues (leaf cuticle, fruit peel), in whole organs, or in the annual or perennial plant as a whole. Since the age composition of the crop and its various organs is not uniform, with older and younger leaves and fruits present at the same time, crops at this stage present a heterogeneous assembly of organs, some near, some still pretty far away from, senescence.

Increasing weakness of the tissues prepares the ground for invasion by facultative parasites of lower virulence, which may have been latent but were unable to attack at an earlier stage. A comparison of market crops and seed crops of the same plant, e.g., cabbage or sugarbeet, will often show this quite clearly, with obligate leaf pathogens (downy mildews, rusts) dominating the market crop, but various agents of rots prominent in the seed crop.

Farkas (1978) cites some examples showing proneness of the crop to certain leaf diseases to be related to levels of sugar: species of *Alternaria* attack yellowed, senescent, low-sugar leaves of sugarbeet and of solanaceous crops much more severely than younger, green leaves. In many other crops attacks of yellowing virus diseases, which have an effect very much like ageing, have been found to heighten proneness to fungal diseases, e.g., incidence of root rots in barley and red clover is raised by barley yellow dwarf and clover yellow mosaic, respectively.

What may be called the strain of senescence may be advanced by environmental strain, may interact with the latter and aggravate it. It can sometimes be delayed by well-timed supply of nitrogen or water, or by artifical inhibition of flowering, e.g., removal of flower trusses from tomatoes.

1.6.2.2 Changes in Crop Climate

The changes occurring in the climate of the foliosphere as the crop passes various stages of growth have been described in the chapter on crop climate (Chap. 1.2) and have also been referred to with regard to disease-proneness of crops in the prematuration stage. These changes are, however, more significant, and more variable, in the mature crop. The reasons are manifold: the susceptibility of the ageing crop to moisture-loving weak parasites, the availability of more numerous points of entry for wound parasites, and the larger inoculum load of strong as well as weak parasites. To give just one example: *Botrytis cinerea*, which can attack senescent tissue best when moisture is high, finds ideal conditions for development in ageing crops with close leaf canopies (see Chap. 1.6.2.8).

However, further changes in the crop climate take place as diseases of all kinds cause many of the leaves or fruits to dry up or shed. Then aeration is improved, inoculum production by pathogens sporulating only on living tissue is reduced, but that by pathogens sporulating on dead tissue increases.

1.6.2.3 Injury and Wound Parasites

The longer the crop grows in the field, the more is it exposed to accumulating external injury from weather factors (wind, hail, frost), from biological factors (animals, insects, other pathogens), and indeed from man (wheel damage, pruning, successive pickings, and cuttings). Since such injuries may be cumulative, they are evidently of greatest potential impact in perennials, e.g., bark injuries to trees.

Injuries are of special significance in the ageing crop because they afford entry to parasites that can invade plants only or preferably through wounds and, once they have penetrated, find senescent tissues easy to colonize.

Hail has already been mentioned as the cause of direct injury, but in senescent crops even minor hailstorms may cause indirect damage. Gaeumann (1951) relates that in Switzerland vintners refer to the cause of white rot of grapes, *Coniella diplodiella*, as the hail fungus, because it is destructive only when it gains entry through hail injury. Incidentally, it is the very unpredictability of direct and indirect hail injury which has made hail the first force majeure against which farmers could take out crop insurance in advanced agricultural economies.

The cases of wind and frost injury, in relation to ageing crops, are similar to that of hail. The greater susceptibility of tissues to pathogen attack renders even limited injury more dangerous than in young crops.

Among the biological agents of injury, insects are the most important. Many insects that neither are directly injurious to the crop nor carry viruses can cause minor injury that provides entry for facultative parasites to invade the ageing plant. In addition, primary pathogens may "open the door" to secondary pathogens.

Finally, man-made injuries are principally related to harvesting operations. This aspect will be treated in the chapter on harvesting practices, both with regard to continued growth of the field or orchard after harvest, and to the effect of harvesting operations on post-harvest disease.

Examples of diseases markedly favoured by wounding of crop tissues are listed in Table 1.14.

1.6.2.4 The Time Element and the Mounting Inoculum

Even rapidly developing pathogens need time to mount an epidemic. As for slower, mostly facultative pathogens, large parts of the crop's growth span may elapse before they cause disease of major proportions. In the ageing crop, inoculum may be heavy, simply because there has been time enough for its formation.

As regards the obligate parasites, mention has already been made of the way inoculum, snowballing to massive amounts, can overcome moderate resistance of

Table 1.14. Diseases markedly favoured by wounding of crop tissues, especially as tissues age

Crop	Disease	Pathogen	Agency causing injury	References
Weather Factors Causing Injury				
Grape	White rot	*Coniella diplodiella*	Hail and other weather factors	Gaeumann (1951)
Cherry	Fruit rot	*Monilia cinerea*	Hail	Gauemann (1951)
Pathogens Causing Injury				
Potato	Dry rot	*Fusarium coeruleum*	*Phytophthora infestans*	Gaeumann (1951)
Apple	Canker	*Nectria galligena*	*Venturia inaequalis*	Gaeumann (1951)
Insects Causing Injury				
Cotton	Boll disease	*Nematospora* (Ashbya) *gossypii*	*Dysdercus* sp.	Gaeumann (1951)
Olive	Bacteriosis	*Pseudomonas savastanoi*	*Dacus oleae*	Gaeumann (1951)
Raspberry	Cane blight	*Leptosphaeria coniothyrium*	*Thomasiniana theobaldi*	Schmidle (1979)

varieties or of ageing plant organs (see Chap. 1.3.1 on the collective approach). Facultative parasites, though they probably have infected the crop later, may also by now have completed several cycles of multiplication of their inoculum, and may take advantage of injured, diseased and/or senescent leaves or fruits to attack the crop severely.

1.6.2.5 Disease in Ageing Tree Crops

In tree crops, ageing is measured by years or even by decades. Most of the factors contributing to age-induced susceptibility of plants therefore act markedly on tree crops, particularly on their perennial components, viz., roots, trunk, limbs, and branches. As old trees lose vigour, and find it increasingly difficult to replace organs affected by environmental or parasite action, they become prone to attack by numerous, especially soil-borne, pathogens, some of which are listed in Table 1.13.

However, as also pointed out in Chap. 2.8.2.2, there are exceptions: the most notable is that of *Verticillium dahliae* found in France to attack olives, almonds and apricot trees when trees are young and vigorous rather than when they are ageing (Vigouroux 1973, 1975; Vigouroux and Castelain 1969).

1.6.2.6 The Ageing Crop as Source of Inoculum for its Neighbours

As shown in the previous sections, the size of inoculum loads derived from many pathogens rises in the ageing crops and this endangers neighbouring crops. More important than the size may be the composition of this inoculum. As the plant ages, the proportion of inoculum made up by some obligate parasites diminishes, as host-specific parasites (rusts, downy mildews, etc.) run out of tissue they can colonize. Instead, the part of facultative parasites and of viruses increases; with the wide host range many of them possess, they represent a far greater danger to a much greater variety of neighbouring crops. The best example probably is *Botrytis cinerea* (see Chap. 1.6.2.8). One field heavily infected with *B. cinerea*, such as a seed crop of lettuce, may be the source of severe infection to crops as varied as strawberries, grapes, beans, and flowers. The same applies to multi-host viruses, such as cucumber mosaic virus and tomato mosaic virus, if vectors are available.

1.6.2.7 Disease Control in the Ageing Crop

Most of the measures that can be taken to lessen impact of diseases on the ageing annual crop or the maturing yield of perennials must be planned well before this stage of ageing and maturation is reached, indeed in part before the crop is sown or planted. These measures include choice of variety, management of inoculum, restriction of the growing span, avoidance of strain, minimizing contact between produce and soil, timing of the date of ripening and harvesting, termination of irrigation, adaptation of harvesting practices, and chemical control.

Varieties. Where a crop is expected to be subject to severe infection at the maturation stage, choice of an at least partially resistant variety if available, assumes great importance. If attacks of several diseases are likely, resistance to at least one of them, preferably the one more destructive or more difficult to control, is highly desirable, e.g., choice of a potato variety resistant to *Phytophthora infestans*, where both this blight and *Alternaria solani* may affect the crop. Growing varieties with short growing spans and early maturation shortens the time over which injuries may accumulate and inoculum build up.

Aeration. Good aeration is of particular importance in the maturing crop, and is to be sought by all available, and economically feasible, means: choice of site, training of the crop on stakes and trellises, and spacing the crop not too densely, as well as thinning out of dense foliage or fruit, e.g., in bunches of grapes.

Avoidance of Strain. Almost any strain, if at all persistent or severe, will "age" the plants, and conversely, senescence makes plants more susceptible to strain. As mentioned in the chapter on stress, strain and predisposition, ripening plant organs may be particularly susceptible to strain that facilitates entry of pathogens, e.g., potato tubers under heat stress before the crop is lifted may be destructively attacked by *Macrophomina phaseolina* (Littauer 1944), as described in Chap. 1.5.6. Avoidance of environmental and nutrient stress by judicious application of water and fertilizers (mostly nitrogen) is a matter of common sense and farming experience.

Contact with Soil. Many of the age-related pathogens are soil-borne. Various methods are available to minimize contact between such pathogens and the ripening produce. Anything that prevents splashing of soil onto the plants is welcome. Irrigation by methods other than sprinkling is therefore preferable to reduce incidence of pathogens such as *Xanthomonas campestris* on *Brassicae* or fruit-rotting species of *Phytophthora* on citrus. Removal of low-hanging branches of fruit trees also helps, as does planting of vegetables on ridges or staking to prevent fruits, e.g., of eggplants or tomatoes (Sharma et al. 1976), from getting too close to the soil. In valuable crops, such as strawberries, covering the soil by plastic mulch will effect drastic reduction in infection of fruit by moulds.

Timing of Sowing and Sprouting and Management of Ripening. Various practices are available to advance or delay ripening and these can be harnessed to reduce losses of mature produce by disease. Methods of advancing yield include early pruning and breaking of dormancy in fruit trees, and crowding crops in dense stands where each plant is limited in space and nutrients: this induces uniform ripening after shortened periods of flowering and fruiting, and is useful for crops destined for "once-over" harvest, such as cotton or chilli pepper for processing. Such practices as the use of early varieties may take advantage of the fact that there has been less time for injuries, and pathogens have had less time to develop. Some of these practices may, however, well favour diseases in other ways, by producing out-of-season growth susceptible to frost and hail damage, and providing high moisture in crowded stands.

Delaying yield may be advantageous where harvest must be timed for specific market requirements, or where later yield may escape injury or disease. Late pruning and dormancy breaking may serve this end, and so may misting to delay bud-break and avoid frost damage (Mittelstaedt 1979). However, the latter practice is inadmissible where *Erwinia amylovora* threatens apples (Spotts et al. 1976). Maturity can also be delayed in some crops by timely applications of water or nitrogen, but these practices also harbour dangers of pathogen attack. Lastly, ripening of crops such as tomatoes can be delayed by removal of the first few waves of inflorescences, and this will delay age-related leaf diseases, such as *Alternaria solani* (Rotem 1968).

The planning of crop maturity to coincide with dry periods is of special importance in seed crops. In this way pathogen attack on the senescent, seed-bearing plant can be limited, and incidence of seed-borne diseases can be reduced. This will be referred to in more detail in Chap. 2.10.2.

Harvest. The timing of once-over harvest in some, and the length of intervals between successive pickings or clippings in other crops is of evident importance in relation to diseases of the mature plant or fruit. The advantages of collecting produce before it is overripe are self-evident, and also find expression in the rates of post-harvest decay. The effects of harvesting practices on disease in mature crops are described in Chap. 2.11.

Chemical Treatment. The use of pesticides close to harvest may present residue problems, but is often indispensible. Where crops, even fairly low-value crops, are threatened at the maturation stage, when all investments to produce them have already been made, the economically reasonable way to safeguard the investment will often be the use of pesticides.

However, the efficacy of most fungicides is related to levels of disease-resistance and size of inoculum. There are, in particular, thresholds of inoculum beyond which the action of fungicides does not suffice to protect the crop adequately. As described above, such heavy loads of inoculum are to be expected at the crop maturation stage more than at any other stage in crop growth. It is therefore imperative to use all available cultural practices to limit inoculum to quantities the chemicals can handle, and conversely to use chemicals early enough to limit build-up of inoculum to levels that will not overcome varietal or age-related crop resistance.

1.6.2.8 Botrytis cinerea – A Pathogen of Soft and Senescent Tissues

Botrytis cinerea is an extremely widespread fungus, occurring in agricultural soils and in crops wherever they are grown. It is extremely polyphagous, with thousands of hosts in dozens of dicotyledonous and monocotyledonous families (Gaeumann 1951). Instead of specializing in hosts, this fungus specializes in tissues, and develops mostly on soft, immature, fleshy, senescent, and necrotic tissues (Baker 1949; Ellis and Waller 1974).

To become aggressively parasitic, the fungus requires high humidity or free moisture and preferably low temperatures (Baker 1949). Its ability to develop profusely at temperatures of 8°–16 °C gives it an advantage over hosts that experience strain at these temperatures.

The nature of the tissues attacked by *B. cinerea* makes it the cause of the following types of diseases:

a) Seedling rot or damping off, and blossom and petal blight – all diseases of young tissue.
b) Grey mould of fleshy organs before and after harvest, such as moulds of strawberry fruit, grapes, onions etc.
c) Mould of senescent or necrotic organs, such as sepals, ageing leaves, and tissues affected by chilling or frost. Desiccating sepals often constitute the food base the fungus needs to attack fruits. Likewise, drying flowers falling on lower leaves may facilitate entry into the latter, especially as in dense foliage humidity around the low leaves is high. The vulnerability of senescent tissue makes *B. cinerea* an important storage pathogen.

Cultural practices to control *B. cinerea* concentrate on the humidity factor: avoidance of excessive foliage density by careful consideration of the rate of sowing and of nitrogen application; good ventilation by sowing or planting in rows parallel to the direction of prevailing winds, by preferring open sites and windy slopes to enclosed sites or low-lying ground, by exposure of susceptible fruit to the air (e.g., by "fenestration" of grapes); and by suitable irrigation practices (see Chap. 2.7 and 2.8).

An effective means of protecting fruit near the ground from grey mould is coverage of the soil by plastic foil. This is widely practiced on strawberries in Israel, and its success seems to have two reasons: the foil prevents loss of radiation heat at night and thus reduces dew formation, and it shields the fruit from contact with soil-borne inoculum.

In addition, in crops picked successively (strawberries, tomatoes), prompt harvesting before sepals senesce and fruits are overripe reduces the build-up of inoculum that endangers fruit picked later.

1.7 Weeds and Crop Disease

Natural vegetation and weeds in cultivated land constitute one of the most important means of spread and survival of crop pathogens. We shall here mention diseases in natural vegetation only in passing, as, with some exceptions, there is not very much the grower can do about it. Weeds, on the other hand, are amenable to management by a variety of practices. Indeed, in many crops, weed management must be considered an integral part of disease management.

1.7.1 Which Pathogens Spread and Survive Through Weeds?

The pathogens chiefly spreading and surviving by aid of weeds are those with wide host ranges and little specialization. Table 1.15 presents a number of examples of pathogens affecting hundreds of plant species in dozens of families. Some of these pathogens are specialized to a certain extent, but most of their strains affect numerous hosts. In each case the host range comprises very many weeds.

The pathogen groups exemplified in the table may be classified as follows:

1. Vector-Borne Viruses, many of which can be seed-borne. In many cases weeds have also been found to be symptomless carriers of such viruses. Some virus diseases and weeds known to spread them are listed in Table 1.16. Thresh (1980) has graphically represented the modes of virus spread from weeds within or alongside the crop or from more distant weeds or wild plants (see Fig. 1.14a).

2. Obligate, Air-Borne Fungal Pathogens. Many such pathogens are highly specialized; their spread by weeds is then often of lesser importance, but there are some important exceptions. In Table 1.17, a list is presented of a number of obligate air-borne pathogens and some of their weed hosts. One group of fungi, the powdery mildews, stands out among these obligate pathogens by the width of their host range. These obligate parasites are not generally spread by weed seed.

The role of wild grasses in the spread of the stem rust of cereals *(Puccinia graminis)* has been studied extensively in Israel (Gerechter-Amitai 1973). The stem rusts of wheat and oats proved to have particularly wide host ranges, and to comprise numerous races. The grass phase of such rusts is important above all in the

Table 1.15. Examples of pathogens with wide host ranges including numerous weeds and wild plants

Pathogen	Disease	No. of Host species	No. of Host families	References
Cucumber Mosaic Virus	Mosaic	775	85	Douine et al. (1979)
Sphaerotheca fuliginea	Powdery mildew	570	24	Hirata (1966)
Leveillula taurica	Powdery mildew	710	59	Hirata (1966)
Sclerotinia sclerotiorum	Cottony rot, stalk rot	361	69	Purdy (1979)
Botrytis cinerea	Grey moulds	Several thousands		Gaeumann (1951)
Phymatotrichum omnivorum	Root rot	1300	85	Taubenhaus and Ezekiel (1936)
Verticillium, albo-atrum, V. dahliae	Wilt	ca. 300		Wilhelm (in Kranz et al. 1977)
Macrophomina phaseolina	Charcoal rot	ca. 300		Filho and Dhingra (1980)
Sclerotium rolfsii	Stalk rot, white rot	ca. 500	ca. 100	Aycock (1966)
Meloidogyne javanica	Root-knot	Over 700		Weischer (in Kranz et al. 1977)
M. incognita	Root-knot	Over 700		
M. hapla	Root-knot	Over 550		

Table 1.16. Some virus diseases known to spread through weed hosts

Crop	Virus	Examples of weed hosts	References
Barley	Barley yellow dwarf	*Lolium perenne*	Anon (1978)
Maize	Maize dwarf mosaic, Maize chlorotic dwarf	*Sorghum halepense*	Damsteegt (1976)
	Maize rough dwarf	*Digitaria sanguinalis*	Bock and Guthrie[a]
Watermelon	Watermelon mosaic	*Momordia dioica*, *M. charantia*, *Coccinia grandis*	Bhargrava et al. (1975)
		Momordia charantia	Adlerz (1972)
Cowpea	Cowpea mosaic	*Phaseolus lathyroides*	Lima and Nelson (1977)
	Cowpea aphid-borne mosaic	*Chenopodium amaranticolor*	Raychaudhuri[a]
Pea	Pea seedborne mosaic	*Vicia villosa*	Stevenson and Hagedorn (1973)
Sugarbeet	Curly top	*Sophia* spp.	Gaeumann (1951)
Carrot	Carrot thin leaf, Carrot motley dwarf	*Daucus* spp.	Howell and Mink (1977)
Various	Cucumber mosaic virus	*Chenopodium* spp., *Amaranthus* spp., *Agropyron repens*	Douine et al. (1979)
Rice	Hoja blanca	*Echinocloa* spp., *Panicum* spp.	Ling (1972)
	Tungro	*Oryza* spp., *Echinocloa* spp., *Leersia hexandra*	Ling (1972)

[a] In Kranz et al. (1977)

build-up and dissemination of immense amounts of inoculum during the short critical period between appearance of primary inoculum and the outbreak of stem rust epiphytotics in grain fields. In a typical case, *P. graminis avenae* was found to infect the grass *Alopecurus myosuroides* Huds. in February, and to build up inoculum which two months later caused an outbreak of rust (of the same race!) in a nearby oat field. Stem rusts have also been found on weeds in Israel in the summer months, when susceptible crops are absent. The epidemiological significance of such oversummering is, however, doubtful. Tho role of spontaneous plants serving as alternate hosts for dioecious rusts is referred to in Chap. 2.2.3.

Species of parasitic plants, such as *Orobanche* spp. and *Cuscuta* spp., also belong here as they are spread by minute seeds to numerous host plants, which also include many weeds.

3. Facultative Air-Borne Fungal Pathogens. The non-specialized species of such pathogens, such as *Botrytis cinerea* and species of *Alternaria*, *Stemphylium*, or *Rhizopus*, spread readily from weeds to crop. *B. cinerea* alone has thousands of hosts (Gaeumann 1951), as described in Chap. 1.6.2.8.

4. Soil-Borne Pathogens. The soil environment presents a physical, chemical, and biological barrier to pathogen dispersal. Fungal soil inhabitants have therefore evolved various methods to overcome their dispersal problems, including systemic invasion to permit a degree of aerial dispersal while within the protection of host

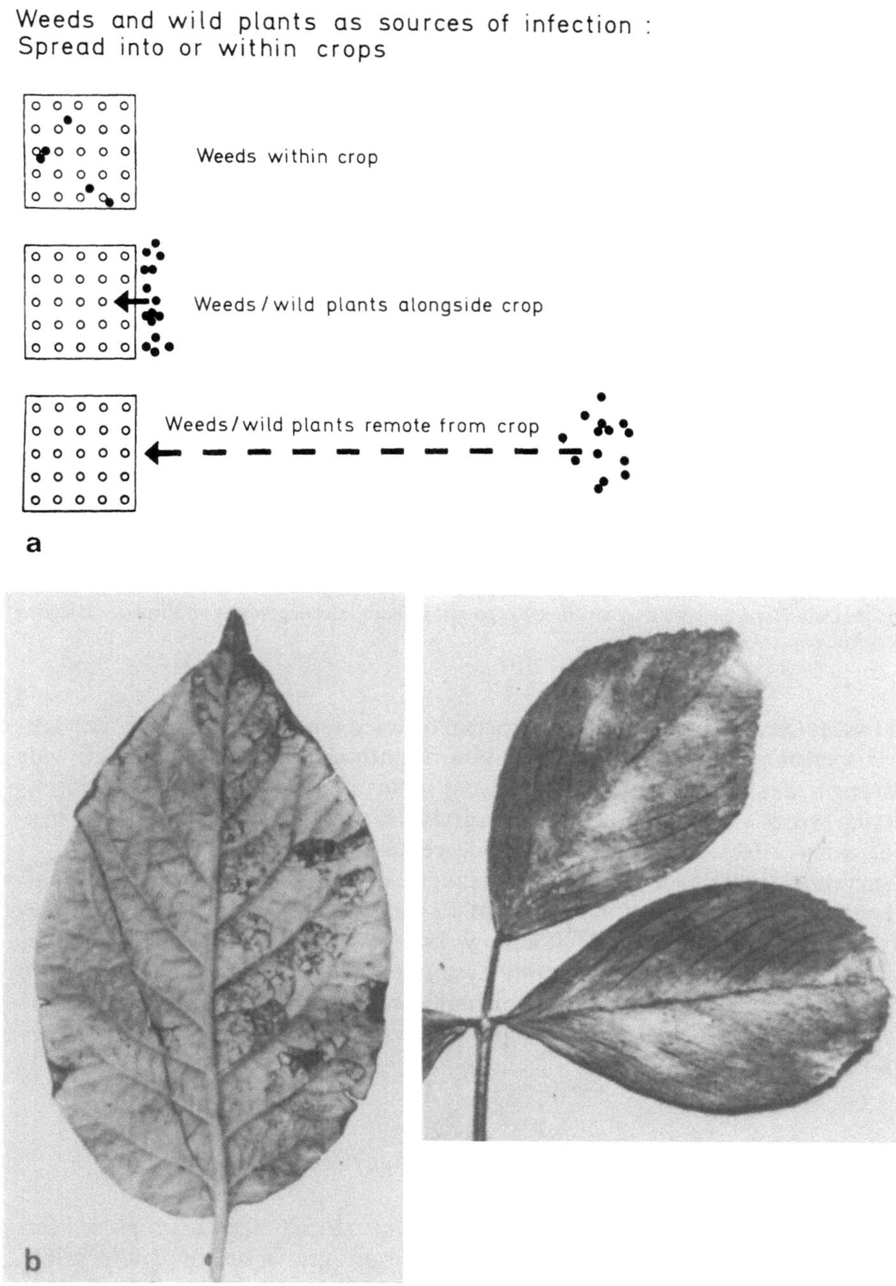

Fig. 1.14. **a** Weeds and wild plants as sources of infection: spread into or within crops (Thresh in press). **b** Symptoms of the powdery mildew caused by *Leveillula taurica* on two of its numerous hosts. *Left* Lower side of pepper leaf – the affected tissue has browned. *Right* Upper side of lucerne leaflets showing light yellow spots. (By courtesy of the Division of Plant Pathology, Agricultural Research Organization, Bet Dagan, Israel)

Table 1.17. Some air-borne pathogens known to spread through wild hosts and weeds

Crop	Disease	Pathogen	Wild or weed hosts	References
Wheat	Stem rust	*Puccinia graminis*	Wild species of *Triticum*, *Aegilops*, *Hordeum*[a]	Gerechter-Amitai (1973)
	Stripe rust	*P. striiformis*	*Agropyron repens*	Hassebrauk and Röbbelin (1975)
	Powdery mildew	*Erysiphe graminis*	Wild species of *Triticum*	Eshed and Wahl (1975)
Barley	Powdery mildew	*Erysiphe graminis*	Wild species of *Hordeum*	Eshed and Wahl (1975)
Maize	Downy mildew	*Sclerophthora rayssiae* var. *zeae*	*Digitaria sanguinalis*	Bains et al. (1978)
	Downy mildew	*Sclerospora sorghi*	*Heteropogon contortus*	Dange et al. (1974)
Cucurbits	Powdery mildew	*Erysiphe cichoracearum*	*Sonchus oleraceus*	Stone (1962)
Tomato, pepper	Powdery mildew	*Leveillula taurica*	*Chenopodium* spp., *Cirsium* spp., *Solanum nigrum*	Palti (unpubl.)
Pear	Fireblight	*Erwinia amylovora*	*Crataegus* spp.	Eden-Green and Billing (1974)

[a] Isolates from *Hordeum* species comprise two groups: isolates infecting wheat and isolates infecting rye, respectively

seed tissues (Sewell, in press). Such dispersal by weed seed presents a very real danger. Dispersal and survival of many soil-borne pathogens is ensured by their wide host ranges, as shown in Table 1.14. A list of some important soil-borne pathogens affecting weeds and aided by them to survive is presented in Table 1.17. These weeds, when affected, all show various degrees of symptoms of disease.

Survival of pathogens by weeds often takes the form of persistence in perennating host organs, as shown for survival of *Verticillium dahliae* in rhizomes of *Cirsium arvense* and other weeds (Skadow 1969). Resistant pathogen structures (sclerotia, chlamydospores, nematode cysts) are frequently released from weed roots into the soil. Many cases are known of weeds acting as symptomless carriers of soil-borne fungi, such as *Fusarium* spp. and *Verticillium albo-atrum*, as mentioned in Chap. 2.3.5.2.

1.7.2 Weeds Particularly Apt to Endanger Crop Health

The greatest danger is constituted by weeds that are closely allied to crops and carry infections in seasons in which crops are either not grown or are ot susceptible. A good example is Johnson grass *(Sorghum halepense)*, a perennial grass weed that harbours many of the viruses of maize and sorghum throughout the winter season (Damsteegt 1976). Control of this grass is essential for the control of viruses such as maize dwarf mosaic.

In semi-arid countries, weeds found in irrigated fields or near their borders serve to carry pathogens through rainless summers. As mentioned above, many grasses

Table 1.18. Some soil-borne pathogens aided in their survival by weeds

Principal host crop	Disease	Pathogen	Weeds affected	References
Various	Wilts	*Verticillium dahliae*, *V. albo-atrum*	44 weeds species in 15 families	Skadow (1966)
Cowpea	Web blight	*Rhizoctonia solani*	*Fleurya estruans*, *Aspilia africana*, *Newbouldia laevis*, *Amaranthus spinosus*	Onesirosan (1975)
Rice	Bacterial leaf blight	*Xanthomonas oryzae*	*Leersia oryzoides*	Mizukami and Wakimoto (1969)
Cotton	Root-knot	*Meloidogyne incognita*, *Hoplolaimus columbus*	*Cyperus esculentus*, *C. rotundus*	Bird and Hogger (1973)
Wheat	Take-all	*Gaeumannonyces graminis*	*Agropyron repens*, *Holcus lanatus*	Curl (1963)
Oat	Cyst nematode	*Heterodera avenae*	*Avena* spp. (wild oats)	Anon (1977a)
Soybean, rice	Rhizoctonia blight	*Rhizoctonia solani*	*Cynodon dactylon*, *Echinocloa* spp.	O'Neill et al. (1977)
Cabbage	Club-root	*Plasmodiophora brassicae*	*Cardamine pratensis*	Gaeumann (1951)

have been found to help rusts oversummer and to attack winter cereals early in their growth (Gerechter-Amitai 1973).

Weeds susceptible to particularly virulent strains of a pathogen may vitiate attempts to eliminate such strains by periodically growing tolerant or resistant crop varieties. This has been experienced in England in hop gardens affected by hop-virulent strains of *Verticillium albo-atrum* (Sewell 1981). It applies equally to USA programmes of instituting zones in which only cereals resistant to certain rust races should be grown in order to impede spread of such races.

Weed tissues also constitute food-sources for many destructive pathogens. Thus zoospores of *Phytophthora cactorum* have been shown to colonize the severed tissues of common weeds (incl. *Dactylis glomerata*, *Lolium perenne*, *Sinapsis arvensis*, and *Stellaria media*) or natural soil surfaces and to effect sporulation within and from such tissues (Dakwa 1970 in Sewell, in press). *Verticillium dahliae* has been found to be preserved in soil for at least one year in residues of the weed *Tripleurospermum inodorum*, and to attack flax or tomatoes subsequently sown (Skadow 1969).

1.7.3 Effects of Cultural Practices on Weeds as Related to Disease Control

In advanced farming weeds are to an ever-increasing extent controlled by herbicides. The effects of such control on crop disease is referred to in the context of integrated control, in Chap. 3.5.1.

Numerous cultural practices affect development of weeds, and thus indirectly the pathogens they carry. These practices are reviewed in detail in Part 2. Only a brief review of their effects on weed and disease development is given here. To ensure absence of weed seeds potentially carrying pathogens from crop seeds and from soil adhering to transplants is an elementary measure of sanitation, but is easier said than done. Of greatest danger is introduction of virus diseases in this way, and all the grower can do is to be most particular in the choice of his seed-suppliers. The "seed content" of agricultural soils can be enormous: up to 39,000 viable seeds have been found per square metre of arable soil, and this only in the top 15 cm (Harper 1974, quoted by Sewell, in press). It follows that, wherever practicable, transplants should be freed from as much soil as possible.

All practices of crop sequence management to control disease must be supplemented by control of weeds that may carry pathogens. Not only weeds on which such pathogens cause disease are to be considered, but also those that may be symptomless carriers.

As regards effects of tillage, which are more fully discussed in Chap. 2.5.4, two aspects seem most important:

a) tillage that inverts soil and buries weeds and their residues to a depth of 20 cm or more is preferred to reduce risks of pathogens affecting subsequent crops;
b) in cultivation for weed control that is carried out while the crop grows, every effort must be made not to throw soil up against the stems of the crop.

The density of the plant cover in fields may be altered extensively by weed populations, which affect humidity relations in the soil and the foliosphere. At early stages of growth this finds expression in increased damping – off and other seedling diseases. At later stages, all diseases that profit from soft growth and prolonged humidity (collar rots, downy mildews) will profit from the conditions created by the weeds. However, in the case of crops affected by partially vector-borne viruses mainly in their young stage, some coverage by weeds may reduce the rate of infection and delay infection until the crop has outgrown the danger.

Irrigation affects weeds and the diseases they carry in various ways. In crops growing slowly, such as onions, irrigation will promote disease, such as *Peronospora destructor*, chiefly by the effects of faster growing weeds on plant density and humidity. In a more general way, however, irrigation acts to preserve growth of pathogen-carrying weeds in the field and at its borders throughout rainless seasons. As mentioned above, such overseasoning is most harmful in the case of weeds carrying viruses, but it is also clearly dangerous where other multi-host diseases (powdery mildews, soil-borne fungi) are carried through crop-less periods. It should also be mentioned that, in one respect, keeping the soil surface free from weeds may present a disease hazard: overhead sprinkling applied to the bare soil surface, rather than soil with some weed cover, causes much more soil to be splashed on to lower parts of trees. This has been found to increase incidence of *Phytophthora syringae* on apple fruits on low-hanging branches (Harris, pers. comm.), and the same is most probably the case also in relation to the brown rot (*Phytophthora* spp.) diseases of citrus.

Part 2 Major Cultural Practices and Their Effect on Crop Disease

2.1 Cost/Benefit and Risk Assessment and the Complexity of Multiple Choice in Pest Control Decisions on the Farm

2.1.1 Cost/Benefit Assessment

Economic criteria are not the only ones to be applied to the choice of agricultural practices, including pest control decisions. Political considerations, such as the effort to reduce imports, social aspects, such as extension of farming to new areas, environmental arguments and health hazards may combine to make governments advocate (or legislate) and farmers adopt practices not otherwise preferable for farm production. However, in a discussion of cost/benefit and risk aspects of pest control, economic criteria generally outweigh all others.

Two of the approaches to disease control, resistance breeding and development of pesticides, do not require the farmer to cover their cost directly. Breeding usually takes place in international, state or university institutions for which many people, besides the farmer, pay in their taxes and contributions. Resistant varieties and hybrids developed by commercial firms are, of course, paid for in the higher price of seed. Pesticides are generally developed by commercial enterprises, who later recover their investment by charging the farmer for their products. The price of a pesticide operation is readily determined. Both the above methods of disease control relate to specific crop-pathogen combinations, their benefits for such combinations are uniform within a fairly narrow range, and cost/benefit relations are not too difficult to assess.

Not so with cultural practices, whether for disease control or for other purposes. Many such practices constitute long-term investments, in general directly at the farmer's expense; they often relate to several successive crops, and their benefits may vary greatly from region to region and even from farm to farm. Moreover, results become apparent gradually, over a period of time, and are not usually as incisive or dramatic as those of some fungicide treatments. As a result, cost/benefit relations are much more complicated to assess.

The concept of benefit requires some clarification. It is not the maximum crop yield that matters, but the economically optimal yield, two concepts that do not necessarily coincide. Thus the yard stick is not the size of the yield, but the size of revenue, be it per unit area, per cubic meter of water or per work day – usually whichever is in shortest supply. The factors to be considered in pre-assessing such revenue comprise:

a) Seasonal changes in the price of a given product, e.g., high prices for early-season fruit, regardless of the fact that such fruit may be more susceptible to disease.
b) Comparative revenue of various crops for a given investment: it may pay more to obtain a moderate yield of wheat, taking into account some losses from disease, than high yields of disease-free barley (Diercks 1975).
c) Comparative revenues expected from entire cropping programmes, as crop sequences comprising several cash crops may pay more than one cash crop and several years' non-cash crops, even if the yield of the cash crop is not very high.
d) Optimal use of available resources of land, water, labour, and machinery.

As evident from the above, farmers are quite likely to adopt, for economic reasons, cultural practices not primarily calculated to exploit the yield potential of a given crop to the full extent that could be realized on their farm. They frequently use practices undesirable from the point of view of crop health, in fact practices known to aggravate certain diseases. This is illustrated by common preference of English cereal growers to sow early in autumn by direct drilling, although this practice heightens the danger of certain diseases (see Chap. 2.5.2.4).

2.1.2 Risk Assessment

An essential element in the farmer's judgement will be the risk he is willing, or can afford, to take in order to obtain the highest revenue. Subsistence farmers often lack marketing and storage facilities, and are in any case not interested in one or several years' top yield achieved at the risk of another year's crop failure, which may mean famine (Harlan 1976). They need steady, reliable yields to feed their families, have selected crops accordingly, and will not readily deviate from practices that have proved themselves in the past (Putter 1980).

Farmers producing yields for sale, however, know they have to face risks: the *market* risk of fluctuating prices; the *weather* risk, especially that of precipitation, which they can sometimes conserve by various practices and which affects successive crops; and the *health* risk of parasite attack on crops managed in different ways (site, season, water etc.). Producers of very high-value crops, such as flowers or glasshouse crops, will be unwilling to take any avoidable risk, and will play safe by using chemicals to prevent pest or disease outbreak. Producers of fairly high-value crops grown in the open – vegetables, spices, fruits – have the sometimes perplexing choice of resistant, but not always high-yielding, varieties, or of chemical control and of the cultural measures discussed in this book. For producers of lower-value field crops – cereals, forage, some industrial crops – the choice narrows and the risk may have to be spread to balance hazards of weather and crop health, e.g., by sowing part of the crop earlier or later than normal, to escape disease.

Various models have been proposed to aid in the assessment of risks due to crop pests (Carlson 1979). These models are, however, evidently only as good as the data available for input. Serviceable input depends on (a) collection of data deriving from previous local experience and the potential damage caused by the combined effect of all major pests and diseases that may attack the crop; (b) availability of methods to monitor pest and disease appearance and epidemic development, e.g., by trapping and field scouting, so that control operations are only carried out if and when necessary (Norgaard 1976, Palti and Ausher, unpubl.); and (c) fairly reliable estimates of the extent to which the loss can be avoided by suitable action.

The last two considerations are more appropriate to chemical control than to control by cultural practices, since the latter are mostly preventive or, in any case, not geared for short-term action.

Most of the models proposed for assessment of risk from disease and of profits from control operations refer to chemical control, with much less attention paid to cultural practices, such as crop sequence (Carlson and Main 1976). Risk assessment in relation to the latter is discussed in Chap. 2.3.2.

Governments, and some commercial enterprises, can do one thing to make risk assessment easier for the farmer: crop insurance. However, up to the present, crop insurance schemes relate mostly to weather risks and other types of "force majeure." Only rarely do they include losses by pathogen attack. Since the attack of certain pathogens or of new strains must still be considered totally unpredictable and/or largely uncontrollable, their inclusion in crop insurance schemes seems warranted.

2.1.3 The Complexity of Multiple Choice in Pest Control Decisions

In modern farming, the alternative routes that lead, hopefully, to the most desirable level of revenue are often so complex that farmers are frequently faced with baffling problems. A German proverb says „Wer die Wahl hat, hat die Qual" (roughly: "multiple choice = painful decision") and that seems singularly appropriate here.

Norton (1976) has pictured the consecutive decisions to be taken by farmers as a "decision tree for crop protection" (Fig. 2.1). In this scheme, the first three steps, all at the pre-sowing stage, are decisions relating to cultural practices, the fourth is choice of the crop resistance level, and all the remaining practices are chemical measures, including all the operations to be performed while the crop grows. To adapt the "tree" for our purposes, we have added examples of in-growth cultural practices for crop protection, viz. irrigation, nutrition, and tillage (Chap. 2.1.3). The fact remains, however, that the impact of cultural practices is to a significant extent made by those adopted in planning the crop.

The complexity of multiple choice may perhaps best be exemplified by a concrete example, such as the choices open to wheat growers in North America, England, and central Europe. Wheat is probably the crop in which most thought and effort has been invested to obtain high yield by all available methods, including resistance breeding, pesticide use and cultural practices.

Let us assume that wheat prices are such that wheat monoculture is economically more attractive than growing wheat in rotation with other cereals or legumes. Introduction of other crops into the rotation (potatoes, sugarbeet) would necessitate additional labour and equipment, which it will pay to secure only if these crops yield well and can be grown fairly often in the rotation. On the other hand,

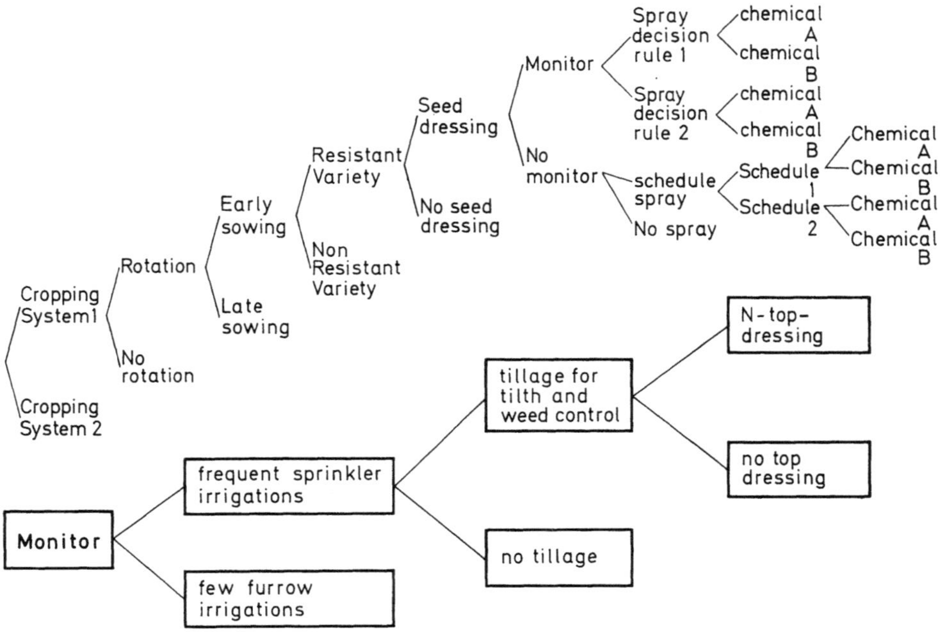

Fig. 2.1. Decision tree for crop protection. *Top* The decision tree as devised by Norton (1976). *Bottom* Adaptation of Norton's decision tree, from the stage of monitoring onwards, to show some alternatives for in-growth protection of crops by cultural practices

maize could be introduced into the rotation without too much trouble, unless that interferes in any way with the high yield of wheat.

The following are some of the health-related questions that then arise, and they are only part of the general complex of questions:

1. Crop Sequence. Wheat monoculture has in many countries proved possible, if a certain reduction in yield is acceptable, but inclusion of other crops may improve overall revenue. As mentioned above, availability of labour and equipment is likely to decide whether additional higher value crops (potatoes, beets) should be grown between the wheat crops (Diercks 1975).

2. Variety. There are varieties resistant or immune to several rusts and *Pseudocercosporella* eye-spot, but not to *Septoria* leaf blotch, *Fusarium* diseases, or take-all (*Gaeumannomyces graminis*). The choice of the variety, from the health angle, depends on the relative prevalence of these diseases, taking into account varietal resistance to moisture stress, the availability and cost of chemical control and auxiliary irrigation, and the benefits of N-fertilization (Leiteritz 1977).

3. Tillage or Non-Tillage? Non-tillage, or direct drilling, has advantages: it helps to control erosion, saves labour and energy, and makes it possible to use short intervals between rains to sow early. However, under successive non-tillage crops the soil tends to compact, which favours take-all, and vigour of growth tends to be reduced, which increases chances of infection by *Septoria* leaf blotch, but lessens those of rust and powdery mildew (Yarham and Norton, unpubl.).

4. *Fertilization.* The fertilization programme will have to correspond to the needs of the variety chosen, with due regard to the facts that (1) different forms of N vary in their effects on the susceptibility of wheat to disease, e.g., NO_3-N tends to favour take-all, but NH_4-N favours *Pseudocercosporella* eye-spot (Huber and Watson 1974; Smiley and Cook 1973), and (2) with N readily available, crops can use the soil's water potential more fully (Papendick and Campbell 1975).

5. *Date of Sowing.* The aspects to be considered are full utilization of labour and equipment, erosion control, the growth stage of the crop and the weather at the time the pathogen presents the greatest danger (Huber 1979), and the chances for chemical treatment at that time.

6. *Moisture Stress.* How well will the variety chosen stand up to moisture stress if rain is deficient? Is auxiliary irrigation available, and at what cost? Is the increase in leaf disease often experienced after sprinkling acceptable or controllable? If, on the other hand, rainfall is above normal, how badly will the wheat lodge? How much should the danger of lodging affect the rate of sowing and the rate and timing of N-topdressings, with special reference to the danger of eye-spot?

The above questions facing the wheat grower have been presented in some detail, because they typify not only what problems the farmer is facing, but also what problems we are facing in this book. Effects of major cultural practices interlock to such an extent with each other and with chemical and genetic methods of disease control, that separate discussions of practices such as crop sequence, tillage, soil amendments, fertilization, and rate and date of sowing run the risk of being somewhat removed from reality. However, every one of these and other practices also have characteristics and effects of their own, which also have to be described. In doing so, after all, in separate chapters, we request the reader constantly to bear in mind that the effect of each operation is largely conditioned by that of others, and that decision-taking in modern farming is commonly a matter of cost/benefit and risk assessment.

2.2 Sanitation

2.2.1 Aims and Limitations of Sanitation

The two aims of sanitation are to prevent the introduction of inoculum into field, farm, or community, and to reduce or eliminate inoculum from diseased fields.

To prevent introduction of inoculum, an attempt must be made to identify all the means by which a given pathogen can be carried from field to field or region to region. This includes seeds and other propagating material, water, animals, plant debris, compost, farm yard manure, implements, wheeled and tracked farm equipment and vehicles, boxes and other packing material in recurrent use and, last but not least man himself. The prevailing direction of winds may also have to be taken into account. Once the potential means of introduction have been thought out carefully, its prevention becomes a matter of common sense, experience, and economic considerations. The fact that it is worth going to a great deal of trouble to keep new pathogens, or races of pathogens, out of farming communities is increasingly realized, and measures to this end have been mentioned in the chapter on the collective approach to disease control. Hanan et al. (1978) have rightly stated that good sanitation requires a special "state of mind" or "awareness." This applies to the individual farmer as well as to the farming community as a whole.

Reduction, or even elimination, of inoculum from field or plantation includes dealing with living plants that carry infection and with debris or dead plant matter that harbours pathogens. Living carriers of infection are additional or alternate plants (crops, weeds or wild plants) that serve as pathogen hosts, and groundkeepers or volunteers that directly transmit disease or serve to "trap" inoculum early in the season. These live carriers may transmit facultative parasites, but their special importance lies in the transmission of obligate parasites, such as mildews, rusts and viruses. Elimination of these plants means killing them by pulling them up, by tillage or by chemical means. This is comparatively simple.

It must be realized, that after some decades of relevant research, all the information now at hand concerning management of crop disease by way of soil microbiota, refers to specific conditions and specific host/pathogen combinations. It does not, and perhaps cannot, lend itself to generalisations. As will also become apparent in the chapter on soil amendments, large-scale field and laboratory research in diverse agroclimates, very much including the tropics and subtropics, is needed to advance this hopeful approach to disease control to the stage of wide, practical use.

Handling dead plant matter, however, is much more complicated. Such matter carriers facultative parasites, often with pronounced saprophytic capability, and is often mixed up with soil. Such plant matter should, if possible, be removed and destroyed by burning or burying it, or it can, with care, be used for composting. Bodily removal of plant debris is, however, frequently uneconomic or impossible. Tillage and soil amendments may then have to be employed to inactivate inoculum, as will be described in the chapters dealing with these operations.

Sanitation in the field or plantation often requires large amounts of labour and is difficult to mechanize. The investment that farmers are willing to make in such operations depends on their assessment of risk and efficacy, and on their abovementioned "state of mind." Availability of labour during the seasons optimal for sanitation measures is also an important consideration.

Sanitation seems of particular importance in the tropics. Here subsistence farmers habitually remove infected plant organs as the crop grows. Such sanitation can be so effective that Putter (1980) in Papua New Guinea could turn taro blight *(Phytophthora colocasiae)* "off" and "on," simply by removing infected leaves daily for a couple of days, or by not removing them. In general, reduction of inoculum by sanitation is far more effective at the endemic or the incipient epidemic stage than at later stages of disease development (Putter 1980).

In tropical plantation crops, the principal agents of root disease, such as species of *Ganoderma*, *Fomes*, *Poria*, and *Armillariella mellea*, are primarily saprophytes that occur in the jungle on decaying timber and can affect a large variety of hosts. To reduce the threat of such diseases, sanitation is the basic, and often the only, method promising success, and has in many cases proved profitable.

2.2.2 Preventing the Introduction of Inoculum

To prevent introduction of a new pathogen, or the new race of a pathogen, into a country is the task of the state, through its quarantine services. This is an essential task, especially in the developing countries; the sins of omission are many, and sometimes the consequences are little short of catastrophic. However, this problem falls outside the scope of this book, which deals primarily with what the farming community or the individual farmer can do.

2.2.2.1 Propagating Material

One of the most frequent and dangerous means of introducing disease is by propagating material, both seeds and vegetative material. No wonder, therefore, that almost all instructions to farmers on how to grow healthy crops, begin with the advice "secure disease-free propagating material."

Seed Transmission. The danger of pathogen transmission by seed is starkly apparent from a study of Richardson's (1979) list of seed-borne diseases. Data culled from this book have been tabulated in Table 2.1 to show transmission of fungi, bacteria, viruses and nematodes by seeds of major crops. Evaluating these data regarding the potential for spread and crop destruction of seed-borne pathogens, the following generalizations (none without exception) may be made.

a) By far the largest number of pathogens carried by seed are *fungi*. These comprise obligate parasites (smuts, rusts, downy mildews) that are highly destructive, but strictly limited in the range of host crops; such pathogens are often carried on the seed surface or in debris attaching to the seed, and can in general be controlled by surface disinfection (an important exception: loose smuts of cereals). The majority of seed-borne fungi are, however, facultative parasites which either cause systemic infection or reach the seed by infection of flower, fruit or grain. Some of these fungi have wide host ranges, e.g., *Fusarium culmorum*, *Verticillium dahliae*, and *Botrytis cinerea*, and thus endanger the crop that carries them in its seed as well as neighbouring crops of other species. The field crops most commonly affected by seed-borne fungi are the cereals, on which smuts and *Helminthosporium* blights are most prominent, legumes (*Ascochyta* blights, anthracnoses) and flax (*Colletotrichum linicola* and *Polyspora lini*).
b) Seed transmission of more than two *bacterial* pathogens occurs in major crops only in the Brassicae, Leguminosae, tomato, and tobacco, and – most importantly – in cereals such as rice and maize. The large majority of these bacterial species, or strains of such species, affect only one or few crops. They are commonly carried within the seed and not amenable to control by surface disinfection.

Table 2.1. Pathogens transmitted by the seed of major crops. (Compiled from Richardson 1979)

Family	Crop	Number of pathogen species transmitted by seed			
		Fungi	Bacteria	Viruses	Nematodes
Alliaceae	Onion, leek, garlic	14[a]	0	2	1
Cannabidaceae	Hop	2	1	3	0
Chenopodiaceae	Sugarbeet	12	2	4	1
Cruciferae	*Brassica* spp.	15	3	1	0
	Radish	9	1	0	0
Compositae	Safflower	10	0	0	0
	Sunflower	10	0	0	0
	Lettuce	11	1	4	0
Cucurbitaceae	Watermelon	5	1	1	0
	Melon	4	0	5	0
	Cucumber	8	1	2	0
	Squash, marrow	5	1	3	0
Gramineae	Oat	27[a]	2	2	2
	Barley	26[a]	2	1	0
	Wheat	32[a]	6	1	1
	Rye	17[a]	1	0	1
	Sorghum	38[a]	4	0	0
	Maize	42[a]	7	4	0
	Millets	33[a]	2	0	0
	Rice	42[a]	5	0	3
	Sugarcane	11	0	1	0
Leguminosae	Groundnut	13	0	5	0
	Chickpea	4[a]	0	0	0
	Soybean	26[a]	7	7	0
	Lentil	3	0	1	0
	Lucerne, trefoil	10	2	1	1
	Phaseolus beans	26	6	10	0
	Pea	14	3	7	0
	Clovers	26	4	6	1
	Vetches	13	1	6	1
Linaceae	Flax	17[a]	1	0	0
Malvaceae	Cotton	15[a]	1	0	0
	Okra, kenaf	10	1	0	0
Rosaceae	Cherry, plum, peach	0	2	6	0
	pear	1	0	2	0
	Raspberry	0	0	5	0
Rubiaceae	Coffee	5	0	1	0
Rutaceae	Citrus	3	1	4	0
Solanaceae	Chilli pepper	17	2	3	0
	Tomato	19	5	7	0
	Tobacco	6	6	8	0
	Eggplant	9	0	1	0
	Potato	0	0	5	0
Sterculiaceae	Cocoa	5	0	1	0
Umbelliferae	Celery	10	2	1	0
		9	1	2	0

[a] All species of *Fusarium* counted as one

c) Transmission of *viruses* by seed is considerably more common in crops of the Leguminosae and Solanaceae than in herbaceous crops of other families, and is also common in rosaceous fruit crops. Apart from the losses caused to the crop sown with infected seed, one of the prominent dangers of virus transmission by seed lies in the wide host range of some viruses, e.g., cucumber mosaic virus affecting cucurbits, Solanaceae, Leguminosae, as well as bananas.
d) Seed transmission of *nematodes* is common in a few crops only. *Ditylenchus dipsaci* is carried on seeds of various crops, including onion and garlic, Leguminosae and oats (all affected by the same strain), and affects many additional hosts. The few other nematodes carried by seed are largely host-specific.

Most of the above-mentioned pathogens carried on seed can only be detected in the laboratory, hence the overriding importance of securing seeds from an impeccable and responsible source. The certificates of freedom from disease, which are commonly attached to seed consignments, are unfortunately not always reliable.

Transmission by Vegetative Propagating Material. Vegetative propagating material includes cuttings, stecklings, bulbs, corms, tubers, rhizomes, stem or root suckers,

Table 2.2. Representative viruses of vegetatively propagated food and ornamental crops. (After Thresh 1980c)

Plant part or organ used for propagation	Crop and virus	No. in CMI/AAB lists
Bulb	Onion yellow dwarf	(158)
	Lily symptomless	(96)
	Narcissus mosaic	(45)
	Tulip breaking	(71)
Corm	Iris mild mosaic	(116)
Cutting	Cassava common mosaic	(90)
	Sugarcane Fiji	(119)
	Abutilon mosaic	
	Carnation ringspot	(21)
	Chrysanthemum virus B	(110)
	Pelargonium flower break	(130)
Stem/root suckers	Banana bunchy top	
	Raspberry bushy dwarf	(165)
Rootstocks/scions	Apple mosaic	(83)
	Cherry raspleaf	(159)
	Citrus tristeza	(33)
	Grapevine fanleaf	(28)
	Passionfruit woodiness	(122)
	Peach mosaic	
	Plum pox	(70)
	Prune dwarf	(19)
	Rose mosaic	
Runners	Strawberry crinkle	(163)
Tubers	Potato leafroll	(36)
	Sweet potato mild mottle	(162)
	Dahlia mosaic	(51)

Note: The numbers in brackets refer to the *Descriptions of plant viruses* published by the Commonwealth Mycological Institute and the Association of Applied Biologists. Further details on the viruses listed can be found in these descriptions runners, rootstocks, and scions. Such material is even more suspect in pathogen transmission than seeds, but detection of disease is in many cases easier.

Viruses can be carried by all types of vegetative material: their spread by potato tubers, flower bulbs, and grafting material is particularly notorious, and hard to detect. Thresh (1980c) has drawn up the list of representative viruses of vegetatively propagated food and ornamental crops that is presented in Table 2.2.

Nematodes can likewise be carried by growing plantlets as well as by dormant vegetative material used for propagation; they are often easier to spot than other pathogens.

Where fungal and bacterial pathogens are concerned, a distinction must be made between young plants and dormant material. Provided the seeds or vegetative sources of seedlings, stecklings or cuttings have been healthy, such plantlets are most likely to be affected by pathogens attacking young tissue in crowded stands under high humidity conditions, especially by downy mildew and allied fungi, and by agents of collar rots such as *Sclerotium rolfsii* and *Sclerotinia sclerotiorum*. Many such diseases are easy to discern. By contrast, dormant propagating material is suspect in carrying pathogens acquired in the mother field, such as species of *Fusarium*, *Phytophthora* or *Alternaria*, or bacteria causing vein discoloration or rots. Such diseases can only be detected by careful inspection and sampling.

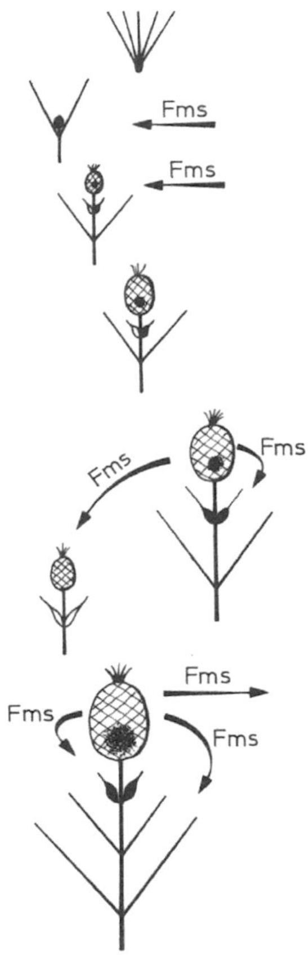

a) Young, healthy offshoot

b) Differentiation of flower parts and bulblets: first vulnerable stage

c) Simultaneous development of fruit, bulblets and the disease affecting them

d) Appearance of external symptoms on and progressive external contamination of bulblets and neighbouring plants

e) Intensification of phenomena described under d; internal contamination of bulblets

Fig. 2.2. Successive stages in the infection of pineapples by *Fusarium moniliforme* var. *subglutinans (Fms)* and in the subsequent transmission of the pathogen to the offshoots commonly used for propagation. (Laville 1980)

An interesting case of propagating material transmitting a disease, which sometimes finds expression only at a late stage of crop growth, has been described by Laville (1980). He investigated the fusariosis of pineapple in Brazil, which is caused by *Fusarium moniliforme* var. *subglutinans* (*F. sacchari* var. *subglutinans*) and induces leaf chlorosis, heart necrosis, and fruit gummosis. Pineapples are propagated by off-shoots developing at various levels below the fruit. By persistent "detective" work, it was found that fruits affected by fusariosis release inoculum which is deposited on the off-shoots and causes brownish necrosis and gummy exudates at their base (Fig. 2.2). Careful elimination of all propagating material showing such symptoms has resulted in reducing fruit rot from 70% to 10%.

Healthy Propagating Material and the Farming Community. Preventing the introduction of inoculum by propagating material is very much the business of the farming community. In many cases the law has stepped in, often at the request of growers, to prohibit spread from region to region, by transport of infected material of diseases such as fireblight *(Erwinia amylovora)* on pears, potato wart *(Synchytrium endobioticum)* and hop wilt caused by *Verticillium albo-atrum* (see Chap. 3.6). All three of these diseases come under regulations governing their transport within the United Kingdom. In China, the transport of seed carrying barley stripe mosaic is similarly regulated (Kelman and Cook 1977). A certain dose of enlightened self-interest should, however, be able to move a farming community to achieve the same end by mutual consent, e.g., by growing regional nurseries under strict sanitary control, or by securing seeds or graftwood from outside sources that are recognized as suppliers of disease-free material, and by insisting on the use of certified seed.

2.2.2.2 Irrigation and Drainage Water That Spreads Inoculum

Irrigation water can be an extremely "efficient" means of spreading inoculum of bacteria, nematodes, and soil-borne fungi. Within the field or plantation, the water applied by flooding or in furrows carries disease from individual plants in the direction of flow to other plants. At the lowest points of the field, inoculum may then accumulate and cause particularly heavy infection, as recorded by Thanassoulopoulos et al. (1980) for *Verticillium dahliae* on olives, and in many other instances.

Of even greater significance for the farming community as a whole is the spread of disease by irrigation or by drainage water to near-by, or even to distant, fields. Water in irrigation ponds, channels, drainage ditches, and lesser or even major rivers has been found to contain many species of pathogens. Most of the relevant studies have been carried out in the USA. These findings, as well as those from other countries, are presented in Table 2.3. The danger of pathogens spreading by means of irrigation or drainage water deserves much more attention, especially in developing countries.

2.2.2.3 Inoculum Introduced in Plant Debris, Compost or Manure

Most crop debris is left in the field in which the crop has grown. The management of such debris is treated in the section on elimination of inoculum in this chapter, and in the chapter on soil amendments.

Table 2.3. Spread of pathogens by irrigation and drainage water

Grop	Disease	Pathogen	Country	References
Cotton	Angular leaf spot	*Xanthomonas malvacearum*		Gaeumann (1951)
Bean	Bacterial blight	*X. phaseoli*	USA	Steadman et al. (1975)
Tobacco	Black shank	*Phytophthora parasitica* var. *nicotianae*	USA	Fenne (1941)
Groundnut, soybean	Black rot	*Cylindrocladium crotalariae*	USA	Krigsvold et al. (1977)
Various	Seedling diseases	*Pythium* sp., *Fusarium* sp., *Phoma* sp., *Rhizoctonia* sp.	USA	Shokes and McCarter (1976)
Various	White mold, stem rot	*Sclerotinia sclerotiorum*	USA	Schwartz and Steadman (1978)
Citrus	Nematode	*Tylenchulus semipenetrans*	S. Afrika USA	Cohn (1976) Thomason and Faulkner (1975)
Olive	Wilt	*Verticillium dahliae*	Greece	Thanassoulopoulos et al. (1980)
Citrus	Brown rot, gummosis	*Phytophthora* spp.	USA	Thomson and Allen (1974, 1976)
Date palm	Wilt (bayoud)	*Fusarium oxysporum* f. sp. *albedinis*	Algeria	Laville (pers. comm.)
Carnation	Wilt	*F. oxysporum* f. sp. *dianthi*	Colombia	Faivre (pers. comm.)
Various	Nematodes	*Meloidogyne hapla*, *Heterodera schachtii*, *Pratylenchus* spp., *Tylenchorhynchus* sp.	USA	Thomason and Faulkner (1975)

However, if the yield removed from the field carries inoculum, either as infected components of the yield itself (rotten fruit) or as infected debris (cereal glumes, fruit stalks, leaves attached to root crops), then the disposal of such debris and culls presents problems of spreading disease.

Some of the post-harvest operations are usually carried out on or near the farm: threshing grain, or grading and packing vegetables or fruits, are all operations that generate considerable amounts of refuse. This must never be redistributed to fields, but must be burned or buried effectively. Nor should straw from fields affected by pathogens such as *Heterodera avenae* be carted to other fields (Anon 1976a).

Many other post-harvest operations are carried out in installations away from the farm, where farmers have no way of supervising the disposal of refuse. Sugar refineries release large amounts of water into sewage systems, and these may well carry sclerotia of *Sclerotium rolfsii* or inoculum of other pathogens, such as *Polymyxa betae*, the carrier of rhizomania (Sutic and Milovanovic 1980). Cotton trash that accumulates in cotton gins is perhaps better known as a source of insect pests, but can carry inoculum of various diseases, including *Xanthomonas malvacearum* and *Verticillium dahliae* (Sterne et al. 1979). Vegetable and fruit canneries are equally suspect in spreading inoculum in sewage water and in boxes re-distributed to farms. Processing plants of all the above kinds can spread diseases to entire communities. Their strict sanitary supervision is essential.

A case of a somewhat different nature has been observed by Golan (pers. comm.) in Israel. White root rot, caused by *Rosellinia (Dematophora) necatrix*, has been found to affect fruit trees particularly in the vicinity of primitive installations for production of earthenware: the furnaces operated for this purpose were heated with the wood of dead trees which often carried the fungus. The latter persisted on weeds and subsequently attacked fruit trees.

A major part of the crop residues removed from fields is utilized on most farms for feeding and bedding livestock. Such residues and the refuse from some processing plants are also composted, to be applied to the soil as nutrients and soil conditioners.

That some sclerotia and other resting bodies of fungi can survive passage through the digestive tract of cattle and sheep is a well-known fact; this includes sclerotia of widespread, polyphagous pathogens such as *Sclerotinia sclerotiorum* (Brown 1937) and *Sclerotium rolfsii* (Leach and Mead 1936; Avizohar-Hershenzon et al. 1969). The percentage surviving is usually not high, and can further be reduced by composting. Fresh manure from livestock fed with leaves of beet crops must not be applied to fields in which such crops are to be grown, as it can transmit pathogens such as *Cercospora beticola* (Anon 1979).

The danger of spreading inoculum through refuse from processing plants can also be limited by composting. Thus, in Arkansas (USA), in a study referred to above (Sterne et al. 1979), composting has been found to kill most of the inoculum of *Verticillium dahliae* and of a species of *Alternaria* in cotton trash, but did not kill *Xanthomonas malvacearum*. In England, careful composting is advocated for hop waste carrying *Verticillium albo-atrum*, that accumulates near picking machines (Sewell and Wilson 1961) and for tomato haulms from crops affected by *Didymella lycopersici* (Phillips 1959). However, undoubtedly, composting as a means of killing inoculum has its problems: in compost heaps in which inner layers reach temperatures of 55°–65 °C, most inoculum may be assumed to be killed in those layers, but the outer layers do not generally reach such temperatures and this is where inoculum is likely to survive. Composting as a sanitary measure can therefore be considered reliable only where this is taken into account and where, if necessary, composts are turned over at least once to expose what was originally its outer layer to higher temperatures.

Introduction of pathogens by sewage sludge has not yet been studied adequately, but the 38 fungal species found by Diener et al. (1976) in activated sewage sludge in the United States included several potential pathogens, such as *Fusarium moniliforme*, other species of *Fusarium*, and also of *Phoma*, *Rhizopus* and others.

2.2.2.4 Inoculum Introduced and Spread by Equipment and Man

Two types of pathogens are the most likely to be spread by equipment and man: soil-borne pathogens and those that are transmitted by sap or mucilage.

The latter pathogens are most readily transmitted by picking and pruning. In both these operations, clippers and hands pass from branch to branch and from tree to tree. Sap containing viruses and mucilaginous secretions caused by bacterial pathogens or by fungi such as *Gloeosporium* spp. are then effectively transferred. So important are pruning implements sometimes considered to be in spreading disease, that in Central America the die-back of cocoa caused by *Ceratocystis fim-*

briata is actually called the "mal de machète" (Schieber 1969). The obvious recommendation is to disinfect all implements used, as well as the hands, at frequent intervals, but this is easier said than done; it requires labourers to be supervised to an extent often difficult to achieve in practice.

Acute danger of virus transmission by sap exists wherever propagating material is prepared by cutting (potato tubers, cuttings of vine, flowers etc.). Bruising and transmission by sap has caused catastrophic outbreaks of bacterial wilt on tomatoes (*Corynebacterium michiganense*) when tomato seedlings were planted by machine (Shoham, pers. comm.).

Soil adhering to wheeled or tracked equipment driving through the fields, or to wheels of trucks transporting the produce, is a potential source of soil-borne pathogens, and so are stakes or boxes after previous use in the field. Driving through the crop also crushes some plants and brushes against others, permitting transmission of sap-borne inoculum, mainly viruses and bacteria. Mowing equipment (see Chap. 2.11) is, of course, even more likely to spread disease, not only virus diseases, but many other pathogens; one disease for which this has been demonstrated most convincingly is the wilt of lucerne caused by *Verticillium albo-atrum* (Heale and Isaac 1963; Haass and Märtin 1977). Machine-hedging of citrus trees has been proved to transmit the exocortis (but not the xyloporosis) disease (Wutscher and Shull 1975), and is certain to transmit many other viruses on fruit trees.

Other equipment that may spread infected soil are trucks bringing bye-products of processing industries to the farm, e.g., molasses or oil-cakes, or cultivation machinery used on two crops susceptible to the same disease. Groundnuts and soybeans, for instance, are both affected by *Cylindrocladium crotalariae* and tillage equipment used on both crops readily spreads the disease (Krigsvold et al. 1977).

Plastic foils used for mulching or covering crops are certainly suspect in carrying inoculum. As rising prices for such foils make their re-use more attractive, the sanitary aspect must not be neglected. Even wires used for trellising tomatoes have been found to carry a pathogen, the tobacco mosaic virus, but are easily disinfected (Palti et al. 1962).

A famous case of man transmitting disease concerns the labourers on tobacco plantations and their chewing of tobacco. This carried tobacco mosaic virus, and by spitting on tobacco plants the labourers spread the virus. The transmission of virus through clothes of labourers working in dense fields and constantly brushing against plants is also well-known. Labourers are also principal agents in spreading some bacterial and fungal diseases, such as the coffee berry disease (*Colletotrichum coffeanum*). They do so while picking the berries, and also by brushing against the berries while weeding the ground below and between trees (Gassert 1977). The spread of *Pseudomonas lachrymans* in cucumber fields can be greatly hastened by manual transmission while the fruit is picked.

A few rules of more or less general validity can be applied to avoidance of spread of inoculum by equipment and man:

a) All in-growth treatments in the crop (tillage, pruning, weeding, picking) should first be carried out in the less infected, usually the younger, crop, and only after that in more severely infected crops.
b) To lessen adherence of foliage-borne inoculum to equipment and hands, no treatments should be performed when the foliage is wet.

c) Wheeled or tracked equipment should not be driven from potentially infected to other fields when the ground is wet.
d) All packing equipment or stakes that have been used before must certainly be cleaned from soil and debris, and should preferably be disinfected.

2.2.3 Elimination of Living Plants That Carry Pathogens

2.2.3.1 Additional and Alternate Hosts

Additional hosts are crops, weeds or wild plants susceptible to the same disease as the crop we are concerned with; alternate hosts are plants which the pathogen (rusts) requires to complete its life-cycle.

The principal danger of additional hosts is that they assist the overseasoning of pathogens, and provide a base from which inoculum multiplies and reaches epidemic proportions early in the season. In climates in which certain crops can be grown all year round, air-borne pathogens with wide host ranges often persist on minor crops to launch massive attacks on more important main season crops. Examples are many: powdery mildews, such as the powdery mildew of cucurbits (*Sphaerotheca fuliginea*) which in Israel persists on winter squash and then endangers the far larger areas of spring-sown melon and cucumber crops, or virus diseases affecting widely disparate crops. Examples of crops that can furnish inoculum for each other have been cited in the chapter on the communal approach to disease control (Chap. 1.3). These examples include viruses transmitted from peaches to sugarbeet and from cucurbits to banana, and a fungal disease common to coffee and cocoa. Rode (1979) has compiled lists of ornamentals that serve for

Table 2.4. Alternate hosts on which some rusts complete their life-cycle

Crop	Rust pathogen	Alternate host	References
Field Crops			
Oat	*Puccinia coronata*	*Rhamnus* spp.	Gaeumann (1951)
Wheat	*P. graminis*	*Berberis* spp.	Gaeumann (1951)
	P. recondita	*Thalictrum* spp., *Isopyrum* spp., *Anchusa* sp.	Saari and Joshi (in Kranz et al., 1977)
Barley	*P. hordei*	*Ornithogalum* spp.	Saari and Joshi (in Kranz et al., 1977)
Sorghum	*P. sorghi*	*Oxalis stricta*	Zogg (1948)
Cotton	*P. stakmanii*	*Bouteloua* spp.	C.M.I.
Fruit Crops			
Plum	*Tranzschelia discolor*	*Anemone coronaria*	Viennot-Bourgin (1967), Rode (1979)
Cherry	*Puccinia cerasi*	*Eranthis* spp.	Viennot-Bourgin (1967)
Pear	*Gymnosporangium* spp.	*Juniperus* spp.	Gaeumann (1951)
Red currant	*Cronartium ribicola*	*Pinus* spp.	Viennot-Bourgin (1967); Rode (1979)
Gooseberry	*Puccinia caricis* var. *grossulariata*	*Carex* spp.	Anon (1948)

the overseasoning of air-borne pathogens of fruit trees in Germany, including hosts for *Erwinia amylovora*, *Nectria cinnabarina*, and *Coryneum carpophilum*.

The role of weeds in serving as additional hosts for spreading crop disease has been discussed in Chap. 1.7.

As regards control of alternate hosts of rust diseases, eradication of barberries, the hosts for aecial stages of *Puccinia graminis*, is a classic of phytopathological literature and practice. Table 2.4 presents a list of some weeds and wild plants that serve as alternate hosts for rusts affecting field or fruit crops.

2.2.3.2 Groundkeepers and Volunteers

The living, individual plants of crop species that may be sources of disease from within the field or from its close neighbourhood are:

a) groundkeepers, plants left over from previous crops; and
b) volunteers that have grown from seeds or other propagating organs of preceding or neighbouring crops.

Groundkeepers, most frequently of root, bulb or tuber crops, can transmit a wide range of obligate and facultative parasites. Since these are usually senescent plants they are most liable to carry inoculum of diseases associated with the ageing plant.

Volunteer plants can become sources of disease in two ways:

a) as direct carriers of all types of pathogens that adhere to or penetrate seeds and vegetative organs of propagation, including numerous bacteria and fungi and some viruses and nematodes; such volunteers also serve to maintain specific races of pathogens.
b) as "trap" plants that catch, early in the season, wind- or vector-borne inoculum which readily infects young tissues, so that the onset of infection cycles is advanced and disease foci may form early in the growth of the crop. Table 2.5 presents a list of some crops in which disease can be greatly furthered by unchecked growth of volunteers.

Although destruction of groundkeepers and volunteers before a new crop is sown is often relatively easy, and can be done by tillage or chemicals, this is not always the case. Thus, in many European countries, transmission of potato viruses through volunteers continues to present serious problems.

2.2.3.3 Roguing

Roguing of diseased individuals in a crop not as yet widely infected may benefit many crops, but is economically justified in particular in plantation crops and in seed crops.

In perennial plantations, which after heavy investment must be maintained over long periods, roguing can slow down the rate of disease spread and add years to the life of the plantation. In fact, in plantation crops in the tropics, where labour is available for this purpose, roguing is standard practice to slow down the spread of viruses such as the cassava mosaic virus (Chant, in Kranz et al. 1977) and the yellow crinkle virus of papaw (Adsuas, in Kranz et al. 1977).

Roguing of infected stools or shoots is recommended most particularly to keep sugarcane diseases down; it is effective against downy mildew *(Sclerospora sac-*

Table 2.5. Some diseases significantly aided in overseasoning by volunteer plants or groundkeepers

Crop	Disease	Pathogen	References
Fungal Diseases			
Wheat	Stripe rust	*Puccinia striiformis*	Line (1976)
Wheat, oat	Stem rust	*P. graminis*	Gerechter-Amitai (1973)
Pearl millet	Downy mildew	*Sclerospora graminicola*	Kenneth (in Kranz et al. 1977)
Tobacco	Blue mould	*Peronospora tabacina*	Lucas and DeLoach (1964)
Soybean	Various	*Macrophomina phaseolina* *Phomopsis sojae*, *Sclerotium rolfsii*	Fett (1978)
	Downy mildew	*Peronospora destructor*	Viranyi (pers. comm.)
Potato	Late blight	*Phytophthora infestans*	
Bacterial Diseases			
Cotton	Angular leaf spot	*Xanthomonas malvacearum*	Sabet (in Kranz et al., 1977)
Virus Diseases			
Sugarbeet	Virus yellows, curly top		Anon (1965b) Brodbent (1969)
Carrots	Carrot thin leaf, carrot motley dwarf		Howell and Mink (1977)
Groundnut	Rosette		Gibbon (in Kranz et al., 1977)
Barley	Yellow dwarf		Bayon and Ayrault (1977)
Nematodes			
Potato	Cyst nematode	*Globodera rostochiensis*	Anon (1977a)
Onion, rhubarb	Stem eelworm	*Ditylenchus dipsaci*	Anon (1975)

chari), smut *(Ustilago scitaminea)*, red stripe *(Xanthomonas rubrilineans)*, and sugarcane mosaic viruses.

In seed crops, virus diseases are also the chief objective of roguing, particularly in crops such as potatoes and beets. In these crops the freedom from disease is often more important than the size of yield. This justifies the expense of roguing.

Roguing of virus-infected plants is a wide-spread practice in Chinese agriculture, where one of the diseases checked in this way is the rosette dwarf virus of wheat (Kelman and Cook 1977).

Roguing should, of course, always be carried out at the earliest possible moment after symptoms become apparent. Timing of the operation is of special significance where pathogens produce resting bodies as the crop matures; roguing must then be completed before this stage is reached, as mentioned by Kenneth (in Kranz et al. 1977) for roguing of setaria before *Sclerospora graminicola* forms oospores.

Roguing is also widely recommended to reduce infection by nematodes and other pathogens on valuable crops. Infection by *Ditylenchus dipsaci* on narcissus (Anon 1974) and tulip (Anon 1972), and by species of *Aphelenchoides* on strawberry (Anon 1973a), can be delayed in this way. Careful and constant roguing (every second day) has proved successful in eliminating the dreaded *Pseudomonas woodsii* from carnations in Israel (Brosh, pers. comm.).

2.2.3.4 Removal of Diseased but Living Parts from Trees

A number of important pathogens of fruit are capable of over-wintering on dormant parts of their hosts, mostly on bud scales. This includes powdery mildews of apple, rose, grapevine, gooseberry, raspberry and hop, apple scab, and leaf curl (*Taphrina deformans*) on peach. Where such infection is recognizable during dormancy, infected parts must then be pruned with great care, so that none are missed. Pruning operations will be referred to in greater detail in the next section and in Chap. 2.13.1.

Periodical removal of infected leaves of fruits involves a great labour outlay. It is practiced in subsistence farming (Putter 1979) and is recommended for some tropical plantation crops. On cocoa, for example, weekly picking of infected pods, before these are mummified, has proved worthwhile to reduce incidence of black pod caused by *Phytophthora palmivora* (Muller 1974) and of Monilia pod rot caused by *Monilia roreri* (Ampuero, in Kranz et al. 1977). On bananas, removal of flowers just before they drop naturally has been advocated to control cigar-end rot caused by *Trachysphaera fructigena* (Beugnon et al. 1970), and removal of distal buds, when all "hands" have opened, to control *Colletotrichum musae* (Feakin 1971).

In climates with mild winters, such as the Mediterranean climate, old leaves, still green, frequently persist until the following spring on deciduous fruit and vine. Such leaves may carry viable spores of diseases such as rusts (*Tranzschelia* spp.), apple scab (*Venturia inaequalis*) and downy mildew of grapes (*Plasmopara viticola*). In this climate, stone and pip fruit trees are often sprayed with oil to break their dormancy, and this treatment is apt to remove such persistent leaves. However, where such sprays are not applied, and in vineyards, these leaves should be removed by hand where strong attacks of such diseases are expected.

Prophylactic removal of clusters of summer bloom, and of shoots arising from scaffold branches, is recommended for apples and pears to avoid their becoming infected by *Erwinia amylovora*, the fireblight bacterium (Anon 1969 b). Removing the flowers at the pink bud stage (by spraying with diquat), or even grubbing visibly diseased peach trees, has been successful in reducing incidence of peach rosette and decline virus in Australia (Smith et al. 1977).

Removal by hand of all basal growth of hops up to 4 ft in height, including infected as well as susceptible but uninfected growth, has been highly effective in reducing infection by downy mildew (*Pseudoperonospora humuli*) in the United States (Romanko 1965).

2.2.4 Destruction or Inactivation of Inoculum Deriving from Dry Plant Matter and Debris

Effects of cultural practices on inoculum derived from dried-up plant organs and crop debris will here be dealt with under two headings:

a) *Parts of trees* that have dried up as a result of parasite action, such as twigs, branches or limbs of trees or mummified fruit, and *entire trees* that have succumbed to disease, or their stumps. The techniques employed to remove such dead matter include pruning, excision or heat treatment of affected parts on the tree, and grubbing. Special techniques have also been developed to treat stumps which it is difficult or uneconomical to remove. The dead matter removed is usually burned or buried.

b) *Crop residues* on the ground that have to be treated in situ. Such residues, which are often mixed with soil, can be dealt with by incorporation in the soil, by burning or flaming, by flooding, and by soil disinfection by heat and other means.

2.2.4.1 Sanitation of Diseased Trees

Pruning is not primarily s sanitary measure. It has important functions in shaping the tree, in removing excess growth and in regulating periods of leafing and fruiting. These functions in themselves have an indirect bearing on disease. A well-shaped and properly aerated crown will be a factor restricting outbreaks of pathogens with high moisture affinities. Bud-break may be regulated to lessen dangers of late frost killing tender organs, with all that this implies for pathogen control. However, what concerns us here is the direct effect of pruning in removing dead wood to reduce the spread of infection.

Pruning as a sanitary measure is applicable to all the numerous diseases that cause blighting, drying up, and die-back of twigs or branches, such as *Botryodiplodia theobromae* on citrus, *Glomerella cingulata* on apples, and many others.

Since many of these pathogens penetrate branches beyond the area visibly affected, the usual practice is to prune several centimetres into the healthy-looking wood.

Pruning is mostly carried out after harvest, and can then be combined with the removal of mummified fruit. This is of great sanitary significance in many crops, such as grapes, for the control of black rot caused by *Guignardia bidwelli* (Stevens and Stevens 1952) and on apples to control brown rots caused by *Sclerotinia fructigena* and allied species (Anon 1965a; Merrill and Petersen 1979); but it is perhaps most important on some tropical crops, e.g., removal of coffee berries affected by *Colletotrichum coffeanum* (Gassert 1978) or of cocoa pods affected by *Phytophthora* (Muller 1974).

Infection on limbs too large to be pruned, or on tree trunks can be excised or sometimes subjected to heat treatment. This is laborious and expensive, but worthwhile especially where first infections in a plantation can be eradicated. Excision of cankers under such conditions is practiced on all types of fruit, e.g., on apples against *Schizophyllum* sp., on citrus against gummosis *(Phytophthora citrophthora)*, on coffee against *Gibberella stilboides* and on all the numerous crops affected by *Gloeosporium perennans* and species of *Ganoderma.*

Heat treatment, or cauterization, of cankers on the bark of fruit trees can be more rapid and therefore cheaper than tree surgery. Hough et al. (1979) have reported complete killing of *Phytophthora parasitica* on citrus bark when lesion temperatures reached 60 °C. Hawkins (1976) has had similar success in cauterizing cankers caused by *Pseudomonas syringae* on stone fruit trees.

If the entire tree succumbs to disease, and if the number of such trees in the plantation is not high, it has to be removed or made innocuous as a focus of infection.

This is of particular importance where the cause of death is one of the multihost fungi that progress by active proliferation of rhizomorphs or similar organs. Good examples are *Stereum purpureum*, the cause of silver leaf disease of fruit trees, and *Armillariella mellea* which causes root rot on a wide range of trees in temperate climates and especially in the tropics. To eliminate trees, they are usually first cut down, and their stump may then be treated in various ways. It may be removed by grubbing, made innocuous by poisoning or be subjected to the action of saprophytes, e.g., by cultivating a carpet of creeping legumes (Wastie 1975). If the latter is planned, it is often advised to ring the trees at an earlier stage in order to prevent carbohydrates from reaching their base, and thus to starve the pathogen. This has been shown to be a good way of subduing *Armillariella mellea* in oilpalm plantations (Turner 1970).

All the above treatments for elimination of stumps as sources of crop disease must also be considered where a crop is to be planted on cleared woodland or in place of hedges, windbreaks etc. (Anon 1961). Combined treatment of tea stumps by fumigation and subsequent mulching and fertilization has proved very effective in suppressing the root rot pathogen *Ganoderma* sp. by increased saprophyte activity (Varghese and Chew 1973).

2.2.4.2 Removal of Crop Debris or its Incorporation in the Soil

Handling crop debris is one of the principal tasks of sanitation. This comprises collecting the debris and removing it from the field, incorporating it in the upper soil layers, burying it more deeply, burning it, or activation soil flora to overcome the inoculum in the debris. The first three aspects will be discussed here, the last two in Sections 2.2.4.3 and 2.4.1.2.

Removal of Debris. Some crop residues are of considerable value: green matter, such as foliage of root crops that can be used for feeding livestock, or dry matter such as straw or cotton stalks, used for bedding and lately also for silage. Removal of such residue is dictated by economic considerations. From the sanitary point of view, it can only be hoped that manure from cattle and sheep fed such residues will be used with care (see below).

Removal of debris is of sanitation value chiefly where the major part of the inoculum is located in aboveground parts of the crop residue. Downy mildews, such as those of lettuce (*Bremia lactucae*), peas (*Peronospora pisi*) and sorghum crops (*Sclerospora sorghi*) form most of their oospores in leaves and stalks. *Verticilium dahliae* produces a profusion of microsclerotia in dead potato and tomato stalks. The species of *Sclerotinia* and *Sclerotium* that affect so many hosts form their sclerotia chiefly in above-ground organs and root collars. In all these and many other cases, the debris is highly infective, and should be removed from the field, buried in pits or burned, if this is economically feasible. Such removal can obviously never eliminate all the inoculum, but can still be well worthwhile.

Incorporation of Debris in Upper Soil Layers. Most crop residue that is not of direct use is ploughed into the soil and enriches it both in organic matter and in pathogens. This raises matters of crop sequence, and of soil amendments and moisture management to activate soil saprophytes. These measures are discussed in the appropriate chapters.

How important is the size of the pathogen population which crop debris can do so much to enlarge in soils in which it has been incorporated? An answer to this question, relating to the various types of fusarial plant pathogens, has been given by Nash-Smith (1970). There is a difference between those *Fusaria* that cause cortical rots (CR) and those that cause vascular wilts (VW). In the CR types, the infection court is fixed, whereas in the VW types it is motile, near the root tip. As the rootlets of the crop move through the soil, they will meet fusarial propagules even if these are present in low, undetectable levels and infection will take place, especially in crops that occupy the soil over long periods such as bananas. For the VW types the size of the pathogen population is therefore of little importance, while it is highly important for the CR types.

Burying Inoculum Deeply. Burial of inoculum-carrying debris below the layers commonly ploughed does not necessarily kill it, but removes it from the site of action. Sclerotia and other resting organs of many pathogens can survive such burial for long periods: they are, as it were, in "deep freeze." Subsequent deep ploughing may again turn them up, viable and ready for infection. It is frequently better policy not to bury such inoculum beyond the layer in which it can germinate, and in which it is subject to attack by other microorganisms. Indeed, aiding it to germinate in the absence of hosts is often a better way to rid soil of inoculum, as will be discussed in the chapter on soil amendments.

2.2.4.3 Burning or Flaming Crop Residue

Burning inflammable plant residue has its problems. Where residues are dry and inflammable enough to be burned without fuel, this can be cheap and fairly cost-effective, but such burning hardly ever destroys all the infective material in the field. It is also objectionable in the more environment-minded countries, because the smoke constitutes an environmental nuisance, and burning stubble on slopes may enhance danger of erosion. In addition, where organic matter is low in the soil, burning stubble is the waste of a resource which should, if possible, be conserved.

Flaming by flame-thrower, or by other devices using fuel, can be much more reliable, as the heat and rate of flaming can be controlled. However, with rising fuel prices, flaming ceases to be economic in many cases.

Table 2.6 presents a list of diseases which have been combatted, with at least partial success, by burning or flaming. Much of the data in this table have been collected by Hardison (1976). Most of the crops to which fire can be applied are graminaceous: cereals in which inoculum can be destroyed after harvest, or pasture grasses which can periodically be freed from inoculum before they make new growth. Most of the remaining crops in the table, such as mint, lucerne and lupin, are also harvested in successive cuttings. Hardison (1976) has pointed out that the increasing success of non-tillage in some major crops, and the resultant problems

Table 2.6. Some diseases at least partially controlled by fire and flame

Crop	Disease	Pathogen	References
Field and Vegetable Crops			
Wheat	Take-all	*Gaeumannomyces graminis*	Hardison (1976)
	Eye-spot	*Pseudocercosporella herpotrichioides*	Hardison (1976) Georgiev et al. (1976)
	Leaf blotch	*Septoria nodorum*, *S. tritici*	Hardison (1976) Dinoor (1980) Saad (in Kranz et al., 1977)
	Flag smut	*Urocystis agropyri*	Hardison (1976)
Wheat and barley	Snow mould	*Gerlachia nivalis*, *Typhula idahoensis*	Huber and Watson (1970)
Barley	Leaf blotch	*Rhynchosporium secalis*	Yarham (1979a)
Oat	Leaf blotch	*Septoria avenae*	Hardison (1976)
Rice	Stem rot	*Sclerotium oryzae*	Webster et al. (1976)
	Sheath blight	*Corticium sasaki*	Hardison (1976)
	Stem nematode	*Ditylenchus angustus*	Sturhan (in Kranz et al., 1977)
Sugarcane	Leaf blight	*Leptosphaeria* sp.	Hardison (1976)
	Fiji disease	virus	Hardison (1976)
Lupin	Brown spot	*Pleiochaeta setosa*	Hardison (1976)
Lucerne	Dodder	*Cuscuta* spp.	Hardison (1976)
Potato	Wilt	*Verticillium dahliae*	Ausher (pers. comm.)
Peppermint	Rust	*Puccinia menthae*	Horner (1965)
	Wilt	*Verticillium dahliae*	McIntyre and Horner (1973)
Asparagus		*Puccinia asparagi*	Rieuf (pers. comm.)
Lowbush blueberry	Die-back	*Diaporthe vaccinii*	Hardison (1976)
(*Vaccinium* spp.)	Canker	*Godronia cassandrae*	Hardison (1976)
Grasses			
Lolium perenne, *Festuca arundinacea*	Blind seed	*Gloeotinia temulenta*	Hardison (1976)
Lolium perenne, *Festuca arundinacea*	Ergot	*Claviceps purpurea*	Hardison (1976)
Paspalum dilatatum	Ergot	*Claviceps paspali*	Hardison (1976)
Festuca rubra	Seed nematode	*Anguina agrostis*	Hardison (1976)
	Stress eyespot	*Phleospora idahoensis*	Hardison (1976)
Bromus inertis	Leaf spot	*Selenophoma bromigera*, *Pyrenophora bromi*	Hardison (1976)
Lolium rigidum	Seed nematode	*Anguina* sp.	Hardison (1976)
Poa pratensis	Leaf mould	*Drechslera poae*	Gray and Guthrie (1977)
	Leaf rust	*Puccinia poae-nemoralis*	Hardison (1976)
	stripe rust	*P. striformis*	Hardison (1976)
	stem rust	*P. graminis*	Hardison (1976)
Cynodon transvalensis	Leaf-gall nematode	*Anguillulina tumefaciens*	Hardison (1976)

of debris management, make burning or flaming an attractive proposition for inoculum control, if it can be carried out at sufficiently low cost.

2.2.4.4 Flooding Fields and Orchards

Flooding fields to rid them of inoculum is an ancient practice, probably first developed for crops grown in rotation with paddy rice. Such flooding has been held to be one of the key factors for the low incidence of soil-borne diseases in present-day Chinese agriculture (Kelman and Cook 1977). In regions with intensive vegetable culture, in the Lower Yangtze and South China, the beds on which vegetables are grown are broken down periodically, and the land is converted to one or two crops of rice or planted to water spinach, water chestnut, water bamboo or lotus (Williams 1979).

In western agriculture, flooding has been adopted for disease control largely as a result of the comprehensive studies by Stover (1954, 1955) on flood-fallowing for the eradication of the *Fusarium* wilt of bananas (*F. oxysporum* f.sp. *cubense*), and of research on control of vegetable pathogens in Florida (Moore 1949; Johnson and Berger 1972). Studying every detail of the effects of flooding on the banana fusariosis, Stover concluded that the key to success lies in restricting the supply of oxygen to the soil as drastically as possible. Thus the fungus survives least in stagnant water, while flowing water bringing in oxygen lengthens its survival. Another factor of importance is temperature, survival under water being 10–20 times greater at 13 °C than at 24°–34 °C. Under ideal conditions, the *Fusarium* did not survive more than 25 days under $1^1/_2$ to 3 ft of water. However, use of flooding for control of banana wilt has now been superseded by use of resistant varieties.

A list of pathogens controllable, at least to a significant degree, by flooding is presented in Table 2.7, and this list will be seen to include nematodes as well as fun-

Table 2.7. Pathogens controllable wholly or partly by flooding

Crop	Disease	Pathogen	References
Fungal Pathogens			
Banana	Wilt	*Fusarium oxysporum* f. sp. *cubense*	Stover (1954, 1955)
Cotton	Wilt	*Verticillium dahliae*	Butterfield et al. (1976), Ionannou et al. (1977)
Vegetables	White mold	*Sclerotinia sclerotiorum*	Moore (1949)
Tobacco	Black shank	*Phytophthora parasitica* var. *nicotianae*	van Schreven (1948)
Nematodes			
Celery	Root-knot	*Meloidogyne* sp.	Johnson and Berger (1972)
	Stubby root	*Trichodorus* sp.	
	Stunt	*Tylenchorhyncus* sp.	
Banana	Burrowing nematode	*Radopholus similis*	Stover and Ostmark (1979)
Rice	White tip	*Aphelenchoides oryzae*	Anon (1968)
Parasitic Plant			
Various	Broomrape	*Orobanche* spp.	Kasasian (1971)

gi. In the work carried out in Florida, it was found that *Sclerotinia sclerotiorum* could be eliminated effectively by flooding and draining the soil alternately at intervals of three days (Moore 1949), and that nematodes were reduced to low levels by two cycles of flooding for two weeks and then drying for two weeks (Johnson and Berger 1972). In California, growing cotton in one-year rotation with flooded rice has proved most effective in controlling cotton wilt caused by *Verticillium dahliae* (Butterfield et al. 1976).

Flooding for a few days also reduced the incidence of safflower rust (*Puccinia carthami*) (Klisiewicz 1977).

In the use of flooding, as of all other methods of soil disinfection, it has to be borne in mind that the soil microflora antagonistic to, or competitive with, the pathogen may be adversely affected – more so than the pathogen at which the operation is aimed. Stover (1959) has, in fact, noted just such an occurrence for the *Fusarium* of bananas: in soils markedly conducive to rapid development of the wilt, flood-fallowing largely eliminated the suppressive flora, and surviving inoculum quickly caused most of the re-planted bananas to wilt.

In countries in which fish are grown in ponds over large areas, cropping of the soil presumably cleaned of pathogens by long submersion deserves to be considered.

2.2.4.5 Disinfestation of Soil by Heat

Three practices of applying heat to the soil to rid it of pathogens are currently employed in the intervals between successive crops:
tillage, steam disinfestation, and solar heating.

Tillage. Exposure of top layers of soil to strong radiation, and thus heating and drying it thoroughly, has long been practiced in the warmer climates as a means of disinfestation, chiefly for the control of nematodes. If the soil is ploughed carefully and repeatedly, this has proved fairly effective against root-knot and free-living nematodes (Weischer in Kranz et al. 1971). It is still the only method cheap enough to be applicable to low-value crops.

Steam. Soil disinfestation by steam is chiefly practiced in greenhouses, but also in nurseries and market gardens. Techniques of applying steam have been reviewed by Newhall (1955). To permit the steam to penetrate soil properly and thus achieve optimal effects in the control of pathogenic soil fungi, bacteria, nematodes as well as pests and weeds, the soil should be loose, free from lumps and fairly dry. However, this method of soil disinfestation has the disadvantage of creating a biological vacuum and involves a fairly high capital outlay and considerable fuel consumption. It is therefore unlikely to be adopted for any but the most high-value crops.

Solar Heating. The utilization of solar heat, applied to soil prepared for sowing, irrigated, and covered by polyethylene mulch (Fig. 2.3) was developed in the early 1970s by Katan and his co-workers (Katan et al. 1976; Grinstein et al. 1979, Katan 1980) an has also been investigated by Pullman et al. (1979) in California. It has been found effective against a variety of soil-borne fungi, nematodes and parasitic

Fig. 2.3. Disinfestation of soil by solar heat under polyethylene mulch; one strip of soil has been left uncovered to serve as a control plot. (Courtesy J. Katan, Faculty of Agriculture, Rehorot, Israel)

Table 2.8. Pathogens controlled by solar heating in Israel and the USA. (According to Katan 1980)

Crop	Disease	Pathogen
Tomato	Wilt	*Fusarium oxysporum* f. sp. *lycopersici*
Tomato	Corky root	*Pyrenochaeta lycopersici*
Tomato, eggplant	Wilt	*Verticillium dahliae*
Potato	Free nematode	*Pratylenchus thornei*
Potato, onion	Root rot	*Rhizoctonia solani*
Onion	Pink root	*Pyrenochaeta terrestris*
Onion	Wilt	*Fusarium oxysporum*
Carrot, eggplant	Broom rape	*Orobanche aegyptiaca*
Groundnut	Pod spot	Various soil organisms
Groundnut	Collar blight	*Sclerotium rolfsii*
Cotton	Wilt	*Fusarium oxysporum* f. sp. *vasinfectum*
Various	Root rot	*Thielaviopsis basicola*
Various	Root rot	*Pythium* spp.
Pistachio nut	Wilt	*Verticillium dahliae*

plants, which are listed in Table 2.8. The depth to which the treatment is effective depends on the intensity of radiation, the duration of the treatment, and the pathogen. At monthly maxima of 49°–52 °C soil temperature at 5 cm, and of 42 °C at 15 cm depth, two weeks under polyethylene sheets sufficed to eliminate *Verticillium dahliae* to a depth of 25 cm, and effected reduction of the population of

Fig. 2.4. Effect of solar heating of the soil on the growth of onions in a *Pyrenochaeta terrestris*-infested soil. Forefield, beds of an untreated plot (drip irrigation) with stunted plants severely affected by the pink root disease; mid-field, the beds treated with solar heat. (Katan et al. 1980)

Fusarium oxysporum f. sp. *lycopersici* by 94% at 5 cm, by 68%–100% at 15 cm, and by 54%–63% at 25 cm depth.

Some evidence has been adduced that solar heating achieves its effects not only by direct thermal control, but also by biological control (Katan 1980). Pathogen populations in deeper soil layers, where relatively low temperatures prevail, were also reduced. Such biological control, which may have prolonged effects, may operate through (a) reduction of pathogen populations or activity by antagonistic microorganisms whose effect is enhanced by heating, and (b) a shift in the biological equilibrium of the soil in favour of microorganisms that prevent re-infestation by the pathogen and renewed build-up of its population, a point of crucial importance.

An added benefit of solar heating is the control of many weeds, with numerous Gramineae especially sensitive to this treatment. Moreover, solar heating of the soil frequently improves crop growth even in non-infested soil.

2.3 Crop Sequence

The term crop sequence is here preferred to that of crop rotation, because the latter implies repeated sowing of the same crop at periodical intervals. This applies much more to extensive farming economies than to intensive cropping, and does not cover monoculture.

2.3.1 Aims of Crop Sequence Management

Regular crop sequences, or what was in the past aptly described as crop rotation, used to be one of the mainstays of agricultural practice. It here denotes sequential sowing of certain crops with intermittent periods of fallow. The general acceptance of this practice then had diverse reasons, which are listed below.

A. General Agronomy.
a) Better use of nutrients.
b) Desirable effects on soil texture, with deep-rooted crops alternating with crops with shallow roots.
c) Water economy, in particular conservation of water in years of fallow.
B. Weed Control. Row crops, in which in-growth tillage for weed control could be practiced, alternated with crops sown by broadcasting; crops likely to stifle weeds by their rapid growth and dense foliage (beets, potatoes) alternated with crops making slow and sparser growth (onions).
C. Soil-Borne Pathogens were held to be amenable to "starving-out" by depriving them of their hosts for several years in succession.

Most of the above agronomic considerations no longer apply. Nutrient supply by fertilizers is relatively cheap. Soil texture can be maintained in a favourable state by modern practices of tillage in conjunction with soil amendments. With the advent of selective herbicides for most major crops, weed control has lost greatly in importance in considerations of crop sequence.

That leaves the aspects of water conservation in dry farming and of disease control, to which the proper sequence of crops and of fallow periods can still make important contributions.

In the context of disease control, in particular, great care must be taken not to upset traditional crop sequences unless careful research has shown this to improve crops. Thurston (1980) has described a case in point. In Peru, where potato soils are commonly infested by the cyst nematodes *Globodera pallida* and *G. rostochiensis*, the Incas passed a law that potatoes must not be grown on the same land more than once in 7 years. It so happens that this is the period required for nematode populations to diminish to a level which makes it possible to produce another potato crop. The Spanish conquerors of Peru, contemptuous of a law that to them seemed senseless, repealed it with disastrous results.

2.3.2 Economic Considerations

The considerations applying to every choice of crops, whether annual or perennial, are also those applying to the choice of crop sequence. They will be listed here with special reference to aspects of crop protection.

a) Assessment of income expected from alternative crops, and of the risks run in sowing each crop. As detailed in the chapter on cost/benefit and risk assessment (Chap. 2.1), these risks include those of the weather and the market, and certainly those of pest and disease development, with its significance to both the individual farm and the farming community. Frequently considerations of income decide farmers to grow cash crops, although these risk disease. Thus, in Arkansas (USA), in soybean fields infested with the nematode *Heterodera glycines*, sowing of lupins every other year has been demonstrated to give best soybean yields. However, lupins not being a cash crop, soybean growers prefer rotations with sorghum or cotton, although these reduce nematode populations less than the lupins (Welch et al. 1977).

b) Availability and cost of labour in general, and particularly in seasons in which the yield is picked. Availability of labour for weeding and sanitation work, and of skilled or semi-skilled labour for spraying, must also be taken into account.

c) The extent to which a crop can be farmed by machinery, again with the accent on collecting the yield, e.g., by cotton-picking or groundnut lifting combines. Is suitable equipment available on the farm, can it be secured by co-operative action of farmers, or are contractors at hand, at reasonable prices, e.g., for applying sprays by ground equipment or aircraft? If expensive machinery, adapted to a specific crop, has been acquired, it has to be used fully for better amortization, and this is a strong incentive for growing that crop as frequently as possible. Monoculture may then be chosen, although it may present disease hazards.

d) Limitations of chemical pest control, including environmental limitations and health hazards to labourers and consumers. Thus crops with produce to be consumed fresh and requiring constant chemical protection (lettuce, tomato) run up against reulations limiting the amount of residue permissible at harvest. They may have to be replaced by crops in which this is not a problem (potatoes, carrots).

As mentioned above, in modern farming investment in farm machinery and installations is often so high that farmers have to grow cash crops at frequent intervals, and cannot afford to take heavy risks. What constitutes a cash crop and which risk is deemed heavy, clearly depends on the farmer's economic status. Advanced farmers with adequate capital may take the long view and choose crop sequences that involve some risk but bear reasonable promise of greater revenue. Subsistence farmers, however, as detailed in Chap. 2.1, cannot afford to run avoidable risks, otherwise they risk famine. They are also strictly limited in resources and, perhaps equally important, bound by tradition. They may therefore stick to a given farm routine as long as this provides the minimum they need. This, for example, is the position described by Glaser and Ogbogu (1974) for cassava growing in Nigeria. Although monoculture of cassava increases incidence of bacterial blight (*Xanthomonas manihotis*), the scarcity of land and low cost of producing the crop makes farmers grow it continuously.

Carlson and Main (1976) have reviewed the cost involved in choosing crop sequences, in terms of income given up by inability to produce the single highest-value crop continuously. This cost may be small, when alternative crops for a given area also have high net returns, and will be high where alternative crops are much less valuable. Linear programming evaluation of cropping patterns can help to assess such costs and indicate crop sequence restraints, i.e., the maximum proportion of total land area a grower can sow or plant with any crop at one time without reducing crop yield or quality. The size of such restraints and their costs have been calculated by Shumway (as quoted by Carlson and Main 1976) for ten crops in California (Table 2.9). He found a 33% restraint on sugar beet acreage, which

Table 2.9. Effects of crop sequence restraints for pest control on crop income per acre in 95 areas of California. (Data in columns 2 and 3 from Shumway, as quoted by Carlson and Main 1976)

Crop	Crop sequence restraint [a]	Range of cost of crop sequence [b]	Interpretation
Maize	80	0	A large proportion of the available area can be sown with these crops, and equally profitable alternative crops are available
Grain sorghum	80	0	
Tomatoes for processing	67	0	
Potatoes	50	0–12	These crops may be grown every second year; in the off-years losses from alternative crops will only be slight
Barley-grain sorghum (double crop)	50	0–14	
Safflower	50	0–17	
Cotton	33	0–88	These crops should not be grown more than once in 3 years. Profitable alternative crops are available in some areas, but in others considerable loss in income is to be expected in years in which alternative crops have to be grown
Sugar beet	33	0–37	
Dry beans	33	0–35	

[a] Maximum percentage of the acreage in local homogeneous production areas that can be sown to the specified crop without reducing crop yield and quality
[b] Per acre reduction in net returns from reducing acres in host crop by rotation restraint. Range from the various production areas surveyed

means that under the conditions studied, growing sugar beets more frequently than once in 3 years in the same field would reduce yields due to the build-up of nematodes, weeds, or disease. The range of costs due to crop sequence restraints varied greatly: in many cases it was zero or very low, because under Californian conditions a great number of profitable alternative crops can be grown. Where the upper limit of the cost range is shown to be high, as in the case of cotton, this indicates that in some of the 95 areas surveyed no alternative crops with incomes comparable to that from cotton were available.

2.3.3 Choice of Crop Sequence in Relation to Other Farming Practices

Decisions on crop sequence, whether relating to the aspect of diseases or not, have to be taken with due regard to a multitude of other cultural practices, and frequently also to chemical control. As Curl (1963) has stated: "Disease control benefits derived from rotating crops can seldom be attributed to that practice alone, since the other necessary practices will affect perpetuation of the pathogen and physiology of the host plant, thereby influencing success of control." The most important practices with which regulation of crop sequence interacts are listed below.

1. Tillage, which redistributes inoculum in the soil, buries some of it, exposes some to drying, and in general affects the amount and location of inoculum the next crop has to contend with.

2. Soil Amendments, because of their effects on the host and their biological effects on soil flora antagonistic to, or competitive with, pathogens; green manure is an essential element in many crop sequences.

3. Fertilization, which can accentuate susceptibility of the crop, but can also help it to outdistance pathogens and "grow away" from them.

4. Water Management, especially the decision whether or not to irrigate a crop in an arid country, evidently constitutes one of the essentials of crop sequence considerations. Many soil-borne pathogens, such as *Sclerotium rolfsii*, will not affect crops in arid countries unless the crop is irrigated.

5. Sowing Seasons. Some important pathogens are limited to certain seasons, and susceptible crops can be grown with impunity in other seasons (see below).

6. Flooding or Soil Disinfestation by heat or by chemicals largely eliminates certain pathogens, changes the balance of forces between pathogens and other soil microflora, and may thus affect "disease expectancy" of several subsequent crops.

7. Weed Control. Weeds susceptible to soil pathogens can ruin the most carefully planned crop sequence (Chap. 1.7.1). This also applies to symptomless carriers of pathogens (see below).

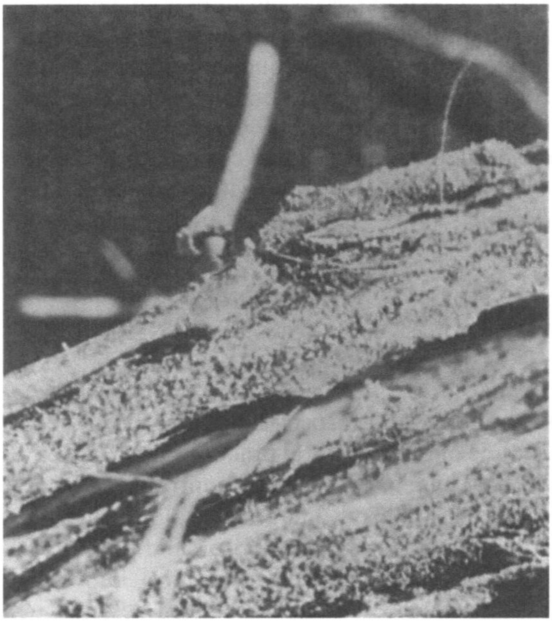

Fig. 2.5. Microsclerotia of *Verticillium dahliae* formed on a potato stalk. Because of the large number of hosts on which this pathogen causes wilt diseases, and the formation of microsclerotia on some of these hosts, populations of *V. dahliae* in the soil can be reduced only by a combination of practices: well thought-out crop sequence, careful weed control, certain soil amendments and, where possible, judicious choice of sowing seasons and varieties. (Photograph by courtesy of the Division of Plant Pathology, Agricultural Research Organization, Bet Dagan, Israel)

It is by now common knowledge that altering the sequence of crops is not, by itself, capable of achieving adequate reduction of diseases caused by polyphagous pathogens, especially those with marked saprophytic ability or equipped with highly persistent organs of survival (chlamydospores, sclerotia etc.). Nevertheless, the combination of planned crop sequence with the other practices enumerated above has led to notable success in managing some of the more destructive soil-borne diseases, such as those caused by *Sclerotium rolfsii* on groundnut (Garren 1961) and *Verticillium dahliae* on cotton (Wilhelm et al. 1966), as explained in Fig. 2.5. Well-planned crop sequences will in many cases slow down pathogen build-up in the soil, and thus delay the necessity for thermal or chemical disinfestation treatments.

2.3.4 The Benefits of Fallowing

That fallowing is beneficial has been known since ancient times. In fact, in one of the cradles of cropping, the Near East, fallowing has been considered not only beneficial but mandatory, so much so that a sabbatical year, in which the entire farming community was not to sow anything, has been prescribed in Jewish religious law.

A distinction must be made between fallowing out of sheer necessity, usually because of limited water supply, and fallowing by choice in order to reap specific benefits. As for fallowing out of necessity, it may suffice to say in the present context that it should be used to best advantage in keeping the field free from disease: repeated tillage is helpful in exposing pathogens to drought, and weed control makes it possible to make the fallow a period truly free from hosts for pathogens. What has been called "ecofallowing," a system of controlling weeds and conserving soil moisture in winter wheat – grain sorghum – fallow rotation with reduced tillage (Doupnik and Boosalis 1980), is on the borderline between fallowing out of necessity or from choice. Its effect on cereal disease largely depends on suitable stubble management.

Fallowing by choice is largely an economic consideration, taking into account revenue from crops that could be grown during the fallowing period, alternative methods of reducing pathogen populations in soil, and benefits of fallowing for succeeding crops. Even the income from reduced yields may be better than no revenue, especially where investment in the field is heavy. Crowther and Cochran (1942) have stated that in Egypt fallowing on irrigated cotton land, although known to promote yields, is seldom practiced, because the expenses incurred in digging irrigation channels are too heavy to forego the income of crops grown every year.

Dry fallowing for pathogen control is generally restricted to extensive farming, where other methods of disinfestation are not economic; it is combined with specific crop sequences, soil amendments, and fertilizer and tillage practices (see chapters on these practices). Such fallowing frequently aims at nematode control. Combined clean fallow and growing barley has reduced populations of *Meloidogyne incognita* in cotton after 9–10 months (Carter and Nieto 1975). One fallowing period between rice crops sufficed to reduce the root lesion nematode *(Pratylenchus indicus)* to acceptable levels (Prasad and Rao 1978).

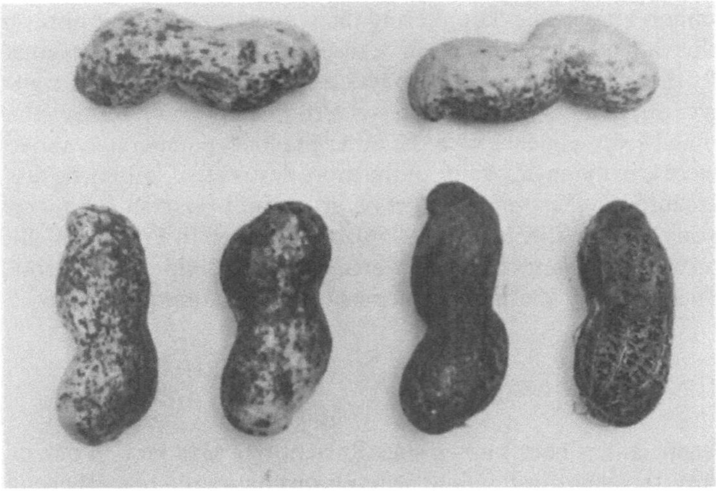

Fig. 2.6. Groundnuts affected by pod rot caused by *Pythium myriotylum*. The disease is markedly affected by crop sequence and irrigation practices. In infected fields, recommendations are for intervals of 3 years between successive crops of groundnuts, wet fallowing, and irrigation intervals as long as practicable (Frank, in Kranz et al. 1977). (Photograph by courtesy of the Division of Plant Pathology, Agricultural Research Organization, Bet Dagan, Israel)

Even a fallow period as short as 6 weeks, intervening between removal of tomatoes and sowing maize, sufficed to suppress *Pratylenchus brachyurus*, which was not suppressed where maize was grown without such a period of fallow (Brodie and Murphy 1975). Dry season fallowing has also proved successful in African regions with dry seasons in controlling brown rot *(Pseudomonas solanacearum)* in solanaceous crops (Waller, pers. comm.).

Wet fallowing, i.e., leaving land free from crops while irrigating it intermittently, is usually practiced for weeks rather than months. The purpose of such fallowing is to make pathogen propagules in or on the soil germinate, spend themselves or become susceptible to attack by saprophytes. This strategy works with pathogens whose propagules can be activated in the absence of root exudates from their hosts. It may be considered for pathogens forming sclerotia, such as *Sclerotium rolfsii* (Smith 1972a) or *Verticillium dahliae* (Menzies and Griebel 1967). Sclerotia of species of *Sclerotinia* may be reduced in numbers by wet fallowing (Smith 1972b), while they will retain viability in dry fallow for many years (Adams 1975). However, as will be mentioned in the next section, caution is needed in applying this practice to land harbouring species of *Fusarium* capable of forming chlamydospores.

Wet fallowing is also partly successful in reducing populations of *Pythium myriotylum*, one of the causes of pod rot in groundnuts (see Fig. 2.6), provided pathogen populations are not too high (Frank 1972). The method is also applicable to pathogens persisting in crop debris on the soil surface. Under arid conditions, *Alternaria solani* populations on potato and tomato debris were greatly reduced by wet fallowing (Rotem 1968).

Flood fallowing, which also comes under the heading of wet fallowing, has been referred to in Chap. 2.2.4.4 as a means of disinfesting soil from some fungal pathogens and nematodes.

Cultivating soil for one or a few crops and then abandoning it to revert to its natural state is the essence of shifting cultivation. This is practiced in the tropics on a very large scale and may also be considered as an extreme case of fallowing. There are many variations of shifting cultivation, e.g., clearing the tropical rain forest, then growing four or five crops of rice, sweet potato and cassava, and subsequently letting the forest take over for 10–20 years. Disease problems encountered in shifting cultivation are rarely as serious as in fields under constant cultivation.

2.3.5 What Are the Chances of Improving Crop Health by Management of Crop Sequence?

The factors to be assessed in weighing the chances of success in bettering the health of crops by optimal arrangement of their sequence are primarily the following:
a) The heritage of pathogens and debris left by the preceding crop.
b) The potential for survival of the residual pathogens, including survival or even multiplication on non-hosts.
c) The fullest, if intermittent, use of varietal resistance in crops grown in sequence.
d) Exploitation of the seasonal limitations of some root-infecting fungi and bacteria to grow susceptible crops in the pathogen's "off-season."

2.3.5.1 The "Heritage" of the Preceding Crop

The root-infecting pathogens likely to affect a crop are in the first place those left over from the preceding crop. If the same crop species, or perhaps a closely allied one, is sown once more, this applies to all kinds of pathogens that may have affected the first crop, specific as well as unspecific pathogens. However, if the second crop belongs to another species, the dangerous "heritage" narrows down to the more strongly saprophytic pathogens, those with wide host ranges, and those otherwise best fitted for survival (see next Chap.).

As regards the host range, three groups of pathogens may be distinguished:

a) Truly polyphagous species, in which there is no host specialization or, if several pathotypes exist, each one will affect a fair number of hosts. Such pathogens are *Pythium aphanidermatum*, *Verticillium albo-atrum*, *V. dahliae*, *Sclerotium rolfsii*, and *Phymatotrichum omnivorum*. They are the most difficult to combat by crop sequence, and often also affect a wide variety of weeds. Many nematodes also belong to this group, especially root-knot nematodes (*Meloidogyne* spp.).

b) Species with polyphagous as well as host-specific strains. Examples are *Rhizoctonia solani*, isolates of which from potato may affect various hosts, but not cereals which are attacked by other strains of the fungus; *Fusarium solani*, with many strains with wide host ranges, but host-specific ones on peas, beans, and some other crops. Several cyst nematodes (*Heterodera* spp.) have to be included here.

c) Pathogens specific to one host or a group of related hosts, such as the many formae speciales of *F. oxysporum*, or the causes of cereal foot and root rot, and the causal agents of most bacterial wilts and root rots.

What the crop leaves behind in the field is not only pathogens, but also its debris, the former often carried by the latter. The extent to which the preceding crop endangers the next crop depends greatly on the amount and nature of such crop debris, and on the length of time pathogens are permitted to persist in it or attack it. What is popularly known as "soil sickness" sometimes turns out to be due to direct effects of the accumulation of toxic decomposition products on the succeeding crop, or to the action of such products on the soil microbiota.

Some root-infecting or soil-borne fungi survive in the debris as dormant mycelium, as does *Gaeumannomyces graminis* in wheat stubble. Many more such fungi produce the dormant propagules to be discussed in the next section. In some cases, such as *Sclerotium rolfsii* and *Verticillium dahliae*, the quantity of propagules formed in crop debris and then released into the soil is so vast, that growing further susceptible crops in the field becomes impossible. For how long? That differs greatly and may range from no more than 1–2 years of non-host growth for *Sclerotium rolfsii* to many more years for *Verticillium dahliae*. Crop debris constitutes by far the most widespread means for persistence of bacterial pathogens in the soil.

It is manifestly impossible to review here the voluminous literature on the pathogen residues left by crops to their successors, or of the effect of the succeeding crop on such residues, and vice versa. Curl (1963) has attempted to do this, limiting his survey largely to data published in North American and English journals. Numerous long-term rotations (3–6 years) have been proposed to eliminate or reduce populations of polyphagous pathogens in the soil. Such recommendations are often of doubtful value in modern farming. However, short-term crop sequences markedly benefiting crop health have also been suggested in certain instances. As these are of more practical interest, some relevant cases are presented in Table 2.10. The crops recommended are all non-hosts of the pathogens concerned. Some of their effects on the pathogens are evidently physical in character, e.g., barley crops reducing nematode populations by drying out the soil, or paddy rice eliminating *Verticillium dahliae*, by submerging soil under water. However, in many other cases the special benefits derived from the crops listed, exceeding the effects of other non-host crops, are clearly to be attributed to biological effects on pathogen or soil flora.

No less important is the particularly bad heritage left by some crops for those succeeding them. Crops that are themselves susceptible to a large number of multi-host diseases are evidently the most dangerous predecessors. Among them, soybeans seem especially undesirable, as they are apt to leave residues of almost all major soil-inhabiting pathogens with wide host ranges. More surprising is the effect of some crops not affected by certain pathogens in raising the level of disease in the succeeding host crop of that pathogen. Thus application of green manure before sowing crops of cotton, potatoes, and various vegetables, has increased incidence of pathogens such as *Rhizoctonia solani* and *Pythium* spp. This will be more fully discussed in Chap. 2.4.1.2.

Special importance attaches to inoculum from preceding crops in orchards planted on land previously cropped with vegetables, or in arid countries, with al-

Table 2.10. Short-term rotations (1–2 years): Some crops exerting specific beneficial effects on pathogen populations left by preceding crops

The beneficial crop	Action exerted and pathogen	The preceding crop	Country	References
Fungal and Bacterial Pathogens				
Paddy rice	Eliminates *Verticillium dahliae*	Cotton	California, USA	Butterfield et al. (1978)
Pea	Eliminates *Gaeumannomyces graminis*	Wheat	France	Lemaire (pers. comm.)
Sudangrass	Eliminates *Pseudomonas solanacearum*	Tomato	Antilles	Messiaen (pers. comm.)
Nematodes				
Barley	Reduces *Meloidogyne incognita*	Cotton	Arizona, USA	Carter and Nieto (1975)
Legumes	Reduce *Pratylenchus brachyurus*	Pineapple	Puerto Rico	Guerout (1975)
Maize	Eliminates *Rotylenchulus reniformis*	Sweet potato	Trinidad	Brathwaite (1973)
Groundnut	Reduces *Meloidogyne incognita*	Tomato	USA	Belcher and Hussey (1977)
Sesame, wheat, *Phaseolus* spp.	Reduce *Pratylenchus indicus*	Rice	India	Prasad and Rao (1978)
Beet	Reduces *P. penetrans*	Cereals	France	Anon (1970b)

most any type of irrigated crop. Olives, almonds, peaches, avocado, and mango are among the crops in danger of being affected by *Verticillium albo-atrum* or *V. dahliae* if planted after eggplants, tomatoes, potatoes or any of the numerous other crops affected by these fungi. Similarly, *Sclerotium rolfsii* may severely attack apples planted soon after land has been cleared of groundnuts or the many other hosts of this pathogen. *Agrobacterium tumefaciens* has been found to attack severely vines planted on soil previously sown with carrots (Golan, pers. comm.). In such cases, the value of the orchard crop to be planted will almost certainly warrant pre-planting disinfestation of the soil by solar heat or chemicals. The same considerations apply where orchards susceptible to root-knot nematodes are planted on old vegetable land.

2.3.5.2 Survival of Pathogens Outside Crop Debris

Dormant propagules constitute the most effective mode of survival by root-infecting fungi. They comprise dispersal spores, resting spores, chlamydospores, and sclerotia (Garrett 1970). Characteristics of these spores are described below.

Dispersal Spores, including ascospores of *Gaeumannomyces graminis*, basidiospores of *Fomes annosus*, and conidia of *Cochliobolus sativus*, are the more delicate forms of dormant propagules; they are not usually capable of remaining viable in the soil for more than a few months, and are therefore most dangerous to susceptible crops sown in close succession.

Resting Spores may first be exemplified by the oospores of *Phytophthora cactorum*. These vary greatly in their period of dormancy, which may extend over many months. They ensure longevity of the population by the fact that, under any given set of environmental conditions, no more than a minor proportion of spores will germinate at any one time. This makes it so difficult to assess the practical significance of contamination by this and similar fungi, from the aspect of crop sequence.

Another type of resting spore is that produced by the cause of club-root of crucifers *(Plasmodiophora brassicae)*. These spores are produced in vast numbers, are fairly resistant to direct environmental influences, but a high proportion sometimes germinates spontaneously in unplanted soil. This rapidly reduces populations, especially in wet, slightly acid soil. Another part of the population is frequently wasted by germination provoked by root-exudates from non-hosts, including grasses.

Chlamydospores are produced by fungi such as *Cochliobolus sativus* and *Thielaviopsis basicola*, and are important in the genus *Fusarium*. In the latter, many species, and many formae speciales of *F. oxysporum* and *F. solani* form chlamydospores, but other pathogenic species *(F. avenaceum, F. moniliforme)* do not. Some fusarial chlamydospores possess one characteristic that other dormant propagules do not possess: if they are induced to germinate in response to nutrient stimuli from fresh plant tissue incorporated in the soil or from root exudates of non-host as well as host roots, they are capable of forming secondary chlamydospores. They can thus actually increase propagule population without benefit of host. This is of the greatest importance in all considerations of crop sequence management and use of soil amendments.

Sclerotia. Two types of sclerotia have to be distinguished. The first type is constituted by sclerotia produced by air-borne fungal pathogens that disperse their spores from fruiting bodies, as do the species of *Sclerotinia*, or from stromata, as in the case of the ergot fungus, *Claviceps purpurea*. Here we are more concerned with the second type, which includes the sclerotia produced by root-infecting fungi. These sclerotia are smaller than the first type, more uniform in size and shape. They are usually produced in large numbers, and function as single infective propagules in the soil. Examples are the sclerotia of *Rhizoctonia solani*, *Phymatotrichum omnivorum*, and *Sclerotium rolfsii*, and the microsclerotia of *Verticillium dahliae*. Most, though not all, of the fungi producing such sclerotia have a wide host range, and can survive in infected host tissue as well as in the soil. These pathogens are among the most difficult to control by cultural practices.

Bacteria. With rare exceptions, pathogenic bacteria survive in the soil, outside hosts and host debris, to a much smaller extent than fungi (Gray 1976). *Agrobacterium tumefaciens*, however, survives in many soils for long periods (see Fig. 2.7), provided temperatures are suitable, the soil does not dry out and the pH level stays close to neutral. On the other hand, bacteria such as *Xanthomonas citri* and *Corynebacterium insidiosum* have been found to survive in field soil for only a few days. When such soil was, however, sterilized before inoculation, these bacteria survived very much longer. This indicates that in ordinary field soil their survival is curtailed by microbe action.

Fig. 2.7. A rose stock affected by crowngall *(Agrobacterium tumefaciens)*, a pathogen capable of surviving in the soil over long periods. (By courtesy of the Division of Plant Pathology, Agricultural Research Organisation, Bet Dagan, Israel)

2.3.5.3 Survival and Multiplication of Inoculum by Aid of Non-Hosts

There are two ways in which non-host plants can aid pathogens of other plants to survive or multiply: by their root exudates, promoting development or facilitating persistence in the soil, or alternatively by having their roots support superficial growth of the fungus or bacterium, without apparent ill-effect on the plant, i.e., as symptomless carriers.

Two examples may here be given for the effects of root exudates of non-hosts. Schroth and Hendrix (1962), investigating *Fusarium solani* f. sp. *phaseoli*, determined that the fungus not only perpetuates itself in soil free of its host, but actually increases the number of its propagules in the rhizosphere of tomato, lettuce, and maize (but not of onion). Survival of the fungus was enhanced by the temporary supply of nutrients by diffusates from non-susceptible plants, which enabled it to vegetate briefly and form new chlamydospores. The other example is *Sclerotium rolfsii*, found in container experiments to maintain the level of its sclerotial population as well in soil in which non-hosts (sunflower, maize) were grown, as in soil with highly susceptible beans (Avizohar-Hershenzon 1962).

Table 2.11. Some major soil-borne pathogens surviving in *symptomless* carrier crops or weeds

Crop chiefly affected	Disease	Pathogen	Symptomless carrier	References
Bean	Root rot	*Fusarium oxysporum* f. sp. *phaseoli*	Sweet potato	Hendrix and Nielsen (1958)
Tomato	Wilt	*F. oxysporum* f. sp. *lycopersici*	Weeds [a]	Katan (1971)
Sweet potato	Wilt	*F. oxysporum* f. sp. *batatas*	Cabbage, tobacco, soybean, bean, watermelon, maize, cotton, cowpea	Hendrix and Nielsen (1958)
Tomato	Wilt	*F. oxysporum* f. sp. *lycopersici*	Sweet potato	Armstrong and Armstrong (1948)
Cotton	Wilt	*F. oxysporum* f. sp. *vasinfectum*	Sweet potato	Armstrong and Armstrong (1948)
Tobacco	Wilt	*F. oxysporum* f. sp. *nicotianae*	Cotton, cowpea, okra, soybean	Armstrong and Armstrong (1978)
Cotton, eggplant	Wilt	*Verticillium dahliae*	Wheat, weeds	Krikun (1976)
Potato	Wilt	*V. dahliae*	Many weeds [b]	Busch et al. (1978)
Cotton	Wilt	*V. albo-atrum*	Barley, sorghum	Benson (1974)

[a] *Oryzopsis miliaceae, Digitaris sanguinalis, Amaranthus graecizans, Malva silvestris*

[b] The weeds include *Amaranthus retroflexus, Chenopodium album, Thlaspi arvense, Solanum nigrum, Portulaca oleracea*

Survival of pathogens on roots of non-hosts has been proved in several important instances. Kraft et al. (1974) attribute the re-appearance of *Fusarium* wilt (*F. oxysporum* f. sp. *pisi*) on peas in Washington (USA) after an absence of decades, to the survival of the fungus on non-hosts. Busch et al. (1978) found in Canada that *Verticillium dahliae* is carried without symptoms by numerous weeds, frustrating attempts to reduce pathogen populations enough to grow potatoes; and Krikun (1976) in Israel similarly proved wheat to carry microsclerotia subsequently proved pathogenic to cotton and eggplant. Table 2.11 lists some major pathogens surviving on symptomless crops or weeds. There are probably few soil-borne pathogens that cannot be transmitted by some weeds. The presence or absence of symptoms means very little. The role of non-hosts in the survival of a pathogen can be assessed reliably only if it is proved that they neither carry the pathogen nor stimulate it by their root exudates. Furnishing such negative evidence is a tough proposition.

Non-host crops sown for the purpose of reducing propagule populations of root-infecting fungi, bacteria or nematodes are called decoy crops, and will be mentioned in Chap. 2.3.8.

2.3.5.4 Use of Varieties Resistant and Seasons Unfavourable to the Pathogens

Proper use of crop sequence as a tool of disease management does not necessarily mean exclusion of a crop species that has in the past suffered from a given disease, until the pathogen is eliminated. Varieties resistant or immune to important soil-borne pathogens are now available for many crops. There are tomato and pea varieties resistant to their respective *Fusarium* wilts, and wheat varieties resistant to *Pseudocercosporella herpotrichioides*, the cause of eye-spot. As regards nematodes, rotation of potato varieties susceptible and resistant to *Globodera rostochiensis* can go a long way to keep the crop healthy (Jones et al. 1967), and there are varieties of cowpeas resistant to *Rotylenchulus reniformis* (Gilman et al. 1978). Even if the size or quality of their yield may be somewhat inferior to that of susceptible varieties (otherwise nobody would continue to use the latter), occasional use of resistant varieties may be justified, because it enables farmers to utilize specialized equipment, know-how, and installations. In such cases one might speak of varietal sequence rather than crop sequence. There is, however, one condition: if the resistant variety, though not suffering from disease, is conducive to pathogen multiplication in the soil (see preceding section), then its use as a factor stabilizing crop sequence loses all merit.

Many pathogens are limited to definite seasons and will neither multiply in the soil nor cause damage to susceptible crops at other times. Thus, snow-mould of wheat, caused by *Typhula idahoensis* and *Gerlachia (Fusarium) nivalis*, can be devastating to winter wheat in the North-West Pacific regions of the United States, but does not affect spring wheat; if combined with fallow, sowing of spring wheat may thus result in a 2–3 years' period without hosts the snow-mould can attack, and this will reduce populations of the pathogen to harmless levels (Baker and Cook 1974).

Another example of using their seasonal limitations to cheat pathogens of their prey is that of *Sclerotium rolfsii*. Under mediterranean conditions, this pathogen infects its most susceptible hosts, such as carrot or tomatoes, only in irrigated summer crops. Crops sown in winter in the open are not affected if they ripen their yield before the late spring months (mid-March to April), although the soil may be heavily infested (Avizohar-Hershenzon and Palti 1962). The same applies to hosts of the root-knot nematode *Meloidogyne javanica*.

2.3.6 Monoculture

The term monoculture will here be used in its restricted sense, meaning cultivation of single or closely allied crop species in annual or seasonal succession, with interruptions only by fallow or intermittent growing of green manure, or by application of soil amendments, not necessarily after each crop.

It is generally accepted that in each agro-ecosystem diversity is a stabilizing factor, and prevents excessive pressure on any part of the system (Browning 1974). This certainly applies to crop/disease relationships. From the general biological aspect, monoculture is therefore clearly dangerous; catastrophic crop losses from disease are much less likely to occur in mixed cropping. Monoculture may also exert selection pressure on pathogens, resulting in the emergence of new pathotypes.

However, for the farmer monocropping has important advantages. Shipton (1977), in his review of monoculture and soil-borne pathogens, has defined these as follows: monoculture *simplifies* cropping by permitting full use of equipment for the crop chosen; and it leads to *specialization* by the farmer, and thus, we may hope, to better know-how, including – it might be added – more knowledgeable handling of disease problems. Shipton also distinguishes between two patterns of disease and pathogen development associated with monoculture: *irreversible* increase in the incidence of disease until a certain level is reached and maintained, or the *reversible* phenomenon of disease first mounting, but subsequently declining to a more or less fixed level.

2.3.6.1 Irreversible Disease Pattern

According to Shipton (1977), the pathogens in this group tend to possess high competitive saprophytic ability and/or persistent survival structures, and to be relatively tolerant of other soil organisms. Among the pathogens that increasingly attack crops in monoculture, the most important are the wilt pathogens of the genera *Verticillium* and *Fusarium*. The question is, of course, at what level this steadily increasing attack will eventually stabilize. If *V. albo-atrum* or *V. dahliae* are present in the soil, incidence of wilt may reach so high a level that cotton monoculture is ruled out. Ashworth (1974) found that *Verticillium* in fields continuously cropped with cotton builds up its inoculum at such a tremendous rate (600% per year) that eventually a level is reached where even interruption of cotton growing for 2–4 years does not succeed in reducing inoculum to a level that will make its resumption possible.

With *Fusarium* the position is less clear. On potatoes, disease incidence in monoculture may not exceed levels also obtained in rotation with winter cereals, and this, in Yugoslavia, has also been found to be true for the *Fusarium* stalk rot of maize, caused chiefly by *F. moniliforme* (Balaz et al. 1978). On peas, melons, and other crops, monoculture will not increase incidence of *Fusarium* wilts if the soil is "suppressive" (see Chap. 1.4.3). Even if this is not the case, *F. oxysporum* wilts appear in general to increase at a relatively slow rate, although they eventually reach devastating levels of incidence.

In 9 years' barley monoculture in Sweden, *Drechslera sativa* continued to increase steadily in frequency (Stetter and Leroul 1979), and thus clearly belongs to this group.

Nematodes in field and vegetable crops show likewise irreversible increase in monoculture, which narrows the community spectrum of parasitic nematodes to those species most favoured by the host (Nusbaum and Ferris 1973). This is most pronounced with the species of *Globodera* and *Heterodera* that are host-specific, such as *H. schachtii* on sugarbeet, *G. rostochiensis* on potatoes and *H. glycines* on soybeans. Presence of these cyst nematodes in the field rules out monoculture of their host crops, unless they are valuable enough to warrant disinfestation of the soil by solar heat or chemicals. Populations of root-knot nematodes (*Meloidogyne* spp.) also rise steadily under continued culture of their many host crops, until sowing of the latter becomes uneconomic (Anon 1968). As regards migratory root nematodes, monoculture trials carried out in Germany over 16 years by Roessner (1979) have shown that populations of *Helicotylenchus* spp. and *Tylenchorhynchus* spp. rose tremendously in plots of winter wheat, and especially of rye, while populations increased much less in oats.

2.3.6.2 *Reversible Disease Pattern*

Shipton (1977) states that pathogens causing disease following this pattern generally have low ability for saprophytic survival. The group includes *Phymatotrichum omnivorum* with its numerous hosts, and the potato scab pathogen *Streptomyces scabies*, and important members are the causes of take-all (*Gaeumannomyces graminis*) and eye-spot (*Pseudocercosporella herpotrichioides*) in cereals. The discovery that incidence of these diseases is reversible, first at Rothamsted, England, and soon after in other countries, has encouraged practical farmers to adopt cereal monoculture even before its effects on pathogens were properly understood.

The general pattern of the rise of take-all and eye-spot in cereals in the early years of monoculture, and of its subsequent decline is illustrated in Fig. 2.8 from Cunningham's (1975) study of take-all in wheat and barley. This author also states the interesting fact that the *virulence* of isolates of *G. graminis* increased up to the time when peak disease levels were approximated in the field, and diminished thereafter; there also was evidence that competitive saprophytic ability decreased in isolates from long sequences of cereals, as against those taken from the first to third crops of cereals following permanent pasture. Studying the decline of take-all in wheat in Switzerland, Zogg and Amiet (1980) found the rate of decline to depend on temperature: it was pronounced at 18 °C, decreasing with temperature, and practically nil at 0 °C. This means that warm spells after harvest may reduce the amount of take-all inoculum left for the next cereal crop, but only if the field is kept free from the many cereal weeds that can transmit the disease.

Most research on cereal monoculture indicates that in fields affected by foot and root rot, maintenance of high yield levels depends on ample supply of nitrogen; this applies equally to crops affected by *G. graminis* (Lockhart et al. 1975) and by *P. herpotrichioides* (Kupers 1976; Obst et al. 1977).

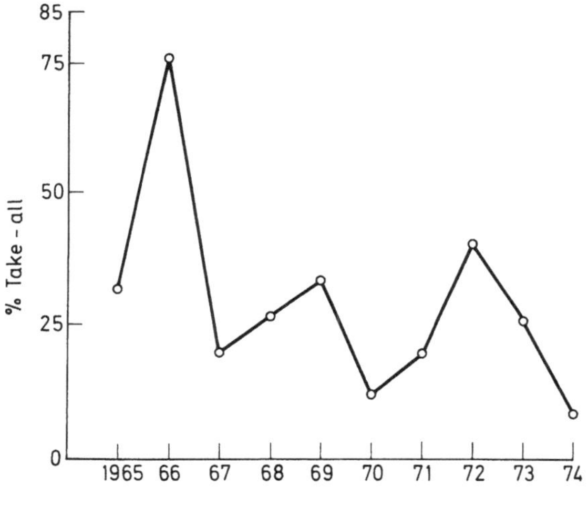

Fig. 2.8. Fluctuations in the incidence of take-all (*Gaeumannomyces graminis*) in a 10-year barley succession. (Cunningham 1975)

2.3.7 Multiple Cropping

2.3.7.1 The Benefits of Multiple Cropping

Multiple or mixed cropping is the rule in native subsistence farming in the tropics, and in other economies in which land or water, or both, are the factors limiting agricultural production, and capital is scarce, while labour is plentiful and cheap and organic manure is available to keep up soil fertility. Multiple cropping plays an important role in labour-intensive farming in China and Japan.

Perrin (1977) has recently reviewed insect pest management in multiple cropping systems, and much of his review is equally applicable to disease management. He enumerates the advantages of multiple cropping, which lead to greater biomass production per unit area, as follows: (a) more efficient use of solar radiation, due to better interception of light by foliage; (b) increased availability of nitrogen for crops mixed with legumes; (c) shading effects, as obtained, for example, where young cocoa trees are planted under oil palms; (d) better use of soil moisture at various depths; (e) smothering of weeds under dense foliage canopies; (f) improved erosion control; and (g) reduction of the risk of crop failure, as one crop species may compensate in growth for another affected by unfavourable environmental conditions. Perrin closes his list with the statement that multiple cropping generally changes the incidence of pests for the better, and this is what we are going to examine here in relation to crop disease.

Multiple cropping may be divided into several practices. There is the simultaneous sowing of field and vegetable crops for either simultaneous or consecutive harvest, e.g., sowing of mixtures of cereals and legumes. Sowing may also be staggered, so that the growth periods of two crops overlap, but do not coincide, as is the practice in some parts of Asia with rice crops into which grain legumes are sown. Finally, in orchards and plantation crops, intercropping may be practiced with annual crops, or two fruit crops may be grown side by side.

2.3.7.2 Multiple Cropping of Field and Vegetable Crops

If two crops are to be sown simultaneously, different types of plant species are usually chosen for the purpose, one crop growing fast with another growing slowly, or a deep-rooted crop with one developing a shallow root system, or a legume crop with another that can benefit from the nitrogen fixed in the legume crop. From the point of view of disease management, such mixing of crops is usually advantageous. There are only few foliage pathogens, notably some powdery mildews, that will attack crops belonging to different botanical families. Each of the two crops thus, to some extent, shields plants of the other from the impact of air-borne pathogens, as well as from vectors of stylet-borne viruses (see below). This should more than offset the microclimatic advantage some fungi and bacteria may derive from the dense foliage of mixed crops. Soil-borne diseases may also be reduced by root exudates of one crop, e.g., onions, being detrimental to root pathogens of the other crop, and for other seasons discussed in the section on decoy and trap crops.

In mechanized farming, crops are chiefly mixed if they can be sown and harvested simultaneously, or if machine harvesting of one does not harm the other crop. Cereal-legume mixtures are the most important of such crops. They are either grown for simultaneous harvesting for forage, or the undersown legume is left after cereal harvesting, and is then repeatedly cut for forage or ploughed-in as soil amendment. In either case, the health of the crop is not adversely affected. Whether such soil amendment will reduce take-all in wheat is still in dispute (Zogg 1963).

Staggered sowing of crops to grow in overlapping periods is an important tool for optimal use of soil water. Where rainy and rainless seasons alternate, crops needing plenty of water, such as rice, can be intersown with grain legumes that will grow on the winter rain but mature as the rains cease. Under similar conditions cotton can be sown into winter wheat, 1–2 months before the latter is harvested. Such practices often involve mowing the earlier crop at a greater height than usual. Due attention must then be paid to the management of stubble, which may carry pathogens able to survive dry seasons, such as *Septoria tritici* on wheat or species of *Ascochyta* on legumes.

A dangerous practice of double cropping is the sowing of summer cereals (maize, sorghum, perhaps rice) in between rows or strips of wheat or barley. Kelman and Cook (1977) have reported that this practice, as observed in China, promotes the incidence of Barley Yellow Dwarf Virus and of *Fusarium graminearum*, which produces its *Gibberella* stage on wheat, maize, and rice and then spreads by ascospores.

A special type of multiple cropping is the sowing of a taller crop to protect a crop of lesser height from virus vectors. Growing of mustard or barley, which grew to heights of 66 and 41 cm, respectively, together with sugarbeet stecklings, has been reported to lower incidence of Beet Mild Yellowing Virus in the stecklings (Heathcote 1968). Radish Mosaic in Japan decreased when radishes were sown between rows of rice or trefoil (Broadbent 1969), and pigeon pea in Haiti was completely protected from virus diseases when grown between rows of tall sorghum (Messiaen, pers. comm.).

A variation of this technique consists in growing several rows of tall crops, such as maize or sunflowers, at intervals of several metres, as barriers in low crops. This has been used to protect sugarbeet stecklings, as above, and to shield seed beds of Brassicae from vectors of the Cauliflower Mosaic Virus. Broadbent (1969), who reported this, also remarked that such barrier crops "are likely to be more effective in protecting plants from stylet-borne than from more persistent viruses, because aphids will often lose the non-persistent virus if they probe the barrier plants. Normally they will fly away over the crop plots when they leave the barrier, because they fly upwards towards the brightest light, but if they should subsequently land within the barriers, there is an increased chance of their being non-viruliferous."

2.3.7.3 Intercropping in Orchards and Plantations

Interplanting or intersowing in orchards is practiced chiefly where the permanent crop is not yet in bearing, and interim revenue is required. This does not usually present a health hazard to the temporary crop, but may have negative aspects for the permanent crop. If irrigation is practiced, this must be adapted to the needs of the temporary, and usually more shallow-rooted crop, i.e., irrigation intervals may be short, which is not what the permanent (orchard) crop needs. In Morocco, the irrigation of lucerne and vegetables intercropped with oil palms has been shown to accelerate the passage of the bayound pathogen, *Fusarium oxysporum* f. sp. *albedinis*, in the palm's stem vessels (Laville, in Kranz et al. 1977). Another important consideration is that polyphagous pests attracted by the temporary crop may spread to the permanent crop.

Sowing or planting of temporary crops susceptible to multi-host pathogens, such as *Sclerotium rolfsii* or *Verticillium dahliae* must, of course, be avoided at all costs, if the permanent crop can be affected by this pathogen. This also applies to nematodes: in France, intercropping of young peaches with melons and other vegetables has led to severe attacks of the root-knot nematode *Meloidogyne arenaria* on the peaches (Scotto de Massese, pers. comm.).

Interplanting in tropical crops may be needed, where shade trees have to be raised to shield plantations from excessive radiation, as in the case of cocoa in Uganda, where radiation stress predisposes the crop to attack by *Verticillium dahliae* (Trocmé 1972). Similarly, in all climates, interplanting of various varieties of the same fruit species, for pollination or other purposes, constitutes a type of multiple cropping which may involve a health hazard. If, for example, the flowering of an early apple variety falls into a period favourable to fire-blight (*Erwinia amylovora*), this variety may be a source of infection for the other varieties in the orchard. As mentioned in Chap. 1.3.3, the apple variety Jonathan, which has widely been used in England as a pollinator, but is highly susceptible to powdery mildew (*Podosphaera leucotricha*), is being replaced widely by the more resistant Worcester Pearmain variety (Butt, pers. comm.).

2.3.8 Decoy and Trap Crops

Decoy crops, sometimes also called hostile crops, are non-host crops sown with the purpose of making soil-borne pathogens waste their infection potential. This is effected by activating dormant propagules of fungi, seeds of parasitic plants, or larvae of nematodes in the absence of hosts which would enable them to continue their development. A list of pathogens that can be decoyed in this way is presented in Table 2.12.

Some of the decoy crops are in themselves fairly valuable, e.g., some of those hostile to broomrape or the groundnuts that can be sown to lessen, though not eliminate, infestation by the root-knot nematode *Meloidogyne incognita*, or the flower crops (hybrids of *Gaillardia* and *Hellenium*) that will reduce populations of *Pratylenchus penetrans*. Most of the decoy crops, however, are of moderate to low value. Their use is thus "expensive," because a season for a more valuable crop is lost. This must be weighed against the cost of sowing in sequence one or more crops which are not attacked, but do not cause quick reductions in pathogen populations. Decoy crops can be used to improve the health of perennials, as in the case of populations of *Verticillium albo-atrum* being reduced in olive plantations by growing *Tagetes* (Baker and Cook 1974), and probably by the use of *Tagetes*, *Crotalaria* and other plants in plantations affected by nematodes.

Trap crops are host crops sown to attract nematodes, but destined to be harvested or destroyed before the nematodes manage to hatch. This has been advo-

Table 2.12. Decoy crops for the reduction of pathogen populations

Crop	Pathogen	Decoy crops	References
Fungi			
Brassicae	*Plasmodiophora brassicae*	Rye-grass, *Papaver rhoeas*, *Reseda odorata*	Garrett (1970)
Potato	*Spongospora subterranea*	*Datura stramonium*	Garrett (1970)
Olive	*Verticillium albo-atrum*	*Tagetes minuta*	Baker and Cook (1974)
Parasitic plants			
Tomato, tobacco	*Orobanche* spp.	Sunflower, safflower, lucerne, flax, chickpeas and others	Kasasian (1971), Krishnamurty et al. (1975)
Various	*Striga asiatica*	Sudangrass	Curl (1963)
Nematodes			
Eggplant	*Meloidogyne incognita*, *M. javanica*	*Tagetes patula*, *Sesamum orientale*	Varma et al. (1978)
Tomato	*M. incognita*, *Pratylenchus alleni*	*T. patula*, castor bean, chrysanthemum	Hackney and Dickerson (1975)
Tomato	*M. incognita*	*T. patula*, groundnuts	Belcher and Hussey (1977)
Narcissus, tomato, okra	*Meloidogyne* sp.	*T. patula*	Hirling (1977)
Soybean	*Rotylenchulus* sp. *Pratylenchus* sp.	*T. minuta*, *Crotalaria spectabilis*	Yoshii and Agudelo (1977)
Various	*Pratylenchus penetrans*	*T. patula*, hybrids of *Gaillardia* and *Hellenium*	Hijink and Suatmadji (1967)
Various	*P. neglectus*	Oil-radish (*Raphanus oleiferus*)	Hirling (1977)
Oats	*Heterodera avenae*	Maize	Hirling (1977)
Various	*Trichodorus* sp.	Asparagus	Hirling (1977)

cated for *Globodera pallida*, using somewhat resistant potato varieties sown early and lifted before temperatures are warm long enough to permit nematode populations to advance beyond larval development, and also for sowing crucifers, to be ploughed-in before the cyst nematodes of beets can develop fully (Weischer 1977). The same objective is achieved in pineapple plantations by planting tomatoes and destroying them before root-knot nematodes can produce eggs (Stevens 1960).

2.3.9 Crop Sequence and Crop Disease in the Future

With the demand for staple crops rising to feed rapidly growing populations, and with farmers everywhere striving to at least equal the income of the industrial population, further intensification of farming is indispensable. This means increased fertilization and irrigation, and rising investment in machinery, storage facilities, equipment to protect crops from weather and parasite etc. To pay for all this, farming revenues will have to rise to levels well above those now attained.

The tendency therefore will be for development in two principal directions: (a) Monoculture of profitable crops, which makes it possible to use expensive machinery to the full wherever labour is a limiting factor but capital is available. (b) Multiple cropping and intercropping where land and/or water are scarce, but labour is available, since crowded and diverse crops on a given area always require more hand labour. The first trend is clearly recognizable in modern farming, especially in temperate regions, the second is that followed by China and is being adopted by many developing countries.

The future of disease control in agriculture developing on these lines lies increasingly in resistance breeding and use of pesticides, but also in the adaptation of cultural practices (e.g., fertilization and irrigation) to monoculture and crowded cropping.

2.4 Soil Amendments and Mulches

Although there is no sharp distinction between them, we shall term soil amendments all the organic matter incorporated in the soil, and mulches all that is left (stubble) or spread on the soil surface. Thus a stubble mulch, when turned in, becomes a soil amendment. Soil amendments have to be considered in relation to the entire microbiota of the soil, both pathogenic and saprophytic. Mulches have additional functions in relation to virus vectors, conservation of moisture, and sometimes in preventing direct contact between the soil and leaves or fruits.

The organic matter available to farmers for incorporation in large fields is chiefly residues of previous crops, such as the stubble of cereals or legumes harvested by cutting, weeds that have come up over a period of fallow, or green manure crops specially sown for the purpose. On a smaller scale, organic matter is available from crop residues, such as foliage of root crops, compost prepared from all manner of organic wastes from farm, town or industry, and farmyard manure. All the above, with the exception of green manure crops, are potential carriers of parasites, mostly of facultative pathogens. Another source of organic waste, which is becoming increasingly important for use in the field, is sewage sludge (Schmid et al. 1973).

Mulches may consist of the stubble of crops, of weeds kept at various heights by cutting or spraying, of straw applied to the soil surface, or of plastic foil spread over the soil. Because of their low cost, or indeed the savings they make possible, only stubble and weeds are commonly used as mulches for large areas.

Except for small fields or highly valuable crops, organic amendments and mulches are seldom used primarily because of considerations of crop health. They are used because they are there anyway (stubble, weeds) or because their effect on the texture and nutrient status of the soil seems worth it (green manure, crop refuse, composts, farmyard manure) or because they offer a solution for environmental problems (sludge). It may even be said that amendments and mulches are often used in spite of their effect on crop health rather than because of it. The pathologist then has to make the best of a situation not of his choice.

2.4.1 Organic Amendments

Amendments to the soil may, of course, be organic or inorganic, but the latter are treated in the section on crop nutrition.

The effect of organic amendments on the soil and on the pathogens and their competitors and antagonists has been reviewed comprehensively by Baker and Cook (1974) in their book *Biological Control of Plant Pathogens*, and in later contributions by Lewis and Papavizas (1975) and Bruehl (1975) to the volume on *Biology and Control of Soil-Borne Pathogens*. The subject is complex, and conclusions drawn from the knowledge available are in many ways not yet ripe for application to the practice of farming. We shall here mention only some aspects of more immediate practical interest, and this includes the effects of soil amendments on nematodes. As pointed out by Katan (pers. comm.), the large body of research containment by soil amendments has been slow in providing results applicable to farming practice. Most of such research has been empirical in nature: screening of materials that were available at low cost. It is to be hoped that future research will aim at analysis of the actions to be expected of specific soil amendments to reduce populations of specific pathogens under well-defined soil and cropping conditions.

2.4.1.1 Effects on Soil and Water Relations

Incorporation of organic matter in the soil is generally considered to have a wholly beneficial effect on the texture of both light and heavy soils. In sandy soils such amendments improve retention of water; this, among other effects, makes it possible to irrigate at longer intervals and reduce the dangers inherent in splashing, lengthening periods of leaf wetness, etc. On heavy soils, organic matter improves percolation of water to deeper layers and makes crusts form less readily in the topsoil. This helps emergence of seedlings and prevents ponding of irrigation water, which causes some non-parasitic diseases such as potato tuber rot (Easton and Nagle 1977) and promotes pathogen infection on underground plant organs. Improved percolation generally improves growth of roots and their resistance to many pathogens, and enables crops to "grow away" from, i.e., win the competition with, various parasites, including nematodes.

2.4.1.2 Effects on Pathogens

Huber and Watson (1970) have divided effects of organic amendments on soil-borne pathogens into effects relating to disease prevention and effect on disease control. In *preventing* pathogens from establishing themselves in new or hitherto arid soil, organic amendments fulfil the primary function of raising the "biological buffering capacity" of these soils, i.e., enriching them with microflora potentially competitive with, or antagonistic to, pathogens that may appear as cropping begins or intensifies. In *controlling* pathogens in the soil, organic amendments act on both host and pathogen, often by their direct effect on the microbiota and their indirect effect on the availability of nutrients in the soil. Harrison (1976) has analyzed results of field tests in which barley straw was incorporated by rototilling in the top 15.2–17.8 cm of soil and reduced populations of *Verticillium albo-atrum* drasti-

Table 2.13. Effects of dry and decayed soil amendments on some diseases caused by soil-borne fungi

Crop and disease	Pathogen	Soil amendment	Effect on fungal population	References
Fungal Populations Reduced				
Potato wilt	*Verticillium albo-atrum*	Barley straw	Reduced by immobilization of N	Huber and Watson (1970)
	V. albo-atrum	Barley straw	Population reduced	Harrison (1976)
Potato black scurf	*Rhizoctonia solani*	Wheat straw	Reduced	Gudmestad et al. (1978)
Bean root rot	*Thielaviopsis basicola*	Oat straw, corn stover, lucerne hay	Reduced by germination and lysis of propagules	Lewis and Papavizas (1975)
Pea root rot	*Aphanomyces euteiches*	Crucifer tissues	Reduced by toxic volatiles	Lewis and Papavizas (1975)
Cotton root rot	*Macrophomina phaseolina*	Lucerne meal, barley straw	Reduced by antagonists	Lewis and Papavizas (1975)
Coriander wilt	*Fusarium oxysporum* f. sp. *corianderi*	Oil cakes	Reduced	Lewis and Papavizas (1975)
Banana wilt	*F. oxysporum* f. sp. *cubense*	Sugarcane residue	Reduced by germination and lysis of propagules	Sequeira (1962)
No crop	*F. oxysporum, F. solani*	Linseed-, cottonseed-meal	Chlamydospore population reduced	Zakaria and Lockwood (1980)
Avocado root rot	*Phytophthora cinnamomi*	Lucerne meal	Reduced by release of NH_3	Lewis and Papavizas (1975)
Root rots of ornamentals [a]	*Phytophthora, Pythium, Thielaviopsis* spp.	Composted tree bark	Reduced by release of chemical inhibitors	Hoitink (1980)
Fungal Populations Increased				
Wheat root rot	*Fusarium solani, F. roseum*	Wheat straw	Increased survival and inoculum density	Lewis and Papavizas (1975)
Wheat snow mould	*Gerlachia nivalis, Typhula idahoensis*	Barley or wheat straw or stubble	Increase	Huber and Anderson (1976)
Sorghum downy mildew	*Sclerospora* sp.	Farm-yard manure	Increased by transmission of oospores	King and Webster (1970)
Potato	*Rhizoctonia solani*	Buckwheat tissue	Increased	Frank and Leach (1976)
No Clear Effect				
Wheat eye-spot	*Pseudocercosporella herpotrichioides*	Straw, farmyard manure	Not clear	Hoeflich et al. (1977)
Wheat take-all	*Gaeumannomyces graminis*	Straw, farmyard manure	Not clear	Hoeflich et al. (1977)

[a] Container-grown azalea, Easter lily, poinsettia, and rhododendron

cally. He concluded that the straw affected neither host resistance nor fungal pathogenity, but had a direct quantitative effect on the pathogen population.

Effects of dry or decayed organic amendments on some major soil-borne fungi are summarized in Table 2.13. Organic amendments consisting of fresh plant material (green manure, weeds, green plant residues) differ from dry or decayed material, as fresh material is not immediately available for colonization by saprophytes. It retains its original host resistance for some time, during which it may be colonized by pathogens rather than saprophytes. Dead plant residues will therefore be safer soil amendments where infection by fungi such as *Pythium* spp. or *Phytophthora parasitica* is to be expected (Garrett 1970).

Some of the effects of green manure on fungal pathogens are listed in Table 2.14. Maturity of the tissues incorporated in the soil must always be taken into account. Lewis and Papavizas (1975) have stressed this in relation to soil amendments by rye and maize. Garrett (1970) has pointed out that incorporation of mature straws or other organic matter with a high C/N ratio at first results in the locking-up of soluble soil nitrogen in the cells of the decomposing micro-organisms. This deprives the crop of available nitrogen and may cause nitrogen deficiency, which needs prompt correction by application of nitrogen. If this is not done, crops like wheat or barley will be rendered more highly susceptible to various disease, including take-all *(Gaeumannomyces graminis)*.

A great variety of effects of soil amendments is mentioned in Tables 2.13 and 2.14, some effects reducing and others increasing disease even if caused by the same pathogen, e.g., *Rhizoctonia solani* (Fig. 2.9). As mentioned above, the way in which amendments act also varies greatly. The most frequent action is on the survival of fungal propagules in the soil, either directly by release of toxic matter or indirectly

Table 2.14. Effects of green manure on some soil-borne fungal pathogens

Crop	Disease	Pathogen	Type of green manure	Effect on fungal population	References
Wheat	Take-all	*Gaeumannomyces graminis*	Rape, pea or mixed grasslegume	Reduction by stimulation of antagonists	Grossmann (1967)
	Eye-spot	*Pseudocercosporella herpotrichioides*	Not stated	Increased by larger spore production	Grossmann (1967)
	Eye-spot	As above	Clover, oil radish	Increased	Schulz (1968)
			grass	Partially reduced	Schulz (1968)
Cotton	Wilt	*Fusarium oxysporum* f. sp. *vasinfectum*	Bean, rape or mustard	Increased by larger spore production	Grossmann (1967)
	Root rot	*Phymatotrichum omnivorum*	Pea, *Melilotus officinalis*	Reduced	Cook et al. (1978)
Potato	Scab	*Streptomyces scabies*	Barley	Increased	Weinhold et al.
			Soybean	Prevented build-up	(1964)
			Pea	No effect	Weinhold et al. (1964)
	Black scurf	*Rhizoctonia solani*	Barley, oats	Slightly reduced	Naumann and Lange de la Camp (1977)

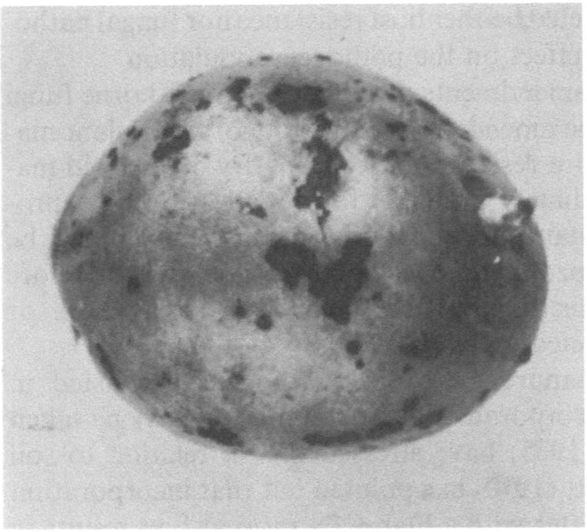

Fig. 2.9. Sclerotia of *Rhizoctonia solani* on a potato tuber. The population of this fungus in the soil has been found to be increased by some soil amendments (buckwheat tissue), but to be reduced by others (barley straw) and by green manure (barley, oats). (Photograph by courtesy of the Division of Plant Pathology, Agricultural Research Organization, Bet Dagan, Israel)

Table 2.15. Soil amendments found to reduce nematode populations

Nematode species	Crop	Soil amendment tested	References
Meloidogyne incognita	Tomato	Sewage sludge	Habicht (1975)
M. incognita	Tomato	Lucerne hay and straw, lespedeza hay, flax hay	Johnson et al. (1967)
M. javanica	Tomato, okra	Sawdust	Singh et al. (1967)
Heterodera marioni	Peach	*Crotalaria spectabilis* in summer, oats in winter	McBeth and Taylor (1944)
H. tabacum	Eggplant	Sawdust, paper	Miller et al. (1968)
Pratylenchus penetrans		Leaf mould + ammonium sulphate	Miller (1977)
Hoplolaimus tylenchiformis, *Xiphinema americanum*		Leaf mould, sewage sludge	Miller (1977)
Pratylenchus penetrans, *Tylenchorhynchus dubius*		Mycelial residues from production of antibiotics, cellulose wastes from the paper industry	Miller et al. (1973)
Helicotylenchus sp., *Tylenchorhynchus* sp., *Meloidogyne* sp.		Mustard oil cake, decayed leaves of *Azidarachta indica*	Saxena et al. (1977)
Pratylenchus penetrans		Oats, sudangrass	Dunn and Mai (1973)
Belonolaimus longicaudatus		Activated sewage sludge	Heald and Burton (1968)
Tylenchulus semipenetrans		Castor pomace (by-product of castor oil extraction)	Mankau and Minteer (1962)

by effects on nitrification processes in the soil or by promoting germination of dormant propagules while no hosts are available to sustain their growth, or by encouragement of competitive or antagonistic microbiota.

We have been unable to find any reference to research carried out on effects on organic amendments on soil-borne bacterial pathogens.

As regards nematodes, addition of organic matter to the soil increases the population of fungi antagonistic to nematodes and apparently also of that of predacious nematodes and of the internal parasites of nematodes (Anon 1968). Table 2.15 lists a great variety of organic amendments that have been found to reduce some species of nematodes of a wide range of taxa. Some of these effects have been obtained by direct action on the nematodes, e.g., the action of sewage sludge on root-knot nematodes (Sayre et al. 1973).

However, in some cases organic amendments have also been reported to have undesirable effects. Thus, incorporation of peat moss in sandy soil accelerated damage by *Tylenchulus semipenetrans* to citrus seedlings (O'Bannon 1968), and cover crops of rye and lupins favoured the stunting of apple seedlings by *Pratylenchus penetrans* (Colbran 1979).

2.4.1.3 Practical Application of Soil Amendments

Available literature shows many positive aspects of the use of organic amendments to reduce the incidence of specific soil-borne diseases. Almost all the relevant research seems to have been carried out in the temperate zone, and cannot be considered applicable to other climatic zones. Yet even within the temperate zone, it appears difficult to generalize from the data obtained: they vary considerably and their applicability to soils and seasons other than those tested is doubtful. It is to be hoped that this promising approach to disease control will receive much more attention in future research in many countries, to match the research carried out, and in progress, mainly in the United States.

2.4.2 Organic Mulches

Until recently, all mulches were organic. However, with the availability of relatively cheap plastic foils these latter are now used increasingly for mulching valuable crops. Their effects are discussed in other chapters [chapters on crop age (Chap. 1.6.2.8) and on physical barriers].

The greatest benefits of organic mulches, especially stubble mulch, are the conservation of precipitation and of organic matter, and reducing dangers of soil erosion. "Stubble mulching" or non-tillage, as discussed in the chapter on tillage, is often adopted by farmers because it enables them to save labour and energy in the preparation of land for sowing, to utilize available labour more efficiently, and to sow crops in quicker succession, or earlier in the season. Among the disadvantages of such mulches are the transmission of certain diseases from season to season, and problems in weed management.

2.4.2.1 Effects on Soil Temperature and Retention of Precipitation

Mulching affects temperature relations of soils and crops in various ways. It reduces the range of diurnal temperatures in the soil, and in temperate climates generally raises the level of temperature to some extent. This is true for stubble mulch and straw mulch, and has been determined for paper and carton mulches by Sumner et al. (1978). The resultant acceleration of root and pathogen growth is often, but not always, in favour of the crop. In climates in which the soil temperature tends to reach higher levels, mulching may have a cooling effect. As already noted, prevention of extreme temperatures in mulched soil prevents *Sclerotium rolfsii* from penetrating roots of apple trees in Israel (Lavee, as quoted by Avizohar-Hershenzon and Palti 1962).

A negative effect of mulching is its interference with the storage in soil of daytime heat which can be released at night. The difference, as compared to unmulched soil, may be as much as 5 °C, and the danger of night frost is thus increased.

Stubble and straw mulch will improve retention of precipitation, often a consideration of overriding importance. Mention has already been made of the welcome effect of stubble in retaining snow, which in some parts of the Great Plains in the USA may make up some 40%–50% of annual precipitation (Greb 1979).

2.4.2.2 Stubble Mulch and Cereal Debris in Relation to Cereal Diseases

Most of the pathogens affecting collar and root of cereals, and some of those affecting their leaves and ears, can persist on stubble and crop debris, but the danger this presents to succeeding or neighbouring crops differs greatly (see Table 2.16).

With the exception of *Septoria* blotch, persistence of pathogens on wheat stubble aggravates incidence of wheat diseases less than might have been expected (Yarham and Norton, unpubl.). With the chemical means now available for cur-

Table 2.16. Effect of inoculum carried in stubble and debris on infection of the succeeding crop of cereals

Crop	Disease	Pathogen	Effect on disease in succeeding crop	References
Wheat	Take-all	*Gaeumannomyces graminis*	Some increase, specially in early sowings	Yarham and Norton (unpubl.)
	Take-all	*Gaeumannomyces graminis*	No effect	Hoeflich et al. (1977)
	Eye-spot	*Pseudocercosporella herpotrichioides*	No clear effect	Yarham and Norton (unpubl.)
	Snow-mould	*Gerlachia nivalis*, *Typhula idahoensis*	Increase	Bruehl et al. (1966)
	Leaf blotch	*Septoria nodorum*	Increase	Yarham and Norton (unpubl.), Hoeflich et al. (1977)
	Leaf blotch	*Septoria tritici*	Increase	Dinoor (1980)
Barley	Leaf blotch	*Rhyncosporium secalis*	Increase	Obst (1975), Polley (1971)
Maize	Stalk rot	*Gibberella zeae*	Increase	Boosalis et al. (1969)
		Fusarium moniliforme	Increase	Boosalis et al. (1969)
		Diplodia zeae	Increase	Boosalis et al. (1969)
	Leaf blights	*Drechslera maydis*, *D. turcica*	Increase	Ullstrup (1976)
	Yellow leaf blight	*Phyllosticta maydis*	Increase	Ullstrup (1976)
	Brown spot	*Physoderma maydis*	Increase	Ullstrup (1976)
	Ear rot	*Diplodia maydis*	Increase	Ullstrup (1976)
	Bacterial wilt	*Corynebacterium nebraskense*	Increase	Ullstrup (1976)

bing infection by eye-spot, persistence of this and other diseases on stubble is frequently no longer a major consideration, where stubble mulching secures the wheat crop other economic benefits (see Chap. 2.5.2.3).

On maize, however, the picture is quite different. All the numerous diseases that persist on stubble or debris of maize are apt to increase their attack on maize crops subsequently grown in the same or in adjoining fields (Ullstrup 1976). The use of straw mulch for the control of virus vectors and of plastic mulches to reduce infection by soil-borne pathogens will be described in Chap. 2.14, which deals with physical and optical means of lessening incidence of disease.

2.5 Tillage

2.5.1 The Aims of Tillage

Tillage operations may be divided into those carried out between successive crops and those performed while the crop is growing. The former are obviously more important in annual crops, the latter in orchards and other perennial crops. The principal effects of these operations in general, and with special reference to crop disease, are the following.

Tillage Between Successive Crops

a) Preparation of the land for sowing and planting, the major aim of tillage ever since man began to cultivate crops, but undergoing major changes in technique in recent years.
b) Incorporation in the soil of organic matter, including manure, green manure, crop residues and weeds or volunteers.
c) Re-distribution of pathogenic organisms (1) from the topsoil, in which they are concentrated, to deeper layers of soil, or their burial at depths that prevent their attack on crops, (2) from isolated foci of infection over the whole field.
d) Shaping the topsoil into ridges or beds for more convenient drainage, irrigation or production of underground organs.
e) Exposure of pathogenic organisms to the sun's radiation, mostly by several successive ploughings.

In-Growth Tillage

f) Elimination of crusts often formed in topsoil after it has been wetted, in order to promote emergence of crops.
g) Weed control in row crops, especially in plantations and orchards.
h) Earthing-up of row crops.
i) Treatment of seasonal crop refuse, such as burying decayed fruit, leaves, and flowers that have shed, or breaking up roots in infected plantations to promote decay before replanting (Fox 1970).

Weed control by tillage has become less important with the advent of effective pre-emergence and post-emergence herbicides. In perennial crops, in-growth tillage may simultaneously control weeds, bury crop refuse and open up irrigation furrows, but also spread pathogens along the rows.

Incorporation of organic matter in the soil will be fully discussed in the chapters on soil amendments, debris management and sanitation.

We shall here concentrate on the significance for crop disease development of tillage for the preparation of land for sowing, and briefly mention a few of the other functions of tillage mentioned above.

2.5.2 Tillage to Prepare Land for Sowing and Planting

Four basic methods of preparing land have been described by Yarham and Norton (unpubl.) in discussing tillage effects on diseases of cereals in England:

a) Mould-board ploughing, in which the soil is inverted to a depth of about 20 cm and then further cultivated to produce a fine seed-bed.
b) Deep cultivation, in which a rigid-tined implement disturbs the soil to a depth of about 20 cm, again necessitating further cultivation.
c) Reduced cultivation, where a relatively shallow (10 cm) tined implement disturbs the soil just sufficiently to produce a seed-bed.
d) Direct drilling (also referred to as stubble-mulch drilling or non-tillage), in which the seed is sown direct into undisturbed soil.

To this should be added:

e) Subsoiling to depths of 35–45 cm, effective for breaking up plough-soles and disturbing the top soil only slightly.

The effects of these and related methods of tillage on crop growth and disease may be reviewed under the headings of depth of soil layers suitable for root growth, and distribution of inoculum in the soil profile and over the surface soil.

2.5.2.1 Depth of Soil Favourable to Root Growth

Crop growth on soil tilled to suitable depth may be expected to be more vigorous, less susceptible to facultative, but more so to most obligate parasites. The depth of soil available for rooting, or the depth at which seeds can be sown in well-tilled soil, may also affect the growth habit of some crops: very shallow sowing, as often resulting from direct drilling, may produce more prostrate growth of cereals and this, in turn, affects the crop's microclimate (Yarham and Norton, unpubl.).

One of the primary aims of tillage is therefore to correct obstructions to root growth. The most common structural cause for such obstruction is soil compaction. When due to passage of heavy vehicles over wet soil, such compaction extends from the soil surface downwards. Its effect on cereal diseases has been studied in detail in England (Yarham 1973), where it was found to aggravate take-all (*Gaeumannomyces graminis*).

In one well-documented case, heavy equipment driven through a barley field increased severity of the disease not only in the succeeding wheat crop, but also in

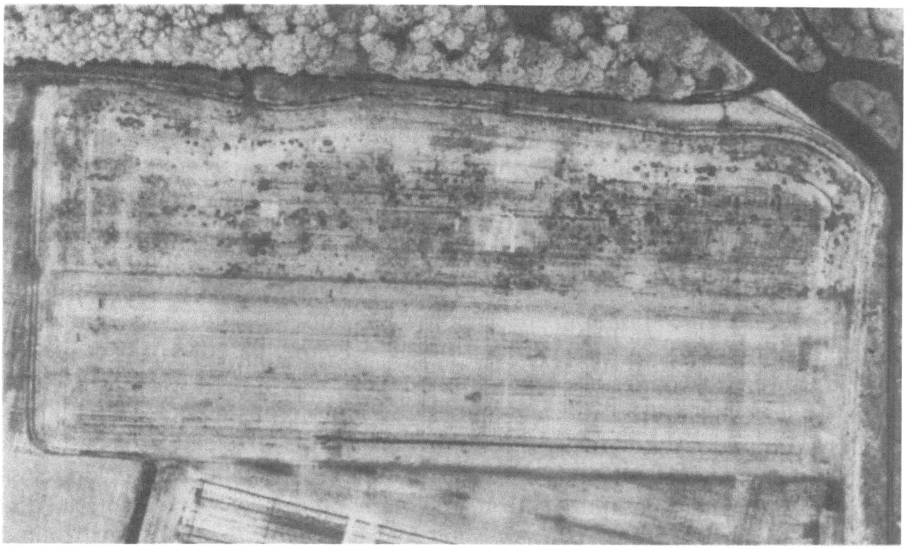

Fig. 2.10. Effect of direct drilling on stunting of barley, associated with *Rhizoctonia* sp., in England. In 1978, the entire field has been direct drilled with barley. In 1979, the upper half was again direct drilled with barley, the lower half was ploughed and then sown with barley. The infra-red photo shows numerous discrete patches of stunting in the direct-drilled, upper half, but not in the lower half that was ploughed (Bell 1980). (By courtesy of the Senior Officer, Aerial Photography Unit, Agricultural Development and Advisory Service, Ministry of Agriculture, Fisheries and Food, Cambridge)

the barley succeeding the wheat. In France, heightened incidence of *Rhizoctonia solani* has been recorded on beans following potato crops harvested by heavy combine machinery (Bouhot, pers. comm.).

Another type of compaction is that associated with plough-soles, commonly due to several successive, shallow cultivations. Plough-soles favour many diseases. A striking example of barley stunt, at least partly due to shallow cultivation favouring infection by *Rhizoctonia*, is shown in Fig. 2.10.

The best way to deal with plough-soles, and in general to loosen up soil in deeper layers and improve percolation, is by subsoiling to depths of 35–45 cm. This facilitates deeper growth of roots and more vigorous development of many crops, enabling them to outpace certain pathogens and increase yield. It has been found particularly effective in reducing loss from nematodes, for a number of reasons: subsoiling may prevent water stress, which renders crops more susceptible to attack by species such as the soybean nematode *(Hoplolaimus columbus)* (Parker et al. 1973), or it may simply allow roots to penetrate to soil depths at which nematodes will not develop. The same considerations apply to some fungal pathogens. Kraft and Giles (1979) have found subsoiling to increase yield of peas in fields with histories of severe root rot due to *Fusarium solani* f. sp. *pisi* and *Pythium ultimum*. Burke et al. (1972) have obtained similar results on beans affected by *F. solani* f. sp. *phaseoli*, especially if subsoiling was carried out under the drill row rather than between the rows.

2.5.2.2 Distribution of Inoculum in the Soil Profile and Over the Field

The benefit, often doubtful, of burying inoculum deeply in the soil has been referred to in the chapter on sanitation (Chap. 2.2.4.2). In many cases the effects of re-distribution of inoculum normally concentrated in the topsoil to depths of 20 cm or even more, as effected by mould-board ploughing, is not clear, at least not where the principal soil-borne diseases of cereals in temperate countries are concerned; but ploughing-in of rice residue to a depth of 25–30 cm renders a high percentage of inoculum of *Sclerotium oryzae* unavailable for infection of rice subsequently sown with flooding (Webster et al. 1976). The effect of direct drilling on distribution of inoculum of cereal pathogens is discussed in Chap. 2.5.2.3.

Spread of inoculum by tillage from isolated foci along the rows or over the entire field is a very real danger for many crops and diseases. A good example is *Verticillium dahliae* which often affects potatoes and other crops in foci. The millions of microsclerotia produced in debris of individual plants affected by the pathogen are easily distributed by all kinds of tillage operations, and render the field useless for further growing of potatoes, cotton, and many other host crops. Bacterial diseases, such as those caused by *Pseudomonas solanacearum*, are likewise distributed most effectively by tillage.

Subsoiling has not been found to affect the vertical distribution of fungi (*Rhizoctonia solani*, *Pythium* spp.) or nematodes affecting cotton, in layers down to 70 cm depth (Hussey and Roncadori 1977).

2.5.2.3 The Diverse Effects of Non-Tillage

"Non-tillage" practices are applied both to annual field crops and to perennial fruit and plantation crops, but details and effects of the practice in the two cases differ considerably.

Annual Field Crops. Non-tillage, which in this context may also be termed direct drilling or stubble mulching, essentially consists in leaving the residue of crops after harvest in the field, and sowing the succeeding crop into the undisturbed soil. The persistence of weeds in such fields is incidental. This practice has the following effects, directly or indirectly affecting disease development:

1. Economic Advantages. Direct drilling of cereals saves labour, makes for better use of available labour and machinery in off-seasons, and facilitates earlier sowing. From the aspect of disease development, earlier sowing does not usually have decisive advantages or disadvantages in cereal growing, but there are exceptions.

2. Survival and Distribution of Inoculum. In non-tilled fields, pathogens capable of aerial distribution enjoy opportunities of spreading spores over long periods. Table 2.17 lists some such pathogens of the cereal crops to which non-tillage is most commonly applied, and shows the considerable risks arising where cereal debris is not incorporated. However, as Garrett (1970) has pointed out, an advantage of growing wheat by non-tillage where take-all is a problem lies in leaving tiller bases infected by *Gaeumannomyces graminis* aboveground, and hence out of reach of roots of the succeeding wheat crop.

Table 2.17. Maize and wheat pathogens spread through inoculum released from stubble in fields under non-tillage management

Crop	Disease	Pathogen	References
Maize	Eye-spot	*Kabatiella zeae*	Anon (1973d)
	Yellow leaf blight	*Phyllosticta maydis*	Anon (1973d)
	Northern leaf blight	*Drechslera turcica*	Anon (1973d)
	Gray leaf spot	*Cercospora zeae-maydis*	Roane et al. (1974)
	Maize Dwarf Mosaic		All et al. (1977)
Wheat	Wheat Streak Mosaic		Wiese (1977)
	Root and foot rot	*Drechslera sativa, Fusarium* spp.	Wiese (1977)
	Leaf blotch	*Septoria nodorum, S. tritica*	Wiese (1977)
	Yellow leaf spot	*Pyrenophora trichostoma*	Wiese (1977)
	Cephalosporium stripe	*Cephalosporium gramineum*	Wiese (1977)
	Head blight, scab	*Fusarium* spp.	Wiese (1977)

3. Soil Compaction. One of the least desirable features of direct drilling is soil compaction. Annual passage of heavy sowing, fertilizing, spraying, and harvesting equipment through the field, without periodical loosening of the soil by cultivation or ploughing, involves considerable risk. The resultant effects on disease, especially on take-all, have been referred to in Chap. 2.5.2.1. A further illustration of such effects is presented in Fig. 2.10.

Perennial Fruit and Plantation Crops. In such crops, a cover of weeds or crops is intentionally maintained, or weeds are eliminated by herbicides. The cumulative effect of non-tillage on disease, sometimes over many years, has to be considered from the following aspects:

1. Economic Advantages. Mowing weed or crop cover is cheaper than ploughing and depends less on soil conditions. Roots of most fruit crops go deeper than those of annual crops, and are not as susceptible to soil surface compaction. In irrigated orchards, water consumption in non-tilled crops tends, however, to be higher than in crops cultivated regularly.

2. Root Injury, Vigour and Disease. All types of cultivation cause some root injury and affect tree vigour to some extent. Increased incidence of *Verticillium dahliae* on cultivated olives and apricots has been reported from southern France (Vigouroux 1975; Vigouroux and Castelain 1969). In Israel, introduction of non-tillage has practically eliminated *Sclerotium rolfsii* from apple orchards, and has reduced incidence of *V. dahliae*, *Rosellinia (Dematophora) necatrix*, basidiomycetes (e.g., *Fomes* spp., *Ganoderma* spp.), and nematodes (*Xiphinema* spp.) in orchards and vineyards (Golan, pers. comm.). The absence of root severance in hop gardens under non-tillage is held to contribute to reduction of inoculum potential of *V. albo-atrum* (Sewell and Wilson 1974a).

3. Non-Burial of Inoculum. In many tree crops, pathogens cause mummification of fruit organs which are often shed. When these persist on the soil surface, they may form fruiting bodies and spread by ascospores. Failure to bury such sources of inoculum has been shown to be one of the disadvantages of non-tillage practices, e.g., in relation to black rot of grapes caused by *Guignardia bidwelli* and brown rot of deciduous fruit caused by *Sclerotinia fructicola* (Merrill and Petersen 1979).

4. Herbicide Treatments in Non-Tilled Plantations. Non-tillage involving removal of weeds by herbicides, leaving a clean soil surface, has affected crop disease in various ways. In hop gardens in England, such practice has been found to increase inoculum of powdery mildew *(Sphaerotheca humuli)* in spring by allowing over-wintering cleistocarps to remain on the soil surface and hibernate in rootstock buds (Royle 1975). On the other hand, non-tillage has proved beneficial in wilt-infested hop gardens, by preventing the distribution of weeds affected by *Verticillium albo-atrum* along the rows by means of cultivators (Sewell, pers. comm.). Application of overhead irrigation to orchards kept weed-free by herbicides facilitates splashing of soil onto the tree trunks and low-hanging branches. This has been found to increase brown rot *(Phytophthora citrophthora)* of citrus fruit, and is bound to favour similarly spread of other soil-borne fungi and bacteria.

2.5.2.4 Tillage Practices as Affecting Cereal Diseases in England

The diversity of tillage effects on diseases, and their interaction with other practices, can be illustrated by reference to cereal crops in England, where this has been studied intensively (Yarham and Hirst 1975; Yarham, pers. comm.). In this work, direct drilling (non-tillage, stubble mulching) of wheat and barley has been compared with cultivation to depths of 10 or 20 cm and ploughing that inverted soil to a depth of 20 cm (see details in Chap. 2.5.2). Some of the results of these studies on major cereal diseases may be summarized as follows.

Take-All (Gaeumannomyces graminis). The change-over from ploughing to direct drilling has had only small effects on the incidence of take-all in wheat and barley, with considerable year-to-year and site-to-site variation. This is readily explained by the multiplicity of factors affected by tillage. Among these, factors likely to reduce the disease in non-inverted soil are the increased activity of antagonists and competitors of *G. graminis* in the top soil and the higher concentration of phosphorus, but lower concentration of nitrogen, in the top soil. On the other hand, non-tillage may favour take-all development where it leads to soil compaction (see Fig. 2.5) by leaving inoculum-carrying debris in the top soil layers into which the succeeding crop's seed is drilled, and by permitting growth of perennial grass carriers of *G. graminis* (Yarham 1979a).

Eye-Spot (Pseudocercosporella herpotrichioides). The effect of direct drilling on the disease in wheat and barley is also slight. Where eye-spot is reduced, this may largely be ascribed to tillage effects on plant habit: in direct-drilled fields, sowing is shallow, the plant habit prostrate (with tillers formed at angles markedly wider than in plants sown more deeply), and the crop climate is accordingly less humid. However, in practice, the dominant effect of all non-plough practices on disease is likely to be the opportunity it gives farmers to sow winter crops earlier. This may make the crop more vulnerable to various diseases, especially to eye-spot, but the advantages of early sowing often outweigh this risk.

Septoria Blotch (Septoria nodorum). The comparative effect of tillage practices on this blotch disease of wheat is chiefly related to the amount of infected debris left on the soil surface. Where this amount is large, burial of debris by ploughing is advantageous (Yarham, pers. comm.).

Rhynchosporium Leaf Blotch (Rhynchosporium secalis). Incidence of this barley disease does not seem to depend primarily on the amount of surface debris left by preceding crops. Although direct drilling leaves more such debris, its effect in producing a more prostrate plant and a less humid crop climate tends to reduce leaf blotch (Yarham, pers. comm.).

Powdery Mildew (Erysiphe graminis) and Leaf Rust (Puccinia recondita). These leaf diseases are more prevalent on ploughed than on direct-drilled fields. The reason seems to lie principally in the reduced mineralization of nitrogen in non-ploughed plots, resulting in less luxurious growth and a less humid crop climate.

2.5.3 Topsoil Management

2.5.3.1 Shaping the Topsoil and Earthing-Up (Hilling)

Shaping the topsoil into ridges or beds is a well-known and useful means of keeping dry the soil that is in direct contact with the root collar or may come into contact with low-hanging fruit subject to infection by soil fungi or bacteria. This seems of special interest in irrigated crops in warm climates, where water is applied at frequent intervals and high temperatures favour rapid infection of soft or senescent tissues. It has proved effective in reducing incidence of *Sclerotium rolfsii* on many vegetables and groundnuts, *Pythium* fruit rot of cucurbits, and *Sclerotinia sclerotiorum* on celery. This and other practices of keeping the topsoil dry, while roots take advantage of the water in underlying layers of soil and on the sides of ridges, is also an accepted means of controlling diseases caused by *Rhizoctonia solani* on lettuce and other crops (Leach and Garber 1970; Pieczarka and Lorbeer 1974). Planting potatoes and tomatoes on raised beds between deep irrigation ditches has permitted production of fairly good crops in soil heavily infected with root-knot nematodes in California (Thorne 1942). However, where somewhat saline water is used for irrigation, sowing crops on ridges may promote accumulation of salts, and thus favour pathogens taking advantage of stress, such as *Macrophomina phaseolina* (Waller, pers. comm.).

Where water is deficient, shaping the topsoil into ridges may have the opposite aim: concentration of water running off the ridges into their bottom, where the crop is sown or planted. With loess and other types of soil that form crusts as soon as wetted, such run-off effectively increases water supply and reduces the danger of water stress, with all its resultant effects on the susceptibility of crops to disease.

Earthing-up potatoes, i.e., throwing soil from between the rows onto the bottom of the potato plants in the row, may facilitate tuber formation and protects the tubers to some extent from late blight infection by spores washed into the soil from blighted foliage (Anon 1976b).

2.5.3.2 Elimination of Soil Crusts

In soils that tend to form crusts, the latter impede emergence of seedlings and may much increase danger of attack by seedling diseases and soil pests. In irrigated crops, dry crusts may be eliminated by frequent, short irrigations, although this increases dangers of seedling diseases, and tillage is preferable. In non-irrigated crops, tillage of the topsoil is so far the only means of helping the seedlings emerge. The use of soil conditioners, such as synthetic polymers, may in future provide an alternative to such tillage, if applied in strips over the rows.

2.5.4 Tillage as a Control Practice

2.5.4.1 Destruction of Weeds and Volunteer Plants Between Successive Crops

Reference to the importance of controlling weeds and volunteer plants has already been made in various chapters. In the present context it should be noted that cereal crops are infested by a number of perennial weeds that carry pathogens and are not readily controlled by chemicals, e.g., *Agropyron repens* which transmits take-all (Yarham 1979a) and Johnson grass *(Sorghum halepense)* that carries virus diseases of maize (Damsteegt 1976). Somewhat contradictory results have been obtained in attempts to show the role played by these weeds in determining the level of disease in subsequent sowings. Where they have been demonstrated to promote disease, especially take-all, mould-board ploughing is certainly the best, and direct drilling the least desirable method of tillage. Fields of tuber- and bulb-sown crops are often plagued by persistence of ground-keepers or germination of volunteers. Many diseases of potato and onion can be transmitted in this way. The effects of tillage in turning in such growth are wholly salutary as measures of sanitation.

2.5.4.2 Weed Control While the Crop Grows

As fas as we are aware, no reference is made in the literature on orchard management to the effect of weed control on the development of diseases, while there are plenty of references to such effects on orchard pests.

In field crops, Garren (1961) has investigated methods of tillage for weed control on the incidence of *Sclerotium rolfsii* on groundnuts. He concluded that tillage that will not throw soil on the plants ("non-dirting" tillage) is acceptable and obviates the necessity for deep incorporation of organic matter in the soil before sowing. Klisiewicz and Beard (1976) have similarly stated that cultivation between rows of sunflowers must not throw soil up against their stems, as this promotes development of *Phytophthora cryptogaea*. The same is certainly true for the cultivation for weed control of all crops susceptible to infection by *Rhizoctonia solani*.

2.5.4.3 Reduction of Inoculum by Tillage

The two methods for reduction of inoculum by tillage are its burial for inactivation by the soil flora or its exposure to heat and drought at the soil surface.

Burial of inoculum has already been referred to in the context of inoculum distribution over the soil profile and of non-tillage. Ploughing-under debris is certainly the basic feature of controlling *Rhizoctonia solani* on its many hosts (Papavizas and Lewis 1979). Even just covering debris with a shallow layer of soil can be effective to prevent obligate parasites from spreading from green debris on the soil surface. In lettuce crops, for example, prompt coverage of debris with soil, just after harvest, reduces danger of spread of both downy mildew *(Bremia lactucae)* and lettuce mosaic virus (Anon 1973 b). Debris containing fungal resting bodies that later liberate air-borne spores should certainly be tilled into the soil; this includes leaves carrying perithecia of apple scab *(Venturia inaequalis)*, oospores of grape downy mildew *(Plasmopara viticola)*, vegetable matter containing sclerotia of *Sclerotinia sclerotiorum*, and many other diseases.

Repeated ploughing during crop-free periods is chiefly known as a measure of some value against nematodes (Anon 1968). It is still widely practiced in lower-value crops, and seems to be effective against root-knot and other nematodes.

Some fungal and bacterial pathogens can be curbed by successive ploughing in dry seasons. To give just two examples: In Texas, mycelial strands and sclerotia of the cotton root rot fungus *(Phymatotrichum omnivorum)* were found to be rapidly killed by such practices (Rogers 1939), and in banana plantations in Central America, disking soil infected by *Pseudomonas solanacearum* five times in the dry season reduced disease significantly (Sequeira 1958).

2.6 Crop Nutrition

Nutrients can affect the relations between crop and pathogen in many different ways. Manipulation of crop nutrition therefore gives the farmer a valuable tool for managing crop health. Fuchs and Grossmann (1972) and Colhoun (1973) have reviewed the effects of nutrients on crop disease, and Ponchet (unpubl.) has emphasized justly that the knowledge available on this subject is inadequate and sketchy.

Nutrient Effects on Host and Parasite

Host Parameters. (a) Vigour of growth, which in its turn may affect the crop climate, and the chances of crops to recover from disease attack. (b) Anatomical and histological characteristics, such as thickness of cuticle and epidermis, or lignification of tissues, and biochemical and physiological reactions, including the formation of corky layers and phytoalexins. (c) Rate of growth, making it possible to shorten stages at which plants are most susceptible (seedlings) or to regulate the period in which such stages occur (bud-burst, flowering), and influencing the time of maturation and senescence. (d) Water economy: proper nutrition may enable plants, under certain conditions, to make fuller use of the soil's water potential.

Pathogen Parameters. (a) Rate of penetration, colonization and reproduction of pathogens. (b) Relative rate of development of pathogens and their competitors and antagonists, especially in the soil. (c) Direct toxic effects of some fertilizers, such as urea, on certain pathogens.

2.6.1 Blanced Nutrition and Nutrition Unbalanced for Economic and Crop Health Reasons

2.6.1.1 Size, Quality and Maturation of Yield

Plant growth is dependent on the supply of many different elements, and each element relates to specific aspects of the plant's metabolism. A proper balance between the elements is needed for the crop to realize its full potential of development. This balance, implying as it does availability of each element in amounts, which together with the amounts of other elements available facilitates maximal yield, greatly depends on local climatic, edaphic and cultural conditions. It is most markedly affected by temperature levels and availability of water, and differs at various stages of crop growth. Balanced nutrition is therefore a relative concept.

As pointed out above, full crop growth may mean neither production of optimal revenue nor particularly healthy growth. Pathogens have nutritional requirements of their own. If these approximate those of the crop, as requirements of many obligate parasites (including viruses) do, the question arises: who benefits more, the crop or the pathogen?

Since nutrients, especially N but also K, affect not only size and quality of yield but also rate of maturation, there is a basic difference in the aim of fertilizing crops producing yields for fresh marketing, and those producing dry produce for consumption or processing. With regard to the latter, which include cereals, nuts, spices, and industrial crops grown for food, fibre or fuel, regulating the rate of maturation may be a minor consideration, except where limited availability of harvesting equipment or fuller use of processing facilities put a premium on harvesting yields successively, to stagger harvest and supply. With most of the dry crops, nutrition aims at maximum size of yield of acceptable quality. Quality may at times even be lower than that expected of the same crop when grown for consumption, e.g., olives or groundnuts grown for oil need not be as free from blemishes as those grown for the market. In dry crops, therefore, the effects of nutrition on disease are of interest chiefly in as much as disease affects the size of the crop.

In crops grown for fresh produce, vegetables, fruits or flowers, the timing of harvest and the freedom of produce from blemish is often as important as the size of yield. Here the principal aim of nutrition, and especially of regulating N-supply, may be to advance or delay ripening, and to produce unblemished produce of good keeping quality, not prone to speedy attack by post-harvest parasites.

2.6.1.2 Crop Nutrition and the Pathogen

The two most immediate objectives of nutrient practices to protect crops from pathogens are:

a) To avoid strain, more particularly nutrient strain, which may directly lower the resistance of crops to facultative parasites. Successful avoidance of nutrient strain may sometimes compensate the crop for environmental strain and thus delay onset of senescence and the strains it brings with it (see Chap. 1.4).

b) To manipulate nutrients to the relative disadvantage of the pathogen: this may involve a direct effect, when either nutrition is balanced at a point better for crop than pathogen (relatively low levels of nitrogen near the maturation stage may benefit cereals, but not their rusts) or nitrogen is applied to shorten periods of special danger. However, as mentioned above, the effect may also be indirect, e.g., stimulation of microbiota competing with or antagonistic to pathogens in the soil.

Grossmann (1970) has placed effects of nutrients on plant resistance to disease in proper perspective. Crops or varieties of very high or very low susceptibility will not be influenced much, but the moderately susceptible crops can be nudged into somewhat better resistance by proper nutrition. High inoculum pressure may at times overcome beneficial effects of fertilization, another case of compensation of one epidemiological factor for another (see Chap. 1.3).

2.6.1.3 Manipulation of Nutrients

By far the most important aspect of the manipulation of nutrients is the rate, form and timing of N-fertilization. This is particularly true in relation to nutrient effects on crop disease. It is therefore the principal subject of the following sections and greatly exceeds in importance anything that can be achieved by varying the supply of other elements.

Manipulation of fertilizer practices strongly interacts with patterns of precipitation and with irrigation practices. In fact, addition of nitrogen or of water to the soil may sometimes produce similar effects on crops. As mentioned before, ready availability of N may assist crops in making fuller use of the soil's water potential (Papendick and Campbell 1975).

Rates of fertilization also have to be adapted to varietal resistance to specific diseases, and to the availability and cost of chemical control. High N-fertilization, leading to heightened danger of rust, may be advantageous and permissible for wheat varieties immune to rust, or where chemicals can be applied promptly and profitably when needed, to control rust. Such fertilization may, however, be utterly out of place with rust-susceptible varieties, or even with moderately resistant ones, if fungicidal protection cannot be secured or timed properly.

Crop nutrition and its effect on disease is a subjet on which a vast amount of research has been performed. Nutrient effects of the application of organic matter to the soil are almost impossible to distinguish from the effects of such applications on soil texture and reaction, and on the attendant effects on the soil microbiota.

Precise information on the effects of various nutrients on crop and disease is almost entirely derived from research with artificial fertilizers, and these will be discussed in the following sections.

2.6.2 Fertilization with Nitrogen

2.6.2.1 General Effects of Nitrogen in Relation to Crop Disease

Nitrogen, more than any other element used in fertilization, affects all the parameters listed in the introductory paragraph to this chapter.

Most pronounced is the effect of nitrogen on the vigour and rate of growth, both factors greatly affecting the level of susceptibility of crops to many pathogens.

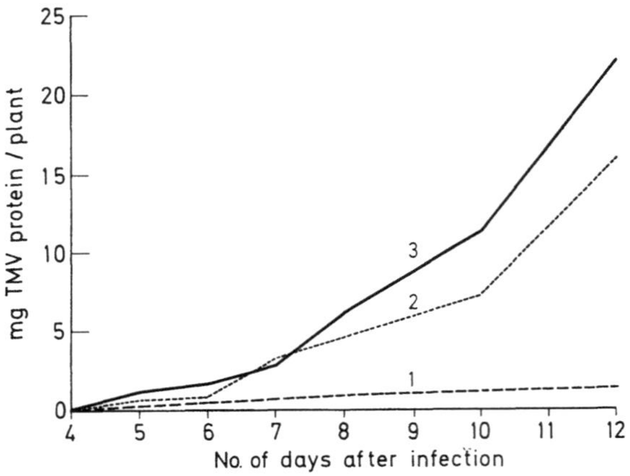

Fig. 2.11. Effects of nitrogen application to tobacco plants on their content of Tobacco Mosaic Virus: *1* plants suffering from N deficiency; *2* plants with normal N fertilization; *3* plants supplied with excess N. (Gaeumann 1951)

It is frequently impossible to distinguish between effects on crop vigour of N and of other invigorating factors, such as irrigation. Vigorously growing crops are, with some exceptions, more highly susceptible to obligate parasites, such as rusts and powdery mildews of wheat and barley, club-root *(Plasmodiophora brassicae)* of Brassicae, and downy mildew as well as virus diseases, such as Tobacco Mosaic Virus (see Fig. 2.11). Some important facultative parasites, however, also profit from high crop vigour, including *Verticillium dahliae* on almond and olives (Vigouroux 1973, 1975), and *Pyricularia oryzae* and *Sclerotium oryzae* on rice (Ou 1972).

The majority of facultative parasites attack low-vigour crops more strongly, and thus profit from low N-levels in the soil. Examples are the wilt diseases caused by *Fusarium oxysporum* strains on tomato, cotton, and pea, and *Alternaria* blights on many crops.

Effects of excess N on host tissues comprise formation of loose tissues, with large thin-walled cells, ample intercellular spaces and wide open stomata – all features that facilitate development of parasites, especially of bacterial pathogens (Grossmann 1970). The increased N-content of host tissues may also benefit pathogens directly, as found by Ou (1972) in the case of *Gibberella fujikuroi* on rice.

2.6.2.2 Nitrate Nitrogen and Ammonium Nitrogen

A number of surveys of the effects of specific forms of nitrogen, especially NO_3-N and NH_4-N, on plant disease have recently been published (Henis 1976; Henis and Katan 1975; Huber and Watson 1974). These have shown that pathogen-crop interactions depend frequently on the form rather than the amount of N available. These interactions are, however, complex in nature, and the effect achieved by particular forms of N in controlling disease may be related to one or more of the following factors:

a) Crop sequence, root exudates and organic amendments: crops such as lucerne, maize or pea, and application of organic manure, stimulate nitrification, and straw decomposes faster, in the presence of NH_4-N (Huber and Watson 1974).

b) The rate and type of fertilization chosen in accordance with the demands of the crop and the desired effect on the rate of its maturation.
c) The rate at which the fertilizer is likely to leach, which in turn depends also on soil type and its organic matter content, on drainage, and on rainfall of irrigation. As NO_3-N leaches more readily than NH_4-N, this is an important consideration.
d) Stabilization of the nitrogen applied by products that inhibit nitrification, such as "N-Serve" [2-chloro-6 (trichloromethyl) pyridine].

The comparative effect on crop diseases of nitrate or ammonium nitrogen has been tabulated by Huber and Watson (1974), and the material in Table 2.18 bases in part on their data. The table shows that wilts and root rots caused by species of *Fusarium* and by *Rhizoctonia solani* are among those favoured by ammonium-N, while wilts caused by species of *Verticillium* and root rots due to *Pythium* spp. are favoured by nitrate-N.

Perusal of this table shows up the farmer's dilemma: all the major crops listed are subject to some diseases favoured by NH_4-N, and to others favoured by NO_3-

Table 2.18. Comparative effects of nitrate and ammonium fertilization on crop diseases

Crop	Disease	Pathogen	Nitrate[a]	Ammonium[a]	References
Cereals					
Wheat	Eye spot	*Pseudocercosporella herpotrichioides*	–	+	Huber and Watson (1974)
	Sharp eye-spot	*Rhizoctonia solani*	–	+	Huber and Watson (1974)
	Take-all	*Gaeumannomyces graminis*	+	–	Smiley and Cook (1973)
	Stem rust	*Puccinia graminis*	+	–	Huber and Watson (1974)
	Stripe rust	*P. striiformis*	+	–	Huber and Watson (1974)
Maize	Root rot	*Aphanomyces euteiches*	–	+	Huber and Watson (1974)
	Root rot	*Pythium* sp.	+	–	Huber and Watson (1974)
	Stalk rot	*Fusarium* sp.	–	+	Huber and Watson (1974)
	Stalk rot	*Diplodia zeae*	+	–	Huber and Watson (1974)
	Northern leaf blight	*Drechslera turcica*	–	+	Huber and Watson (1974)
Sorghum	Stalk rot	*Fusarium moniliforme*	+	–	Erinle and Edmunds (1970)
Rice	Blast	*Pyricularia oryzae*	–	+	Huber and Watson (1974)
	Brown spot	*Drechslera oryzae*	+	–	Dasgupta and Chattopadhyay (1977)

[a] See footnote on p. 138

Table 2.18 (continued)

Crop	Disease	Pathogen	Nitrate [a]	Ammonium [a]	References
Legumes					
Bean (*Phaseolus*)	Root rot	*Fusarium solani* f. sp. *phaseoli*	–	+	Snyder et al. (1959)
	Wilt	*F. oxysporum* f. sp. *phaseoli*	–	+	Huber and Watson (1970)
Broad bean (*Vicia faba*)	Chocolate spot	*Botrytis fabae*	–	+	Huber and Watson (1974)
Pea	Root rot	*Aphanomyces euteiches*	–	+	Huber and Watson (1974)
	Root rot	*Pythium* sp.	+	–	Huber and Watson (1974)
Soybean	Root rot	*Aphanomyces euteiches*	–	+	Huber and Watson (1974)
	Cyst nematode	*Heterodera glycines*	+	–	Huber and Watson (1974)
Various	Charcoal rot	*Macrophomina phaseolina*	–	+	Filho and Dhingra (1980)
Vegetable and Industrial Crops					
Potato	Root and stem rot	*Rhizoctonia solani*	–	+	Huber and Watson (1974)
	Wilt	*Verticillium albo-atrum*	+	–	Huber and Watson (1970)
	Scab	*Streptomyces scabies*	+	–	Huber and Watson (1974)
Tomato	Wilt	*Verticillium albo-atrum* *V. dahliae*	+	–	Smiley (1975)
	Wilt	*Fusarium oxysporum* f. sp. *lycopersici*	–	+	Woltz and Jones (1972)
	Fruit and root rot	*Colletotrichum phomoides*	+	–	Huber and Watson (1974)
	Wilt	*Pseudomonas solanacearum*	+	–	Huber and Watson (1974)
Tobacco	Root rot	*Thielaviopsis basicola*	+	–	Smiley (1975)
Cotton	Root rot	*Phymatotrichum omnivorum*	+	–	Smiley (1975)
	Wilt	*Fusarium oxysporum* f. sp. *vasinfectum*	–	+	Smiley (1975), Henis (1976)
Sugarbeet	Damping-off, black root	*Rhizoctonia solani*	–	+	Huber and Watson (1974)
Other Crops					
Citrus	Brown rot	*Phytophthora citrophthora*	–	+	Boccas and Laville (1978)

[a] + Indicates that disease development is favoured by nitrate or ammonium, respectively
– Indicates that the opposite is the case

N. From the point of view of phytomedicine, the decision which nitrogenous fertilizer to choose therefore has to be based on full information on the pathogens likely to attack the crop, and the relative frequency and severity of the attack to be expected. This is an example of the level of judgement expected of the modern farmer in using cultural practices for disease control.

2.6.2.3 Timing of Nitrogen Fertilization

The principal concern in the timing of nitrogen application is avoidance of excess N at the stage of crop maturation. Such nitrogen may

a) cause soft growth of leaves at a time when foliage is dense and humidity conditions particularly favourable to pathogens;
b) delay maturation (possibly into seasons more favourable to the pathogen) and give pathogens extra time to affect the produce, and
c) prevent full maturation of the produce in the time allotted to the crop.

Much of the research on effects of late applications of nitrogen refers to cereals and their diseases. In wheat, such gifts of nitrogen are apt to favour development of leaf blotch *(Septoria nodorum)* (Oppitz and Hoeser 1978) and of rusts (Anon 1979). Application after emergence of the flag leaf also caused increased incidence of powdery mildew *(Erysiphe graminis)* in late stages of growth (Last 1954).

Similar problems exist with regard to some diseases of rice. In detailed research by Mayee et al. (1976) on timing and splitting applications of nitrogen at the stages of planting, tillering and panicle initiation, complete doses (50 kg/ha) applied at one of the last two stages resulted in an incidence of *Sclerotium oryzae* of about 60%, as compared to only 34%–38% where applications were split between the three stages. Split applications of N are also preferred in order to minimize incidence of *Pyricularia oryzae* (Bernaux 1976) and of *Cochliobolus miyabeanus* (Anon 1970a). As regards the bacterial blight of rice *(Xanthomonas oryzae)*, division of nitrogen top-dressings into several small doses had a similarly beneficial effect (Mizukami and Wakimoto 1969).

On potatoes, restriction of N application to the period up to flowering may be expected to assure proper maturation of the tubers, and thus lessen their susceptibility to the bacterial rot caused by *Pectobacterium carotovorum* var. *atrosepticum* (Ziegler et al. 1976).

A special case, in which N application late in the growing season is beneficial, is that of apple scab *(Venturia inaequalis)*: foliar application of urea to apples before leaf fall has resulted in better microbial deterioration of leaves and less ascospore formation (Butt, pers. comm.).

2.6.2.4 Toxic Effects of Nitrogen on Pathogens

Not all the effects of nitrogen are achieved via its effects on host tissues. Some nitrogen compounds are directly or indirectly toxic to fungi.

Ammonia has been found to affect germinability of the sclerotia of *Sclerotium rolfsii* by increasing the NH_4 content of the sclerotial environment to toxic levels (Henis and Chet 1967). For this reason, ammoniacal fertilizers are widely preferred for application to soils infected with this fungus.

Ammonia also has a direct toxic effect on *Phymatotrichum omnivorum* on cotton, and possibly on club root *(Plasmodiophora brassicae)* on Brassicae and take-all *(Gaeumannomyces graminis)* on wheat and barley (Fuchs and Grossmann 1972).

Calcium cyanamide exerts toxic action on a number of soil-inhabiting fungi, including *S. rolfsii*.

Urea, frequently applied as a foliar spray, has proved directly toxic to various foliar pathogens. It also has a therapeutic action on some *Fusarium* wilts, e.g., that caused by *F. oxysporum* f. sp. *lini* on flax, and may in some cases act as an antidote to the toxins produced by these fungi (Fuchs and Grossmann 1972).

2.6.3 Fertilization by Phosphates, Potassium and Calcium

2.6.3.1 Phosphates

Knowledge of the effect on crop diseases of phosphates per se is fragmentary and in part contradictory. Somewhat surprisingly, since P is involved in metabolic processes vitally important to the plant, its action on disease resistance is variable and does not seem to be very conspicuous (Király 1976).

Of the few research results clearly expressing the influence of phosphates, four cases may be mentioned. In long-term field experiments at Rothamsted, England, Mattingly and Slope (1977) found that repeated applications of phosphate fertilizers delayed the onset, and greatly lessened the severity, of take-all disease *(Gaeumannomyces graminis)* on barley. In the United States, application of triple superphosphate significantly reduced incidence and lesion morphology of potato scab *(Streptomyces scabies)* (Davis et al. 1976). However, incidence of cucumber mosaic virus was shown to increase on spinach as the amount of P supplied was increased (Cheo et al. 1952), and so was that of *Septoria nodorum* on wheat (Cunfer and Touchton 1979).

Most of the cases in which salutary effects of phosphates on plant disease have been observed seem to have been connected with one of two situations: either the crop grew on soil markedly deficient in P, and correction of the imbalance contributed to its health, or maturity of the crop was somewhat advanced by liberal supply of P, and this helped it escape from obligatory parasites preferring younger tissues, such as the downy mildews.

Obviously, the information presently available is of little value as a practical guide to farmers in improving disease control by better use of phosphate fertilizers.

2.6.3.2 Potassium

The effect of potassium on crop disease depends on the nutrient status of the soil, on the rate of application, and largely on its balancing effects of other elements (especially nitrogen) on the plant.

The mechanism by which potassium affects disease caused by pathogens differs greatly, and includes both direct and indirect effects.

Direct effects are stimulation or reduction of pathogen penetration, multiplication and survival, and of its aggressivity and rate of establishment in the host. Vari-

Table 2.19. Some diseases reduced in severity by the effect of ample K-supply on host resistance

Crop	Disease	Pathogen	References
Field and Vegetable Crops			
Wheat	Stem rust	*Puccinia graminis* f. sp. *tritici*	Király (1976)
Wheat, barley	Powdery mildew	*Erysiphe graminis*	Perrenoud (1978)
Rice	Brown leaf spot	*Cochliobolus miyabeanus*	Ismunadji et al. (1973)
	Stem rot	*Helminthosporium sigmoideum*	Ismunadji et al. (1973)
	Narrow brown leaf spot	*Sphaerulina oryzina* (= *Cercospora*)	Ismunadji et al. (1973)
	Bacterial blight	*Xanthomonas oryzyae*	Tagami (1968)
Cotton	Wilt	*Fusarium oxysporum* f. sp. *vasinfectum*	Black (1968)
	Wilt	*Verticillium dahliae*	Garber et al. (1976)
	Root-knot	*Meloidogyne* spp.	Nusbaum and Ferris (1973)
Melon	Wilt	*F. oxysporum* f. sp. *melonis*	Perrenoud (1977)
Tobacco	Leaf spot	*Alternaria longipes*	Last (1954)
Tomato	Early blight	*A. solani*	Király (1976)
Cabbage	Yellows	*Fusarium oxysporum* f. sp. *conglutinans*	Walker and Hooker (1945)
Cauliflower	Downy mildew	*Peronospora parasitica*	Felton and Walker (1946)
Pea	Root rot	*Aphanomyces euteiches*	Perrenoud (1977)
Cassava	Bacterial wilt	*Xanthomonas manihotis*	Adeniji and Obigbesan (1976)
Maize	Stalk rot	*Gibberella zeae, G. fujikuroi*	Shipton (1979)
Maize	Smut	*Ustilago maydis*	Perrenoud (1977)
Maize	Northern corn blight	*Drechslera turcica*	Perrenoud (1977)
Fruit and Tree Crops			
Grape	Downy mildew	*Plasmopara viticola*	Perrenoud (1977)
Grape	Grey mould	*Botrytis cinerea*	Király (1976)
Apricots	Brown rot	*Sclerotinia fructicola*	Perrenoud (1977)
Oil palm	Wilt	*Fusarium oxysporum* var. *albedinis*	Turner (1970)
Rye-grass	Snow mould	*Gerlachia nivalis*	Perrenoud (1977)
Lucerne	Root rot	*Fusarium* spp.	Perrenoud (1977)
Lima bean	Bacterial blight	*Pseudomonas syringae*	Perrenoud (1977)

ous phases of disease development may well be differently affected: in the case of potato late blight *(Phytophthora infestans)*, potassium does not influence the frequency of infection, but it increases development of lesions and intensity of sporulation (Fuchs and Grossmann 1972). The amount of potassium available to the crop generally has little effect on its susceptibility to virus diseases.

Indirect effects of potassium on disease development derive from:

a) Promotion of wound healing, which evidently affects infection by wound parasites, e.g., *Botrytis cinerea* on grape (Király 1976).
b) Heightening resistance to frost injury, and thus lessening losses caused by parasite invasion following such injury.
c) Delaying maturity in some crops, thereby postponing the strain of senescence and its attendant effects, mostly infection by facultative pathogens.

Table 2.19 presents a list of diseases reduced in severity by application of potassium, including a variety of fungal and bacterial pathogens, some obligate and many facultative. Effects on virus diseases are usually not well marked, nor does potassium greatly affect nematodes though there are some exceptions.

Among pathogens favoured by application of potassium, mention must be made of the cause of rice blast *(Pyricularia oryzae)* and of the nematode *Meloidogyne incognita*, which infects numerous crops (Perrenoud 1977).

Of the two most commonly used K-salts, potassium chloride and potassium sulphate, the chloride is generally considered to raise crop resistance to disease more effectively, e.g., in the cases of potato scab *(Streptomyces scabies)* and *Monilia* rot of apricots; it is, however, somewhat less effective than the sulphate in maintaining resistance against certain viruses, such as the leafroll and Y viruses in potatoes (Fuchs and Grossmann 1972).

2.6.3.3 Calcium

The principal effect of calcium on host/pathogen relationships seem to be on the composition of cell walls of the host, making them more resistant to penetration by facultative pathogens. Diseases caused by the latter are therefore prominent among those known to be reduced by high calcium level in the plants. These diseases comprise some caused by *Rhizoctonia solani*, *Sclerotium rolfsii*, *Botrytis cinerea*, and *Erwinia phytophthora* (Kiraly 1976), and wilts caused by strains of *Fusarium oxysporum* on melons and carnations (Blanc et al. 1979). The resistance of lucerne to the stem nematode *(Ditylenchus dipsaci)* is likewise strengthened where the Ca supply is ample (Fuchs and Grossmann 1972).

However, in some cases calcium seems to favour certain pathogens. Thus black shank (*Phytophthora parasitica* var. *nicotianae*) on tobacco is more severe on high-Ca soils (Kincaid and Martin 1972), and high calcium contents of potato tubers increase their proneness to attack by *Streptomyces scabies* (Fuchs and Grossmann 1972).

As mentioned in Chap. 1.4, one of the principal benefits of calcium in the protection of plants lies in its use for the adjustment of soil pH to the detriment of pathogens favoured by acid soils, such as *Plasmodiophora brassicae*.

2.6.4 Minor and Trace Elements

Information on the effects of minor and trace elements on crop disease is fragmentary. A somewhat arbitrary collection of items on this subject is presented in the following.

Iron. Use of iron chelates has had beneficial effects in two cases involving *Verticillium* wilts. In Florida (USA), injection of chelated iron into the soil of a mango grove planted on old tomato land caused a dramatic recovery of the mango from wilt due to *V. albo-atrum* (Goldweber 1975). Similarly, in Israel, the health of groundnuts severely affected by *V. dahliae* on calcareous loess soil was much improved by spreading Sequestrene Fe-183 (Fe-EDDHA) on the soil (Krikun and Frank 1975). Foliar application of iron compounds is also reported to reduce incidence of *Stereum purpureum* in deciduous orchards (Fuchs and Grossmann

1972). On the other hand, low levels of iron were conducive to control of *Fusarium* wilt of tomatoes in Florida (Woltz and Jones 1972).

Manganese. Applied by foliar spray, manganese caused a slight but significant reduction of potato scab caused by *Streptomyces scabies* (Davis et al. 1976). Applications of manganese have also been found to reduce potato tuber susceptibility to late blight caused by *Phytophthora infestans* (Mudich 1967), and of young pumpkin seedlings to *Sclerotinia sclerotiorum* (Abia et al. 1977). However, doses of manganese optimal for growth of tomatoes also stimulated their infection by Tobacco Mosaic Virus (Mudich 1967). Incidence of the stem gall disease (*Protomyces macrosporus*) on coriander was likewise increased by application of manganese (Gupta 1976). As in the case of iron, low levels of manganese had a controlling effect on *Fusarium* wilt of tomatoes (Woltz and Jones 1972).

Molybdenum. Applications of molybdenum, even more than those of manganese, reduced infection of potato tubers by late blight, and also diminished incidence of *Ascochyta* blight on beans and peas (Mudich 1967).

Silicon. Effects of deficiency of availability of silicon on plant diseases have been studied chiefly in relation to the rice crop. Deficiency of this element increased susceptibility to most of the important rice pathogens, including *Pyricularia oryzae* and *Sclerotium oryzae* (Ou 1972), *Cochliobolus miyabeanus* (Anon 1973c), and *Xanthomonas oryzae* (Tagami 1968).

2.7 Moisture Management in Non-Irrigated Crops

The term moisture management comprises here both soil moisture and moisture in the foliosphere. Most of the more important cultural practices affect moisture in one way or another. In non-irrigated crops, tillage and soil amendments have perhaps the most obvious effect on soil moisture, and anything that interferes with air movement in the crop has clear effects on moisture in the foliosphere. However, the topography of a field, the properties of the soil, crop sequence, nutrition, choice of sowing date and other practices also influence moisture levels.

Moisture management practices are chiefly effective where moisture is limited by regular seasonal or daily dry periods, or by irregular dry spells. Good management will then alleviate crop strain. On the other hand, drainage will, of course, be effective where excess rain, or soil texture and topography, cause too much water to be retained too long for the crop's health.

2.7.1 Practices Affecting the Soil Moisture Potential, and Their Effects on Crop Disease

The effects of many cultural factors on conservation of soil moisture are described in appropriate chapters of this book, and will here be recapitulated only in brief.

1. Organic Matter. Anything that affects the organic matter content of the soil affects its water-holding capacity and water potential. This includes soil amendments and green manure, crop sequence, with remnants of roots of various crops distributed variously over the soil profile, and harvest practices, such as mowing cereals high or low. The moisture-conserving effect of organic matter cannot be separated from its biological effect, its influence on microbiota and their relation to plant pathogens. In soils rich in organic matter, water-deficit stress situations are less, or less frequently, likely to arise.

2. Tillage and Non-Tillage. One of the most important benefits of non-tillage is directly connected with water management: the prevention of erosion by water and wind. Moreover, suitable tillage, especially on heavy soil, deepens the soil layer to which water can penetrate, loosens up the soil, and has proved beneficial in controlling diseases such as *Fusarium solani* and *Pythium* root rot of peas (Kraft and Giles 1979). Cultivation interrupts the continuity of capillaries from deeper soil layers to the surface, reducing evaporation. Prevention of crust formation in the top soil facilitates germination and reduces seedling disease. It also restricts run-off of water, and thus both conserves water and diminishes risks of inoculum spread from field to field or to ditches leading to other fields.

3. Crop Sequence. In regions with limited rainfall, a principal aim of fallowing is to conserve moisture, often in conjunction with reduced tillage (Doupnik and Boosalis 1980). Fallowing with repeated tillage may, however, be practiced for the opposite reason: to dry out the soil and reduce the populations of nematodes and of some soil fungi. Growing in sequence crops with root systems developing at various depths also helps to keep the soil percolative, and thus affects its water potential.

4. Drainage. The effects of drainage on crop disease are to mitigate or avoid the stress of excess water, and create conditions in which fungal pathogens associated with high soil moisture (see Table 1.4) and bacteria do not enjoy decisive advantages over the crop. Buchholtz (1944) has described how tile drains laid at intervals of 10–15 m reduced incidence of *Aphanomyces cochlioides* on sugarbeet growing in strips directly over the drains to 25%, while in adjoining 3-m-wide strips 73%, and in the centre between the drains 90% of the plants were affected.

2.7.2 Optimization of the Use of Rainfall as Related to Crop Disease

All the factors mentioned above as affecting conservation of moisture have an obvious bearing on the optimal use of available rain. Another aspect of water management is sowing or planting crops when the rainfall to be expected can serve them best and restrict risk of disease.

This applies predominantly to crops grown in regions with well-defined rainy and rainless seasons. The chief problem is frequently how to grow crops on natural precipitation, but have them ripen in the dry period, so as to preclude attack of pathogens associated with high moisture, including those affecting senescent tissue and seeds (see Chap. 2.10). Experience in Israel shows that crops that can benefit from being grown on rain-fed soil and ripen after rains cease include melons, which thus escape downy mildew and fruit rots, and sugarbeet, which escapes *Sclerotium rolfsii*.

Growing forage crops, such as grains or legumes, in rainy winters and then harvesting them for hay or silage before leaf diseases contribute to their deterioration, is another example of optimal use of rain.

Rainfall periods can also be utilized to the full by giving certain crops an early irrigation, to get them started before the rains come, and then letting them ripen in dry spring weather after the rains have ceased.

Published references to the infection of crops sown at various dates before, during or after the monsoon period are rare. With reference to the angular leaf spot of cotton *(Xanthomonas malvacearum)*, Sahni and Gera (1967) recommend early sowing, so that plants reach maturity before the onset of the monsoon, to reduce incidence of the disease.

2.7.3 Management of Leaf Wetness in Rain-Fed Crops

The moisture level in the foliosphere, and especially the duration of leaf wetness, are basic factors in the development of many leaf and fruit pathogens. In non-irrigated fields management of moisture in the foliosphere means, in practice, creation of conditions conducive to rapid evaporation of precipitation from soil and foliage. All the factors listed in Chap. 2.7.1 as affecting the moisture level of the topsoil, e.g., tillage and drainage, also influence relative humidity and dewfall in the crop. A number of practices have an immediate bearing on the level of moisture in the crop and on duration of leaf wetness:

1. Practices Affecting Air Movement Through the Crop. The density of sowing or planting, direction of rows in relation to that of prevailing winds, and the gradient of slopes are important factors affecting the drying of foliage wetted by rain or dew. Windbreaks have an obvious influence, and the seasonal strength and direction of winds have to be considered in choosing sowing seasons. The most drastic exclusion of air movement is, of course, coverage of crops by plastic or glass.

2. Practices Affecting Shade. These practices, referred to in Chap. 1.2.2.2, include choice of slopes, taking into account the respective rate at which dew dries on northern or western and on southern and eastern exposures; or shading by windbreaks, or by specially grown shade trees, as practiced in some tropical plantations. In tall crops, or crops staked or trellised, the direction of the row determines its shade effect. In Israel, rows of trellised tomatoes running north to south were less affected by *Alternaria* or *Stemphylium* blights than rows running east to west: in the latter the shaded northern part of the plants was severely infected (Rotem and Cohen 1966).

2.8 Irrigation

Most of the cultural practices mentioned in this book aim at improving conditions of growth in locations or seasons naturally more or less suitable to crop growth. Irrigation, however, facilitates cropping where without application of water this would either not be possible at all or would only give meager results. It therefore exerts a fundamental influence on crops and cropping schedules. The effects of irrigation on crop disease interact with those of all other major cultural practices. In describing irrigation effects, some overlapping with descriptions of other practices is unavoidable.

2.8.1 Effects of Irrigation on Soil and Foliosphere Climate

Soil. The most profound effect of irrigation is, of course, its effect on the soil. In a rainless climate, irrigation determines some physical properties of the soil, e.g., formation of crusts, the depth of soil available to root growth, the rate and size of plant development, the composition of the flora and fauna of soil microbiota, and response of the soil to tillage or compaction.

Interaction of irrigation with soil microflora to control disease is exemplified by the case of potato scab *(Streptomyces scabies)*. Lapwood and Hering (1970) have shown that scab attack on potato tubers is prevented by maintaining soil water near field capacity during tuber formation. By timing irrigation in various ways, e.g., beginning it only after tuber formation had already been initiated or stopping it before tubers were completely formed, scab could be induced on certain limited parts of the tuber – those that had formed when moisture was too low (Fig. 2.12a). The effect of irrigation on the scab is ascribed to the increase of bacterial flora antagonistic to *S. scabies*, under high moisture conditions (Baker and Cook 1974).

The effects of irrigation on the soil are longer-lasting than its effects on the foliosphere climate, but they are given to wide variations by action of macroclimatic factors (temperature, radiation, occasional rain) and by crop factors such as root penetration, rate of water removal by the crop, and rate of decomposition of organic matter.

An important aspect of irrigation in relation to soil is erosion. Choice of irrigation technique and rates of irrigation must take this danger into account. On certain soils, furrow irrigation is preferable, as it provides a certain amount of erosion control; trickle irrigation, often preferred on slopes because of its ease of handling, may be undesirable where rainstorms may occur: the soil saturated by trickle irrigation will be unable to hold the rain water, which will start moving downhill (Tulang and Bedish 1975); erosion will then be stronger than in furrow-irrigated plots.

Foliosphere. Under arid conditions, any kind of irrigation induces fuller shoot development, and this in turn ensures more shade, lower temperatures, and longer periods of high moisture in the lower foliosphere and upper soil layers (Rotem and Palti 1969). Lower leaves and root collars may thus be exposed to increased dangers from the attack of fungi and bacteria favoured by such conditions.

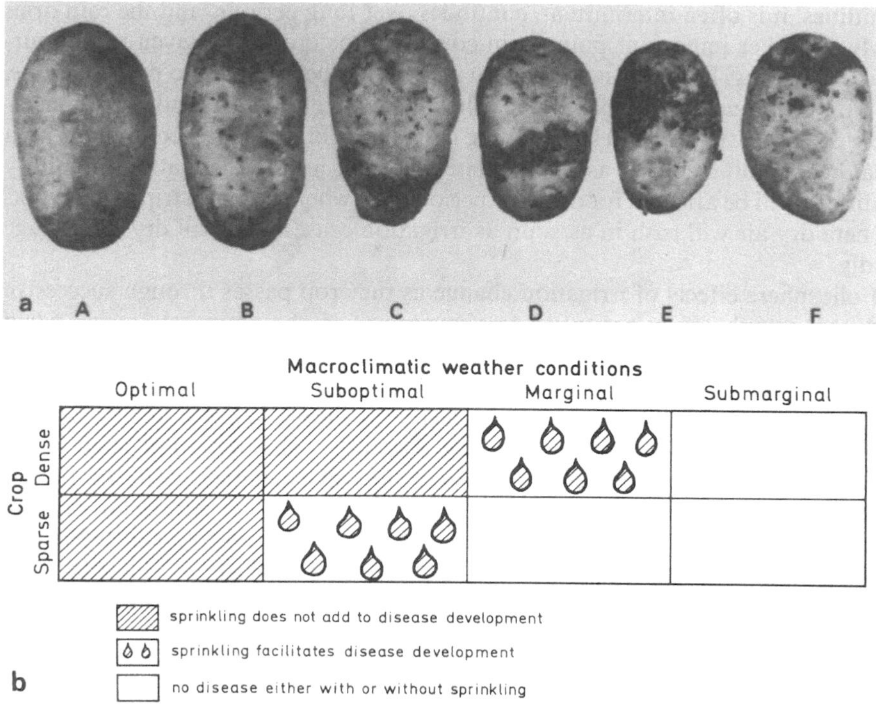

Fig. 2.12. a Relationships between time of irrigation and position of scab lesions on tubers. *A* Control irrigated to maintain soil water potential of 0–0.2 bar; *B, C, D* From plots where irrigation was stopped twice during tuber formation, but progressively later each time. Note that scab is progressively farther away from stolon end on *B–D*, which corresponds to progressively later periods during which watering was discontinued. Note scab at apex end of *B–D*, which corresponds in each instance to the second time watering was discontinued; *E, F* Watering discontinued once: *E* beginning about halfway through the period of tuber formation, and *F* later than *E*. Scab developed when watering stopped. (Lapwood and Hering 1970, as reproduced by Baker and Cook 1974)
b Schematic representation of the effects of sprinkling and of crop density on disease development under various macroclimatic conditions. Sprinkling is apt to facilitate disease development in dense crops mainly when conditions are marginal for the pathogen, but in sparse, well-aerated crops also when conditions are better, though not optimal. (Rotem and Palti 1969)

The effect of irrigation on the foliosphere climate is much more transient than its effect on soil climate. As mentioned in the chapter on microclimate (Sect. 1.2.3), the foliosphere effect is of importance in the development of leaf and fruit diseases where conditions are suboptimal for the pathogen: addition of a few hours of leaf wetness by sprinkling, or impeded ventilation in a dense crop, may then promote disease in the irrigated crop. On the other hand, under extremely dry (submarginal) conditions, the effect of irrigation on foliosphere climate is usually too short to enable development of moisture-dependent pathogens. This is graphically represented in Fig. 2.12b.

The effect of sprinkling on foliosphere climate is essentially different from that of rain. Rainfall generally occurs on cloudy days, or on days with high atmospheric humidities, it is often intermittent, continues over long periods, and the rain dries slowly. Another important point: rain covers all the crops in a given neighbourhood. Sprinkling, however, is applied for a fixed number of hours to parts of crops, often on bright and sunny days. The foliosphere effect then depends to a considerable extent on the size and shape of the irrigated field. If a large, contiguous area is watered simultaneously, as is sometimes the case in flood irrigation, humidity conditions will be affected for a longer period than where narrow strips are sprinkled; here dry air will rush in as soon as irrigation ceases, and will dry the foliage rapidly.

Foliosphere effects of irrigation change as the crop passes through successive stages of growth. They are much less important in the youngest stages, when ventilation is still good, than at later stages, as the crop grows dense. Different techniques of irrigation may therefore suit different stages in crop development.

Under certain conditions, irrigation has been shown to increase dewfall and lengthen hours of leaf wetness in this way. This effect has been implicated in the increased incidence of disease on sprinkled plots of cucumbers affected by the downy mildew *Pseudoperonospora cubensis* (Duvdevani et al. 1946) and of tomatoes affected by *Colletotrichum phomoides* (Raniere and Crossan 1959).

The effects of irrigation, and more particularly of sprinkling, on foliosphere climate complicate attempts at forecating attacks of leaf diseases, such as *Phytophthora infestans* (Easton 1976).

2.8.2 Irrigation Effects on the Host Crop

2.8.2.1 Extension and Manipulation of Growing Season

Extension of the season in which a crop can be grown on normal rainfall can be effected by (a) sowing before the rainy season to give the crop a start before the rains come; (b) sowing in the rainy season to continue growth and ripen after the rains have ceased; or (c) a combination of sowing before and ripening after the rainy season. Irrigation for the above purposes is commonly called auxiliary irrigation.

Sowing in dry seasons to give the crop a start by irrigation before rains begin is a common practice where rainy seasons are short, but irrigation water is available, and where young stages of growth or young organs are highly susceptible to disease attack when it rains. In Mediterranean countries, a single irrigation given to winter cereals before the rains start may greatly advance growth and will start the crop off under conditions unfavourable to early infection by smuts. However, such irrigation risks drought damage to young plants if rainfall begins later than normal. The fuller development of cereals thus irrigated and then grown on normal rain may also at times be conducive to lodging and to attacks of *Septoria tritici* and rusts at the end of the rainy season.

A striking example of the effects of irrigation preceding the rainy period on diseases of a tropical crop, is that of coffee grown in the Cameroons at altitudes of 1,650 m, as described by Muller (1975, 1977). The coffee berry disease *(Colletotrichum coffeanum)* attacks most severely when the early stages of fruit growth coincide with the rainy period. Muller applied 3–4 irrigations in January and Febru-

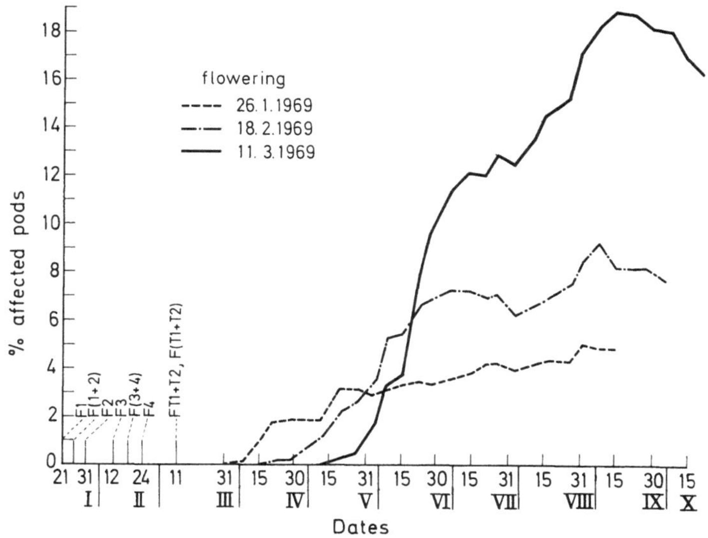

Fig. 2.13. Mean percentage of pods affected by coffee berry disease in plots induced by early irrigation to begin flowering on 26th Januar and 18th February, respectively, and in non-irrigated plots that began flowering on 11th March. (After Müller 1977)

ary, 6–9 weeks before rains normally begin, and induced flowering 5–6 weeks before it took place in unirrigated plots. This resulted in the reduction of the percentage of affected berries from about 18% in the control to about 4% in the plots first irrigated early in January (see Fig. 2.13). The irrigation had additional effects in a 2–3-fold development of the leaf canopy, and his increased the yield potential; but it also increased attacks of rusts *(Hemileia vastatrix, H. coffeicola)* to a level so severe that the yield potential could be realized only if the rust was controlled by spraying. The net result of the early irrigation was therefore that yields increased and spray applications were reduced from five, normally given against coffee berry disease, to three required to control the rusts.

Extension of the growing season of normally rain-fed crops by means of irrigation is of general interest where it meets the demand of the market or ensures continued supplies to processing plants or to feed livestock. From the plant disease aspect, such lengthening of the growing season is of particular interest, where crops can ripen their yields under relatively dry conditions, free from pathogens, especially those that attack senescent tissue. This applies, inter alia, to seed crops of numerous legumes and vegetables.

The manipulation of the seasonal growth of fruit crops in temperate climates by misting, in order to reduce risks of frost damage, is referred to in Chap. 2.9.5.

2.8.2.2 General Effects of Irrigation on Crops and on Stress

The general aim of irrigation is to enable the farmer to grow crops under soil, temperature and light conditions favourable to crop development, where all (or most) of what is lacking is water. This largely applies to warm climates, and in these the effect of irrigation is predominantly promotion of vegetative development, with growth of reproductive organs frequently delayed. This implies that irrigation chiefly promotes vigour, delays senescence to some extent, and produces plant canopies that eventually grow dense and create a crop climate often quite different from that of uncultivated or unirrigated fields. [See chapters on crop climate (1.2.2), crop age (1.6.2.2), and rates of sowing (2.9.1).]

Vigour induced by irrigation can help crops attacked by pathogens that do not develop rapidly, to mature their crops in spite of the attack. This is true for many annual crops attacked by root-knot nematodes (*Meloidogyne* spp.) and other nematodes. Vigour also helps some fruit trees affected by viruses to produce yields before the disease gains the upper hand.

However, not infrequently increased vigour means increased susceptibility to obligate pathogens (see Chap. 1.6.2), as well as to some other pathogens. Thus Vigouroux (1973, 1975) has shown the heightened susceptibility of vigorous stone-fruit and olive trees to *Verticillium dahliae* in France. These diseases are much rarer in non-irrigated groves.

Many crops respond to irrigation applied at longer intervals by formation of flushes of young growth highly susceptible to pathogen attack. This facilitates development of downy mildews, such as *Plasmopara viticola* on vine, of some powdery mildews, such as *Podosphaera leucotricha* on apple, or of rusts, e.g., *Tranzschelia* spp. on stone fruit.

One of the principal benefits of irrigation, as related to plant disease, is that it avoids severe water-deficit stress. This is of particular importance where fruit trees or other perennials have to subsist and be guarded from disease in dry seasons, before they produce their yield in wet seasons. Thus mango trees, when grown in African regions with dry seasons, can be saved by irrigation from the water stress that predisposes them to attack by *Botryodiplodia theobromae* (E. Laville, pers. comm.).

Irrigation, if applied frequently at low rates, may entail a certain amount of temporary water stress, at the end of irrigation intervals. Such irrigation promotes shallow root growth rather than deep penetration of roots. This renders the crop highly dependent on moisture conditions in the upper soil layers, which are those drying most rapidly when moisture supply is delayed. Among pathogens capable of taking advantage of fluctuations of soil moisture, *Fusarium* wilts, e.g., *F. oxysporum* f. sp. *lycopersici* on tomatoes, are the most common.

However, restriction of root growth to upper soil layers sometimes enables the crop to escape attack of pathogens limited to deeper soil layers. Thus, in Florida (USA), the nematode *Radopholus similis* cannot persist in upper layers of sandy soil, where moisture fluctuations are pronounced; frequent light irrigations inducing shallow root growth prevent attack of the nematode on citrus trees (Tarjan and O'Bannon 1977).

Supplementary irrigation is increasingly used in continental climates for intensified production of field crops. When varieties with high yield potentials, supplied with ample fertilization cannot realize their potential because of lack of rain, irrigation often proves rewarding. Research in the German Democratic Republic on irrigation of major crops has shown that diseases of cereals (Schoenrock-Fischer et al. 1976), sugarbeet (Klaus 1976), and potatoes (Ehrenpfordt 1976) are not markedly affected by such irrigations.

A lesser effect of irrigation on hosts is the increase of guttation. This has been found to increase infection of many crops, especially cereals (rice, rye, turfgrass), by a variety of pathogens, e.g., *Drechslera sorokiniana* on bentgrass *(Agrostis palustris)* (Endo and Amacher 1964). Guttation water, sometimes reabsorbed by plants, probably also facilitates leaf penetration by bacteria, such as *Xanthomonas campestris* on Brassicae.

2.8.3 Irrigation and the Pathogen

2.8.3.1 Survival and Management of Inoculum in Irrigated Crops

The most important effect on pathogens of irrigation applied throughout dry seasons is that it does away with the "drought incision" in inoculum development (see Chap. 1.1.2.2). Where two annual sowings of the same crop can be carried out, e.g., in rice fields, or where two crops susceptible to the same disease can be sown in one year, e.g., wheat and maize, inoculum is bound to increase to dangerous proportions. In addition, irrigation promotes growth of weeds in and around the irrigated area. Thus, cereal grasses growing in such habitats have been shown to transmit cereal rusts (Gerechter-Amitai 1973). Clearly, irrigation contributes greatly to the survival of inoculum on living hosts.

When it comes to the management of crop debris carrying inoculum, however, irrigation offers some means of control. Destruction of inoculum by flooding has been referred to in the chapter on sanitation (Chap. 2.2.4.4). Thorough drying of soil and then rewetting it will weaken sclerotia of some major pathogens, such as *Sclerotinia sclerotiorum*, *S. minor*, and *Sclerotium rolfsii* (but not of *S. cepivorum*) until they are colonized and destroyed by soil microorganisms (Smith 1972a, b). Alternate drying and wetting of soil also rapidly reduces the number of viable propagules of *Verticillium dahliae* (Menzies and Griebel 1967). Similarly, *Alternaria solani* persisting in dry debris of potatoes and tomatoes without loss of viability can be eliminated by periodical wetting of such debris (see Fig. 2.14).

Air-borne inoculum can, in certain cases, also be reduced or inactivated by overhead irrigation. Cohen et al. (1974) induced germination of spores of the cucumber mildew *(Pseudoperonospora cubensis)* by short, daily sprinklings – too short to permit the germtube to penetrate the leaf; this controlled the mildew effectively. Although it cannot be termed irrigation, it is of interest to note that Yarwood (1939) reduced infections of several powdery mildews, including *Sphaerotheca pannosa* on roses, *Erysiphe polygoni* on bean, *E. graminis* on barley and *E. cichoracearum* on cucumber, by applying water at 70 lbs pressure to the foliage. R. Tramier (pers. comm.) has confirmed that the rose mildew will not develop on leaves kept wet for 2–3 h a day.

Overtree misting to delay blooming and reduce risks of frost has been found to result in lower incidence of apple scab *(Venturia inaequalis)*, probably because ascospores were released while misted trees were still dormant and/or because the moisture film produced by misting on the leaf surface was thick enough to interfere with ascospore release (Spotts and Ferree 1979). Misting can, however, result in severe development of fireblight of pears caused by *Erwinia amylovora* (Spotts et al. 1976).

Fig. 2.14. Tomato leaf showing symptoms of *Alternaria solani*, a pathogen affected by irrigation in various ways: (a) it persists in dry debris from season to season, but can be eliminated by intermittent wetting of the debris; (b) its spores are extremely drought-resistant and will germinate even where leaf wetting is intermittent, with periods of wetness alternating with periods of dryness. (By courtesy of the Division of Plant Pathology, Agricultural Research Organization, Bet Dagan, Israel)

The extent to which overhead sprinkling washes off spores and thus reduces viable inoculum depends largely on droplet size. R. Cassini (pers. comm.) found that inoculum of *Drechslera turcica* is washed off maize leaves sprinkled by water canons producing large-size droplets, while ordinary overhead irrigation favoured the disease.

2.8.3.2 Dispersal of Inoculum in Irrigated Crops

Irrigation water may carry inoculum to all crops. Conversely, all irrigated crops, excluding perhaps those under trickle irrigation, may be the source of inoculum for other crops reached by drainage water. The danger of pathogens, including nematodes, lurking in irrigation water derived from rivers, ponds, reservoirs, channels etc. has been referred to in the chapter on sanitation (Chap. 2.2.2.2).

As regards soil-borne inoculum already in the field, irrigation by flooding can spread this in all directions; this has frequently been witnessed in flooded rice crops in relation to *Sclerotium oryzae*, whose small, light sclerotia are particularly fit for water dispersal (Bernaux, pers. comm.). Irrigation water has also been found to

Table 2.20. Some pathogens disseminated by overhead sprinkling, by splashing of soil or by splashing of inoculum from parts of the shoot

Crop	Disease	Pathogen	References
Peach, almond	Twig canker	*Fusicoccum amygdali*	C. Grosclaude (pers. comm.)
Apple	Bitter rot	*Gloeosporium fructigenum*	Pappo (1965)
Apple	Fruit rot	*Phytophthora cactorum*	C. Grosclaude (pers. comm.)
Citrus	Brown rot	*P. citrophthora*	
Maize	Eyespot	*Kabatiella zeae*	F. Rapilly (pers. comm.)
Tomato	Anthracnose	*Colletotrichum phomoides*	Raniere and Crossan (1959)
Potato	Late blight	*Phytophthora infestans*	Rotem and Palti (1969)
Bean	Halo blight	*Pseudomonas phaseolicola*	Walker and Patel (1964)
Bean	Grey mould	*Botrytis cinerea*	Bochow (1975)
Celery	Blight	*Septoria apii*	Bochow (1975)
Cucumber	Angular leaf spot	*Pseudomonas lachrymans*	Bochow (1975)
Cabbage	Club-root	*Plasmodiophora brassicae*	Bochow (1975)
Cabbage	Black rot	*Xanthomonas campestris*	Palti et al. (1962)

spread bayoud (*Fusarium oxysporum* f. sp. *albedinis*) from isolated foci over whole oil palm plantations (Laville, in Kranz et al. 1977). In all crops, furrow irrigation will evidently disperse inoculum chiefly along the rows.

Overhead sprinkling disperses inoculum of many soil-borne fungi and bacteria by splashing, and also washes inoculum from higher to lower organs of the plant. A list of some diseases aggravated by such dispersal is presented in Table 2.20.

Rotem and Palti (1969) have described the effects of sprinkling on dispersal of air-borne spores and on the viability of these spores. The following paragraphs are drawn largely from their descriptions.

Many of the most important foliage and fruit pathogens form their spores at night and release them in day-time. Although the amount of spores dispersed varies according to environmental and biotic factors, the diurnal periodicity of their release is quite constant in most fungal species. In the semi-arid zones, in which irrigation is most important, the spores of many species are released earlier in the day than in more temperate zones. Thus, in Israel, about 80% of the spores of *Alternaria solani* are released during 2 to 3 h, around 11 a.m. (Rotem and Reichert 1964); spores of many downy mildews are released between 7 and 9 a.m., and under very dry conditions their release may be limited to half an hour, between 7 and 7.30 a.m.

Sprinkler irrigation affects this situation by precipitating spores from the air, and – as noted above – splashing them about. It also provides the moisture necessary for spore germination. As splash dispersal of spores normally disseminated by wind is conditioned by the presence of spores on the leaf surface, its highest rate will occur when irrigation is applied before these spores are released by natural means. Consequently, early morning sprinkling can potentially disperse, by splashing, spores of most fungi, while noon sprinkling will be effective chiefly for spores whose natural release occurs later. The evening and early night sprinklings will find few spores to disperse, while sprinkling performed just before morning will again meet spores to be splashed about. This has been found to be true for *Phytophthora*

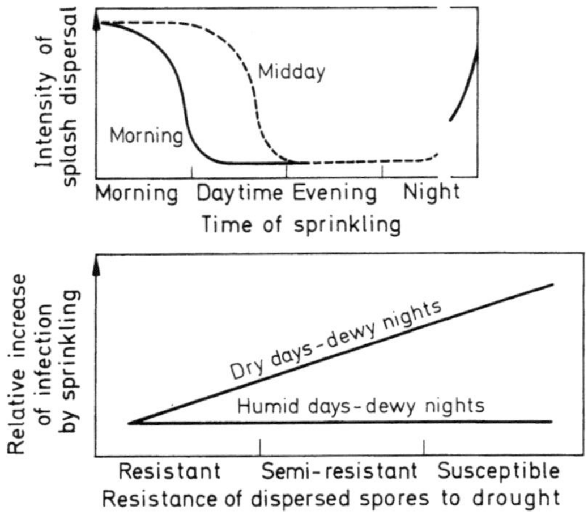

Fig. 2.15. Some effects of sprinkling on dispersal of inoculum. (Rotem and Palti 1969)
Top Effects of sprinkling applied at various times of the day, on the intensity of splash dispersal of spores released chiefly in the morning, or at midday
——— *morning*, ----- *midday*
Bottom Effects of daytime sprinkling on the relative increase of infection caused by spores with various levels of drought resistance, when nights are dewy but days are either dry or humid

infestans in Israel (Rotem et al. 1970), but data for other fungi are hypothetical. The interplay of the factor mentioned above on intensity of splash dispersal is presented schematically in Fig. 2.15 (top).

The viability of dispersed spores of various fungal species under dry conditions often furnishes the key to understanding the effects of sprinkling on disease caused by these fungi. This will be illustrated here by reference to spores of two fungi: the extremely drought-resistant spores of *Alternaria solani* and the highly drought-sensitive spores of *P. infestans*. Both these pathogens form their spores at night, and both attack potatoes and tomatoes. There the similarity ends: the *Alternaria* spores, dispersing at noon, are capable of surviving the dry hours of the day on which they are dispersed, and of many subsequent days. They will germinate as soon as dew falls and sprinkling is, therefore, of little additional benefit to the disease where dew is present (Rotem 1968; Rotem and Reichert 1964). By contrast, the sensitive spores of *P. infestans*, dispersing earlier in the morning, will not readily survive dry days. Surface irrigation will not help them in doing so, but sprinkling performed at or close to the time of dispersal will assist the freshly formed spores to infect hosts before losing their viability. Sprinkling applied later in the day may come too late to rescue spores, if the preceding hours have been very dry, but may still find spores viable and help them germinate, if atmospheric humidity has been at a fairly high level.

Spores of other fungi may be expected to react to sprinkling at various times of the day according to their resistance to drought. This has been expressed schematically in Fig. 2.15 (bottom): where days are humid and nights are rich in dew, sprinkling will not increase infection, which occurs no matter how sensitive or resistant the spores are to drought. However, where dewy nights alternate with dry days, daytime sprinkling may greatly increase infection by sensitive spores which would otherwise not be able to withstand the dry day. To a lesser extent, sprinkling will increase infection caused by semi-resistant spores, and only infection by drought-resistant spores will not be increased, because it will be high anyway (Rotem and Palti 1969).

2.8.3.3 *Attraction of Virus Vectors to Irrigated Crops*

As irrigation is applied when rain is deficient and rain-fed crops as well as wild flora dry up, the irrigated crop becomes a green island in the midst of drying vegetation. This in many cases occasions mass migration of potential or actual virus vectors from such vegetation into the irrigated crop.

The danger of such vector influx into an irrigated crop constitutes an important consideration in the choice of sowing and planting dates of crops to be irrigated. This subject is more fully discussed in the chapter on sowing and planting dates (Chap. 2.10.2). As stated there, it is sometimes better to delay sowing an irrigated field for some time after rain-fed vegetation has dried up, to allow for a period in which vector populations will be reduced by lack of hosts.

2.8.4 Techniques of Irrigation

The four principal techniques of applying water are by flooding, furrows, overhead sprinkling and drip or trickle irrigation systems. A fifth technique, subsoil irrigation, is only used under certain circumstances, where the water table is relatively high and can periodically be raised to various levels near the soil surface.

Table 2.21 presents a comparison of economic, agronomic, and plant protection aspects of four irrigation techniques. It shows that, from the point of view of disease control, overhead sprinkling has many disadvantages. However, the advantages it presents in labour cost and rate of water use explain its wide adoption by growers who can afford the initial outlay. The newer technique of trickle irrigation has numerous advantages, but requires an even higher investment. In the following sections, disease control aspects of overhead and trickle irrigation will be discussed in some detail.

2.8.4.1 *Overhead Sprinkling May Promote Disease*

All that will be said here on potential promotion of disease by sprinkling refers to conditions neither extremely dry nor extremely wet, but to conditions where wetting of soil, foliage, and plant collars for a few hours may have a marked effect on pathogen and disease (see Figs. 2.10 and 2.11).

Effects of sprinkling on dispersal and viability of spores has been considered in Chap. 2.8.3.2. We shall now follow the effect of sprinkling after viable spores have landed on the crop and the factors influencing subsequent infection of the sprinkled crop, as described by Rotem and Palti (1969).

The success of viable spores to establish themselves on the host depends chiefly on the period over which the moisture required for penetration is furnished during the actual period of sprinkling and thereafter, until the foliage dries. The rate of drying depends on the vapour pressure deficit of the air, on wind velocity, and on crop density. In one experiment in Israel, the rate of drying of tomato foliage after sprinkling ranged from 5 min for outer leaves, under a cloudless sky, with strong wind, at 36 °C and 16% relative humidity, to 4 h for inner leaves, under a cloudy sky, with no wind, at 17 °C and 86% relative humidity. In the latter case drops per-

Table 2.21. Comparison of economic, agronomic and plant protection aspects of four irrigation techniques

	Flooding	Furrows	Sprinkling	Trickling
1. Percentage of surface soil wetted	80–90	30–50	100	20–30[a]
2. Economic considerations:				
Price of installation	Low	Low	Fairly high	High
Cost of labour	Low to medium	High	Low	Low
Water consumption per unit area	Medium to high	Medium to high	Medium	Low
3. Agronomic considerations:				
Use on topographically difficult sites	Unsuitable	Difficult	Possible, with some difficulty	No problem
Limitation of irrigation period by wind	None	None	Severe	None
Use of somewhat saline water	Possible	Limited	Limited (scorch)	Possible, preferably under plastic mulch
4. Host aspects:				
Moistening of foliage	Slightly moistened	Not moistened	Moistened	Not moistened
Moistening of collar	Moistened	Not moistened	Moistened	Moistened
5. Pathogen aspects:				
Surface dissemination of inoculum	All over field	Along rows	Limited	None
Splash dissemination of inoculum	None, or little	None	Much	None
Washing inoculum off plants	None	None	Frequent	None
6. Integration with control by pesticides:				
Timing, in relation to wetness of ground	Soil has to dry before pesticides can be applied by ground equipment			No limitation
Washing-off of fungicidal deposits	None	None	Washed off to various degrees	None

[a] This figure applies to crops grown in rows 1–1.5 m apart

sisted until dewfall, and a continuous moisture period of 20 h thus resulted from sprinkling and dew (Rotem and Cohen 1966).

The extent to which sprinkling at various times of the day assists infection also depends greatly on the temperature level favouring spore germination and on the speed with which the infection process is completed. This has been depicted graphically in Fig. 2.16 and is further explained in the legend to that figure.

In rainless seasons, dew and sprinkling constitute the major sources of moisture on leaves, stems, and fruits. How these interact under various environmental conditions may again be demonstrated best by comparing their effects on those two contrasting types, *Alternaria solani and Phytophthora infestans*. This comparison is presented in Table 2.22.

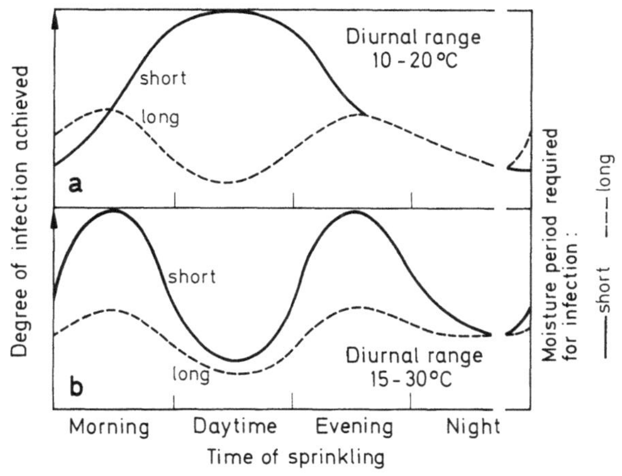

Fig. 2.16 a, b. Schematic representation of the effects exerted on infection by sprinkling, applied at various times of a dry day-dewy night cycle, at two diurnal temperature ranges, and in relation to pathogens requiring short or long moisture periods for completion of the infection process. The pathogens are assumed to produce spores viable throughout the day, and to have infection optima of 20 °C.
a On days with a 10° to 20 °C daily range of minimum and maximum temperature, pathogens capable of completing infection within short moisture periods will achieve maximum infection under midday sprinkling; but pathogens requiring longer periods for infection can only utilize the extended moisture periods produced by dew combined with morning or evening sprinkling, and the amount of infection is then reduced by lower temperatures
b On days with a 15° to 30 °C daily range of temperature, even the pathogens requiring only short moisture periods find conditions in midday-sprinkled plots too hot, and will achieve better infection under morning or evening sprinkling; the pathogens requiring longer moisture periods will also take advantage of morning or evening sprinklings continuous with dew periods. (Rotem and Palti 1969)

Table 2.22. Interaction of dew and sprinkler irrigation in their effect on *Alternaria solani* and *Phytophthora infestans* on potatoes and tomatoes under the environmental conditions encountered in various locations in Israel. (Rotem and Palti 1969)

Environmental conditions	Development of	
	A. solani	P. infestans
Completely arid and dewless oasis of the Negev desert	Slight development only under sprinklers	Complete absence
Daily atm. humidity minima 35% or below, nights rich in dew, no rain, in N and W Negev	Dew alone suffices for infection and epidemic development; sprinkling is without effect	Dew alone inadequate; sprinkling required for blight outbreaks
Daily atm. humidity minima above 60%, night rich in dew, no rain, in the coastal plain	Dew alone suffices for infection and epidemic development; sprinkling is without effect	Dew may suffice, but is materially assisted by sprinkling, in leading to blight outbreaks
Atm. humidity always high, dew plentiful, no rain, in the coastal plain (spring, fall)	Dew alone suffices for infection and epidemic development; sprinkling is without effect	Dew alone can support blight outbreaks; sprinkling irrelevant

The differences in behaviour of the two pathogens are due chiefly to the above-mentioned differences in the drought resistance of their spores. Those produced by *A. solani* will remain viable under practically all the conditions under which tomatoes and potatoes are grown (Rotem 1968). In rainless periods the promotion, by sprinkling, of the disease caused by this fungus is, therefore, limited to conditions where either dew is deficient or temperatures during the dew period are very low. On the other hand, for *P. infestans* and similar types of pathogens, daytime conditions critically affect their reaction to sprinkling; in the absence of rain, dew alone will support full infection only if fairly high daytime humidities enable spores to survive until dewfall. As humidity during the day increases, the effect of sprinkling becomes more and more prominent, but in the extreme drought prevailing in the desert, where nights lack dew, sprinkling alone cannot induce blight development.

2.8.4.2 Trickle Irrigation

Trickle (drip) irrigation has been developed over the past 15 years and presents many cultural advantages not related to plant protection, e.g., economy in water use, use of somewhat saline water, use of shallow soil by simultaneous irrigation and fertilization, and irrigation under plastic mulches. The use of trickle irrigation may be expected to expand greatly where the crop can bear the expense, and its plant protection aspects therefore deserve attention.

The benefits of trickle irrigation for plant protection have recently been reviewed by Palti and Shoham (1979). As regards *crop climate* and *stress*, an important characteristic of trickle irrigation is that it can be practiced on slopes, where other modes of irrigation are difficult to apply. Slopes that present fewer risks of chilling or freezing can thus be planted with crops susceptible to low-temperature stress and frost injury; in addition, the down-draughts common on slopes provide relatively dry foliosphere conditions. No less important is the avoidance of the periodic moisture stress common under other forms of irrigation. Maintenance of a constant supply of water by trickling, on soils of a wide range of permeability, is an effective means of guarding crops from predisposition by water deficit to diseases such as *Alternaria* blights and powdery mildews.

Sztejnberg and Golan (1980) have studied effects of trickle irrigation on diseases of deciduous fruit trees in Israel. Two diseases practically disappeared with this practice: *Glomerella cingulata* was eliminated because inoculum was no longer splashed about, as it is under sprinkling, and *Sclerotium rolfsii* because trickling lowered soil temperatures and increased activity of microorganisms inimical to the pathogen. On the other hand, incidence of white rot (*Rosellinia necatrix*) rose along the trickle lines, as this fungus thrives at lower soil temperatures and constant humidity.

Trickle irrigation under plastic mulches facilitates growing some crops in dry climates, with positive effects on disease control. The mulches can prevent contact between plant organs and soil: *Rhizopus* fruit rot of strawberries has been practically eliminated in Israel in this way, and the incidence of *Botrytis cinerea* rot has been greatly reduced. Black mulches suppress weeds, which could affect the crop climate or serve as carriers of disease. Mulching of plant beds, and consequent prevention of evaporation from the soil surface, makes it possible to irrigate with somewhat saline water, with reduced danger of salinity effects on the crop.

Proper placement of the trickle irrigation line has been found to be of crucial importance to minimize *Phytophthora* diseases of citrus. Timmer and Leyden (1976), in the United States, reported lower incidence of *P. parasitica* foot rot of grapefruit grafted on sour orange, when trickle-irrigated trees were planted on ridges than when planted on level soil. In Corsica, E. Laville (pers. comm.) found an increase in *Phytophthora* gummosis, when trickle lines were placed too close to the trees. If properly managed, the opportunity offered by trickle irrigation to place the irrigation line accurately and to shift it periodically, according to the crop's needs, is useful for disease control. Thus, in tomatoes grown for processing in Israel, most of the ripening fruit lies on the soil; if the trickle line is then shifted from its earlier position on plant beds to run *between* the beds, much fruit rot (such as that caused by *Rhizopus* sp.) can be avoided (Shoham, pers. comm.).

2.8.5 Minimizing Disease by Irrigation Management

Two groups of factors interact to determine the scope and time of disease development in irrigated crops (Rotem and Palti 1969):

a) weather factors (rain, atmospheric humidity, temperature, radiation), through their effects on the crop climate, the formation of dew, and the dispersal and viability of spores; and
b) pathogen and host characteristics, foremost among them the time of spore dispersal, drought resistance of spores, and speed of the infection process.

These factors, considered against the background of capital outlay, and labour cost and availability, have to guide the grower in choosing the irrigation technique best suited to minimize disease, and the timing and rate of irrigation.

2.8.5.1 Choice of Irrigation Technique

Among the factors listed in Table 2.20 as characterizing the various irrigation techniques, and the biological factors mentioned in relation to spore viability and infection, the choice of irrigation techniques has to concentrate on three aspects:

a) Is progress of the disease going to be markedly affected by addition of limited hours of leaf and collar wetness?
b) Does splash dispersal represent real danger?
c) Is water-deficit stress likely to occur and to further disease?

If all the three questions can be answered in the negative, the type of irrigation chosen is irrelevant to disease management. If any one of these questions is answered positively, however, a choice must be made between surface irrigation (flooding, furrows), sprinkling and trickle irrigation.

At early stages of crop growth, dangers of water-deficit stress are particularly acute, so that trickle irrigation, if available, offers a considerable advantage. At these stages, ventilation in the crop is generally good, and foliage will dry rapidly after sprinkling, but since much of the sprinkled water hits the ground, the danger of splash dispersal of soil-borne pathogens is high. The limitations imposed on sprinkling by wind are an added disadvantage, especially where dry winds may blow for longer periods and dry out the surface soil where the young crop's roots are.

As the crop canopy becomes continuous, problems of leaf wetness and ventilation assume major importance. Irrigation techniques wetting neither collar nor foliage are then preferable, and where the crop is sprinkled, the time of day at which this is done may be crucial (see next Chapter). As mentioned above, the continuous moisture provided by trickle irrigation under dense foliage canopies favours collar diseases, such as white rot caused by *Rosellinia necatrix* (Sztejnberg and Golan 1980), and *Phytophthora* gummosis of citrus crops.

In the maturing crop, aeration and the rate at which surface soil dries up after irrigation are among the major factors determining invasion of senescent organs near the ground by facultative parasites. Sprinkling, which wets the entire soil surface, is certainly the least desirable mode of irrigation, from this aspect. As mentioned above, covering the soil by plastic mulch and supplying water by trickling below the mulch can solve this problem.

2.8.5.2 Timing of Water Supply

The principles governing the timing of irrigation, its beginning as well as its frequency, from the aspect of disease management, are:

a) providing the crop with as uniform a supply of water as possible, to obviate stresses of water deficit or excess;
b) timing of giving of water in relation to periods of heightened susceptibility of host to disease; and
c) minimizing hours of contiguous leaf wetness, e.g., avoidance of sprinkling in continuation of dew periods.

Uniform Water Supply is a matter of soil preparation (levelling), as well as determination of suitable irrigation intervals. These intervals have to depend on characteristics of the soil, such as texture and organic matter content, and of the crop, especially depth of root system and size of transpiring foliage. To save labour, promote development of deeper roots, and reduce periods of high moisture at the soil surface and in the foliosphere, fewer applications of larger quantities of water at longer intervals are generally preferable to more frequent irrigations at lesser rates. Restricting the number of irrigations is of greatest importance where sprinkling is practiced. In fact, very frequent sprinkling may create conditions in which certain diseases are no longer controllable. Thus, on cucumbers grown in Israel on sandy soil in mid-summer, when sprinkling 2–3 times a week was necessary to keep the crop growing, downy mildew *(Pseudoperonospora cubensis)* developed so strongly that even frequent fungicide applications failed to control it (Palti, unpubl.).

Periods of Heightened Susceptibility to Disease. One of the periods in which most crops are particularly susceptible to disease is the stage of germination and emergence of seedlings from the soil, up to the time the young stem has hardened to some extent. Irrigation techniques have been developed to reduce incidence of seedling diseases. One such technique consists in irrigating the soil before sowing, letting the top layer dry, and then sowing into the underlying wet layer. This results

Fig. 2.17. Millet affected by grain smut *(Sphacelotheca sorghi)*. Incidence of the disease is affected by the effects of various irrigation practices on soil wetness and aeration when the crop germinates. (By courtesy of the Division of Plant Pathology, Agricultural Research Organization, Bet Dagan, Israel)

in seedlings emerging mostly through dry soil, where damping-off fungi cannot attack them. This is a widely used practice in Israel, e.g., in avoiding damping-off of cotton caused by *Rhizoctonia solani* (Palti et al. 1975). As seedling stands are often crowded, prior to thinning, it is most desirable to delay further irrigations until thinning to final stand has been performed.

In Egypt, two methods of sowing and irrigation are used for cereals. The "herati" method consists of broadcasting seed over moist soil, and ploughing it in to a depth of 8–10 cm, the "afir" method in sowing on dry land, harrowing to about 4 cm depth, and irrigating only thereafter. Jones and Abd-el-Ghani (1940) found the "afir" method of post-sowing irrigation to result in lower percentage incidence of all the smuts they investigated: flag smut *(Urocystis tritici)* and bunt *(Tilletia caries)* of wheat, covered smut of *(Ustilago hordei)* of barley, and grain smut *(Sphacelotheca sorghi)* of millet. They ascribed this success of "afir" sowing to both the shallower depth of sowing and the wetness and lack of aeration in the topsoil.

Mention has already been made of a classical case of protecting the host at a vulnerable stage, the prevention of potato scab *(Streptomyces scabies)* by appropriate timing of irrigation at the stage of tuber formation (Lapwood and Hering 1970; Lapwood et al. 1973). Also on potatoes, continued irrigation, right until the crop is lifted, will prevent charcoal rot *(Macrophomina phaseoli)* of tubers in potato crops grown in warm climates (Littauer 1944) Fig. 1.12, p. 57).

Minimizing Hours of Leaf Wetness. The rationale for reducing, as far as possible, the hours of leaf wetness due to sprinkling, the crop climate conditions and pathogen characteristics which make this important, have been described in Chaps. 2.8.3 and 2.8.4. The critical point is the timing of the period of leaf wetness due to sprinkling in relation to the dew period.

The ideal solution, winds permitting, is sprinkling at night, when plants are anyway wet with dew. Automatic equipment for turning sprinklers on and off is used increasingly for this purpose. If night sprinkling is impracticable, the guiding principle for day-time irrigation is that it should not start before dew on all the foliage has dried off, and should be discontinued in time for leaves to dry before the following night's dewfall. If prevailing winds make it necessary to choose between morning or afternoon sprinkling, the period in which there are least viable spores on the leaves should be chosen.

2.8.6 Irrigation and Crop Disease – Outlook

Irrigation of wide areas not so far farmed for lack of water is one of the concrete hopes of mankind to feed its increasing populations. Where such irrigation will be applied to desert areas free from dew, it will not in itself create severe disease problems. However, where irrigation will serve for crop production in extension of rainy seasons, or in regions rich in dew in their rainless seasons, its interaction with climatic and crop factors must be understood. For this reason, relations between crop disease and mode and timing of irrigation in various climatical set-ups has been described in such detail in the foregoing sections.

In areas newly brought under irrigation in developing countries, flooding or furrow irrigation are likely to be chosen. Fortunately, these techniques if properly performed are least problematical from the point of view of plant health. However, even their use requires study to minimize disease, as shown by the work performed in Egypt on control of cereal smuts (Jones and Abd-el-Ghani 1940).

Where irrigation is used in temperate climates to intensify production, the technique chosen is almost invariably sprinkling, because it saves labour. Problems of increases in disease incidence have been reported from maize and potato fields irrigated in this way. The techniques and timing of water application to crops growing in temperate climates, in relation to disease, deserve to be studied more closely.

Trickle irrigation, with its many technical advantages and its desirability also from the plant health aspect, is likely to be used on a rapidly increasing scale wherever capital is available and crops repay the outlay.

2.9 Rate of Sowing and Planting, and Density of Stand

The density of the stand and of the foliage canopy, as resulting from rates of sowing and planting and subsequent thinning or pruning and water and nutrient supply, can have a profound influence on disease development. The effects of crop density and shade on the crop climate have been discussed in Chap. 1.2. We shall here consider physiological and epidemiological consequences of density in relation to the crop climate it helps to create. However, in doing so we must not lose sight of the fact that the overriding consideration in choosing rate of sowing is the economic one: densely sown and somewhat more disease-prone crops may, under certain conditions, give heavier yields than crops sown less densely, and this may well outweigh all other aspects. Where uneven germination or seedling mortality are to be expected, due to lack of seed vigour, to seed- or soil-borne diseases, or to environmental conditions, sowing more densely than usual can be of help in obtaining adequate stands.

2.9.1 Effects at Various Stages of Growth

The principal physiological results of dense stands are, on the one hand, etiolation of growth and delay in the hardening of young tissues beyond their initial soft state, and on the other hand earlier senescence of individual organs or of whole plants.

In seed-beds, crowded stands of beet, tobacco, and many vegetable and flower crops develop elongated stems that are prone to attack of all damping-off organisms and of soil-borne fungi that can affect crops only before stems lignify. A good example of the latter is *Sclerotium rolfsii* which, in rainless seasons, will attack irrigated cotton only when seedlings are unduly crowded (Avizoher-Hershenzon and Palti 1962). Leaves and petioles of crowded plants are also easy prey to downy mildews, e.g., *Peronospora tabacina* on tobacco and chilli pepper, *P. parasitica* on Brassicae, and *P. farinosa* on beet, and to *Phytophthora infestans*. In fact, the latter will attack moderately resistant eggplants in Israel only when they are grown in crowded seed-beds (Palti and Netzer 1963).

The growth span of annual crops, when grown in dense stands, depends chiefly on the availability of water and nitrogen. If these are supplied in abundance, crowded stands will lead to extensive or excessive shoot growth, dense canopies, delay of maturity, and eventually senescence, under conditions of high humidity. On the other hand, if water is limited, dense sowing may induce stress which favours diseases such as the stalk rots of maize caused by *Diplodia maydis* and *Gibberella fujikuroi* (Nyvall 1979).

The high level of humidity prevailing under dense canopies favours the large majority of air- and sap-borne fungal and bacterial pathogens, e.g., the causes of coffee rust (*Hemileia vastatrix*), coffee berry disease (*Colletotrichum coffeanum*), blister blight of tea (*Exobasidium vexans*), downy mildew fungi of all crops, the species of *Cercospora* causing leaf blights of sugarbeet, groundnut, celery and other crops, and the species of *Alternaria* and *Stemphylium* that affect solanaceous and other crops, as well as bacterial pathogens of leaf and fruit spots and rots on tomato, chilli, bean etc. As regards soil-borne pathogens, density of stand favours in particular those that cause local, not systemic, infections on root and collar, e.g., *Rhizoctonia solani*, and *Sclerotium* and *Sclerotinia* species.

As the crop matures, and water continues to be available, the danger of disease heightens in dense-canopy crops more markedly than in sparser crops. Attacks of

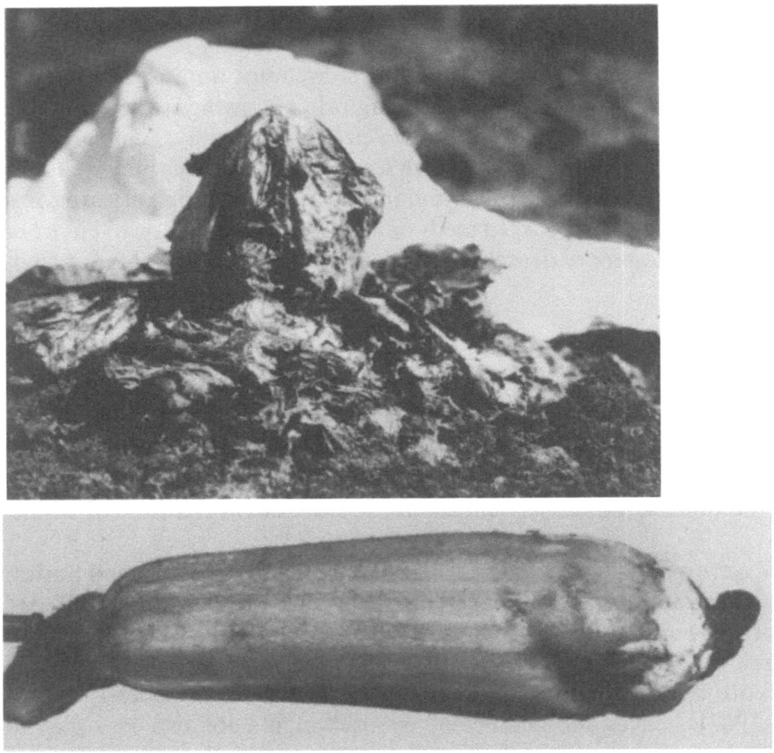

Fig. 2.18. The white mould caused by *Sclerotinia sclerotiorum* is favoured by high humidity in the dense canopy of crowded crops. *Above* Head of lettuce affected by white mould. *Below* Fruit of vegetable marrow affected by the mould. (By courtesy of the Division of Plant Pathology, Agricultural Research Organization, Bet Dagan, Israel)

foliar pathogens favoured by high humidity, and difficult to control by chemicals where foliage is dense, may cause shedding of leaves which then serve as food bases for facultative pathogens with ready saprophytic ability. The classical examples for this chain of events are dense stands of groundnuts affected by *Cercosporidium personatum*, causing massive shedding of leaves in which *Sclerotium rolfsii* multiplies to attack stem bases (Garren 1961).

When crowded crops eventually approach the end of their growth span, high humidity under their dense canopy will facilitate the attack of facultative pathogens on senescent tissues (see Chap. 1.6). In such cases, severe attacks may be expected in particular of pathogens affecting succulent or sugar-rich tissue, e.g., of *Sclerotinia sclerotiorum* on carrots celery, lettuce, and cucurbits (Fig. 2.18) and *Sclerotium rolfsii* on sugarbeet, as well as bacteria on potato tubers.

However, crowded stands may occasionally slow fungal infection. Dinoor (in press) has noted that in Israel the spread of *Septoria tritici* in wheat fields is delayed in erect, dense stands. The reason is that, when infection starts from the lower leaves, water droplets penetrating into the dense stands lose much of their impact, and splash dispersal is more limited than in more widely spaced plants.

In crops grown with a limited supply of water, crowding of annual crops leads to premature ripening. This is a strategy sometimes employed to obtain early crops of limited size, either for considerations of market prices (vegetables, flowers) or to escape pests or diseases that develop only later in the season (cotton, sugarbeet). Although water-deficit stress may then heighten susceptibility of the crop to certain pathogens, only few of the latter are capable of causing losses under dry conditions.

2.9.2 Root and Shoot Contact in High-Density Crops

Roots. Crowded stands imply increased chances of roots of neighbouring plants meeting in the soil, and of passage of pathogens from one plant to another due to the proximity or to actual root contact. This certainly applies to all pathogens that actively search for new roots to infect, such as nematodes or certain fungi (*Armillariella mellea, Stereum purpureum*). Heale and Isaac (1963) have investigated the spread of *Verticillium albo-atrum* in closely spaced stands of lucerne. The fungus spreads underground not only from lucerne to lucerne but also to many common weeds, evidently under conditions in which inoculum was plentiful.

Shoots. Crowding of leaves and fruits raises the chances of their suffering injury from wind, that knocks them against each other. This, and the actual proximity, facilitates contact transmission of viruses that are not solely dependent on vectors, and of pathogens (among them many bacteria) exuded from diseased tissue either in guttation water or in exudates caused by disease. Nematodes affecting shoots are also transmitted by leaf contact in dense crops, e.g., *Aphelenchoides ritzemabosi* on chrysanthemums (Anon 1969a). Many other pathogens, among which *Botrytis cinerea* is prominent, can colonize senescent tissues (flowers) and then spread by contact to fruit they would not otherwise affect at this stage.

2.9.3 Density and Disease in Tree Crops and Vines

In tree crops, density of canopy is determined by spacing of the trees, by their training (cordons, pergolas etc.), and by vegetative development of each individual crown.

The factors affected by density are mainly those relating to humidity. In a dense crop – dew will cover a relatively small proportion of growth; rain will run off the canopy, and concentrate between trees, or will run down limbs and trunks; leaf wetness will persist longer after rains, and periods of high relative humidity will be extended after humid nights. This is also related to wind being impeded by dense foliage.

Thus the density to which canopies are designed to grow is a basic component of moisture management in tree crops. The decision what density is most desirable, from the aspect of disease development, involves the following considerations:

1. Lessening Chances of Disease. Dense canopies protect fruits and trunks from sun scald, which often facilitates pathogen attack.

Dense top foliage shields lower foliage from dew; where dew is a major source of leaf wetness, this reduces danger of attack on lower leaves of pathogens such as downy mildew of vine *(Plasmopara viticola)*.

2. Increasing Chances of Disease. The longer hours of high humidity in and under dense crowns are conducive to increased development of most disease of branches, leaves, and fruits. This may involve more die-back, and consequently weakening of trees, and more pruning.

Spraying dense-canopy trees against parasites is less effective, and also more costly.

2.9.4 Crowded Stands May Mean Lower Incidence of Certain Systemic Diseases

In the development of non-systemic disease, where each propagule or group of propagules effects one localized infection, the greater the number of roots, fruits, and leaves available for infection, the better the chances of pathogen meeting susceptible organ. Therefore, as described above, multitudes of plant organs, under the humidity conditions of the crowded crop, mean plentiful infection. This applies to all levels of inoculum quantity, whether sparse or abundant.

Among virus diseases, those due to mechanically transmitted viruses, such as potato virus X or bean common mosaic, spread faster in crowded than in sparse stands.

However, in the cases of vector-borne viruses and of diseases due to soil fungi spread by discrete propagules, such as sclerotia, the position is basically different. Here a single infection per plant may result in disease in the whole plant. If inoculum or vectors are plentiful and likely to reach every plant anyway, then the density of stand is immaterial. If the amount of inoculum is limited, however, and can be expected to come into contact with only a limited number of plants, a larger proportion of the latter is apt to escape systemic infection in the crowded stand consisting of many individual plants than in sparser stands with fewer, and probably more fully developed plants. As Vanderplank (1947) has put it: the larger the "catchment" area offered by an individual plant, the greater is the probability of infection, and the higher will be the percentage incidence of disease over a given area (Fig. 2.19).

Table 2.23 presents a list of vector-borne viruses found to affect crowded stands less than sparse stands. Investigating groundnut rosette, Farrell (1976) found that in crowded stands not only was the proportion of affected plants lower, but virus incubation in these plants was slower. The fact that of two vectors of the soybean mosaic virus, one caused more infection in crowded, the other in sparser stands, shows how much detailed research is required before the relation between any one vector-borne virus disease and the density of the crop can be determined.

The fungal disease for which the phenomenon of lower incidence at closer spacing of the crop has been studied most profitably is the wilt caused by *Verticillium albo-atrum* and *V. dahliae* on cotton in California (Wilhelm et al. 1976, 1978).

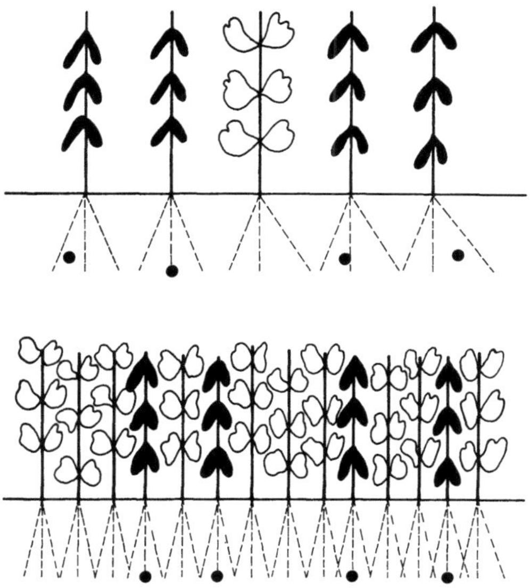

Fig. 2.19. Schematic explanation why a systemic soil-borne pathogen may affect a densely sown plot less than a plot sown at a lower rate, provided the inoculum level is not high. Each black point represents a unit of inoculum. The black plants have been affected, the white plants are healthy. *Above* Widely spaced plants, each with a large "catchment area." Contact of inoculum with each of the plants results in a comparatively large area of affected crop. *Below* Densely sown plants: each affected plant constitutes only a small proportion of the whole crop

Table 2.23. Vector-borne virus diseases that affect crowded stands of crops less than sparse stands

Crop	Virus	Vector	Country	References
Groundnut	Rosette	*Aphis craccivora*	Nigeria, Malawi	A'Brook (1964) Farrell (1976)
Tomato	Leaf curl	*Bemisia tabaci*	India	Sastry et al. (1976)
Soybean	Mosaic	*Aphis citricola* [a]	USA	Halbert et al. (1979)
Cucumber	Mosaic	*Aphis gossypii*	France	Leclant (pers. comm.)
Barley	Yellow dwarf	Various aphids	France	Lapierre (pers. comm.)

[a] But another vector of soybean mosaic, *Capitophorus elaegni*, causes more numerous infections in denser stands

This is ascribed to the reduction of effective inoculum per plant in proportion to the increase in the number of plants per unit area in the densely planted fields. Short-branching, arching phenotypes of cotton have been developed that are ideal for planting of 120,000 to 180,000 plants per hectare, and this combined approach has given outstanding control of the wilt.

On soybeans in the United States, the incidence of brown rot *(Cephalosporium gregatum)* in rows 30 inches apart has been found to be higher when spacing in the row was 6–12 inches than when it was 1–4 inches (Halbert et al. 1979).

Reduction of disease caused by a pathogen of an entirely different type, but also soil-borne and capable of systemic infection, the corn smut *(Ustilago maydis)*, has also been reported from the United States (Wilcoxson 1975). The disease was less prevalent on hybrids, crosses and inbred lines of maize sown at the rate of 18,000 than on those sown at 5,000–7,000 plants per acre.

Present-day knowledge does not permit of generalizations concerning the effects of crowded crop populations on infection by systemic diseases. In principle, there is no reason why the results reported for *Verticillium* wilt on cotton should not apply to many other wilt diseases. Small quantities of inoculum are commonly encountered where a crop has shown limited foci of soil-borne disease, from which cultivation spreads the pathogen thinly over wider areas. To obtain precise information on the value of closer spacing of crops in such a situation would seem to be a worthwhile objective of applied studies.

2.10 Sowing and Planting Dates and Manipulation of Flowering and Fruiting Periods

2.10.1 Date of Sowing and Incidence of Disease

What makes the farmer choose a specific date for sowing or planting his crop?

In subsistence farming there are two considerations: dates that will yield the maximum size of staple crops, and dates that will ensure more or less fresh food for the longest possible yearly period.

When the farmer sows forage crops for use on his own farm, the continuity of fodder supply, whether fresh greenage, hay or silage, is the central consideration in deciding when to sow.

In growing crops for sale, the decision is more complex and is mainly guided by the following considerations:

a) dry crops (cereals, nuts) and non-perishable crops for processing, e.g., fibre crops, are mostly sown when they are expected to give maximum yields, which implies optimal use of precipitation. This only applies, however, if suitable storage facilities are available, otherwise considerations of storage may take precedence.
b) perishable crops grown for processing will be sown either when yields can be high or when the processing plant can absorb them or needs them to keep operating for as many months as possible every year (sugarbeet, canning tomatoes or fruits).
c) perishable crops for local and foreign markets are sown when the periods of ripening meet periods of demand, so as to obtain maximum revenue.

Anticipation of losses from extreme weather conditions (frost, hail, drought, heat) often determines sowing dates, or guides the regulation of flowering and fruiting in orchard crops, but the greater or lesser chances of crops sown on various dates becoming infected by pathogens are rarely the major factor in the farmer's decision when to sow (see also Chap. 2.5 on tillage of cereals). Nevertheless the extent to which variations in sowing dates can affect incidence of disease must be clearly defined, so that the risk of disease can be weighed appropriately.

In regions with well-defined dry and rainy seasons, the onset of rains and duration of the rainy period is the prime consideration in choosing sowing dates, as has been pointed out in the section on optimization of the use of rainfall (Chap. 2.7.2). This applies to cotton, groundnuts, rice, and other high-temperature crops where rains coincide with heat, and to wheat, clover, potatoes, and other crops growing under moderate temperatures in regions with cooler rainy periods.

In climates with more evenly distributed rainfall, levels of temperature, as related to the crop's temperature requirements, will largely determine the date of sowing.

Relevant to the choice of sowing dates of all irrigated crops is the seasonal availability and rate of supply of water. This may be an overriding consideration where water is limited, high-income crops (orchards, cotton) have to be taken care of first, and less profitable crops (forage) have to be fitted in when sufficient water is left over.

Wherever sowing dates are fixed in relation to rainy seasons, considerations of possible water stress (and consequent disease) are of great importance. However, in many crops more room for manipulation is being afforded by breeding of varieties that either stand up to certain amounts of stress, are fairly resistant to stress-induced infectious disease, or have shorter growth periods. The advances in maize breeding deserve special mention in this respect.

The choice of sowing dates in relation to crop disease has one principal aim, viz. to reduce to a minimum the period over which infective agent (propagule, vector) meets susceptible host tissue. This is also the aim of operations to influence the time of flowering and fruiting, especially by pruning and by breaking of dormancy of deciduous crops.

Choice of sowing dates will be consered under three headings:

1. Avoidance of seasons in which vector development peaks, or in which vectors have no alternative (wild) vegetation to feed on.
2. Considerations of the relative growth rates of host and pathogen under various weather constellations, and especially at various levels of temperature.
3. Avoidance of seasons in which massive development of inoculum may compensate pathogens for unfavourable weather or resistance factors.

The decision whether to sow earlier or later in the season may not be easy, even where considerations of only one disease are concerned. The case of the cereal mildew *(Erysiphe graminis)* is a good example. In temperate areas (a) spring-sown wheat or barley usually develops more mildew on crops sown late than on those sown early: but (b) autumn-sown crops have more disease on seedlings, the earlier they are sown; after the winter, however, the earlier crops have less mildew than crops sown late (Jenkyn and Bainbridge 1978).

However, few crops are ever attacked by a single pathogen, and sowing crops on a specific date may help to reduce the risk of one, but heighten that of another disease or pest. Huber (1979) has called this the "multi-adversity" relationship between sowing date and crop health. He exemplified by describing the situation with which growers of winter wheat are faced in Indiana (USA). Early sowing increases the yield potential, but also the severity of take-all, *Septoria* leaf-blotch, powdery mildew, leaf rust, yellow dwarf virus, *Pseudocercosporella* foot rot and the scab, root rot and snowmould caused by species of *Fusarium*. On the other hand, incidence of *Rhizoctonia solani*, *Typhula* snowmould and bunt may be increased by late sowing.

A similar dilemma has been described by El-Behadli and Al-Hassan (1979) with regard to sugarbeet in Iraq. In early sowings (15th September to 1st October), *Rhizoctonia solani* caused more damping-off than *Pythium aphanidermatum* or *Phoma betae*, but in November sowings the reverse was the case. Moreover, in the earlier sowings head rot caused by the three fungi was more prevalent, but *P. betae* killed more plants in the later sowings.

Only local experience in assessing the relative importance (loss potential, frequency of occurrence) of the various pathogens concerned can guide farmers to optimal sowing dates in such cases, which are all too common in agricultural practice.

2.10.2 Minimizing Virus Diseases by Varying Sowing Dates

Successes achieved in reducing incidence of virus diseases by sowing crops when vectors are least likely to harm them, are listed in Table 2.24. In this table, viruses have been arranged in three groups, which overlap to some extent, but indicate the practices that may lessen their incidence.

a) Viruses with vectors restricted to well-defined seasons or with clear seasonal peaks in their development. Losses from many of these viruses can be reduced if care is taken to sow the crop so as to avoid its most susceptible stage coinciding with the vector's peak. Annual occurrence of such peaks varies; in France, a monitoring service has been set up for the months of September to November, to advise growers of the occurrence of aphids on barley and the attendant danger of Barley Yellow Dwarf affecting earlier crops (Lapierre, pers. comm.).

b) Viruses with vectors capable of developing in various seasons, but the disease they cause is economically important only when affecting young crops. This group includes Cucumber Mosaic Virus and other important diseases, from which crops sown in most seasons have to be protected while young.

c) Viruses with vectors that attack crops chiefly when no other vegetation is available. This is probably true for many virus vectors affecting irrigated crops in semi-arid areas after rains have ceased and spontaneous vegetation has died down.

Table 2.24. Minimizing incidence of virus diseases by choice of appropriate sowing and planting dates

Crop and disease	Vector	Country	Effect of sowing date on vector and disease	References
Viruses with vectors restricted to well-defined seasons or with clear seasonal population peaks				
Groundnut rosette	*Aphis craccivora*	Nigeria	Vector population smaller in crops sown in June than in July	Booker (1963)
Tomato spotted wilt	*Thrips tabaci*, *Frankliniella* spp.	world-wide	No sowing or transplanting in warm, humid seasons when vector populations peak	Chagas and Vicente (in Kranz et al. 1977)
Maize rough dwarf	*Laodelphax striatella*	Israel	Sow in late May, after vector populations on grasses have declined	Harpaz (1972)
Carrot motley dwarf	*Cavariella aegopodii*	Australia	Delay sowing until vectors's spring dispersal period has passed	Stubbs (1956)
Barley yellow dwarf	Aphids *Rhopalosiphum* sp.	England, France	On winter cereals sown early vectors reach second seasonal peak; on late (April) sown spring crops, peaks are reached at early growth stages	Yarham and Norton (pers. comm.), Bayon and Ayrault (1977)

Table 2.24 (continued)

Crop and disease	Vector	Country	Effect of sowing date on vector and disease	References
Viruses with vectors restricted to well-defined seasons or with clear seasonal population peaks (continued)				
Broad bean (*Vicia faba*) subterranean clover red leaf virus	*Aulacorthum solani*	Tasmania	September sowings coincide with peak of vector flight; sowings in May, July, or November much less affected	Johnstone and Rapley (1979)
Bean golden mosaic	*Bemisia tabaci*	Brazil	Late dry-season sowings come up when cooler temperatures reduce vector population	Costa (1975)
Broad bean (*Vicia faba*) Sudanese broad bean mosaic virus	*Aphis craccivora*, *Acyrthosiphon sesbaniae*	Sudan	Sow in mid-October to produce well-developed plants less attractive to vectors, when vector populations peak in January	Abu Salih et al. (1973)
Viruses destructive chiefly when affecting young crops				
Tomato cucumber mosaic	Aphids	France	Plant as first trusses form, as crop susceptibility then diminishes	Lot et al. (1979)
Tomato yellow leaf curl	*Bemisia tabaci*	Israel	Spring-sown crops affected only slightly, as vector populations reach high levels only in summer	S. Cohen et al. (1974)
Sugarbeet yellows	*Myzus persicae*	world-wide	Sow early enough for crop to reach less susceptible stage as vectors multiply	Klinkowski (1967)
Broad bean strain of cowpea mosaic	*Sitona* sp.	France	Sow in winter, so that plants grow resistant by April–May, when vectors fly	Ferault (pers. comm.)
Viruses borne by vectors that visit crops chiefly when no spontaneous vegetation available				
Cassava mosaic	*Bemisia tabaci*	Togo	During rainy season, much wild vegetation attracts the vector; in dry seasons, vectors concentrate on irrigated cassava crops	Dengel (1978)

2.10.3 Relative Growth Rates of Crop and Pathogen, and Levels of Crop Susceptibility, at Various Sowing Dates

Since most crops grow under ranges of temperature and humidity wider than those favouring pathogen development, sowing or planting at seasons which give the crop an advantage is obviously good strategy. This has already been referred to in the chapter on crop sequence (Chap. 2.3.5.4): soils infected with pathogens limited to warm seasons (*Sclerotium rolfsii*, root-knot nematodes) or to cold seasons (snowmoulds) can be used for susceptible crops in the pathogen's "off-season."

A certain amount of such seasonal management is possible for a large majority of host/pathogen combinations in all climates with more or less marked seasonal changes in weather. A few cases, selected for the variety of factors and considerations involved, are presented in Table 2.25, and one example in Fig. 2.20. These examples refer to field crops to which changes in sowing dates to lessen risk of disease are most relevant. This, of course, does not mean that they are not used for vegetable and other crops as well, but, as mentioned above, with these crops market considerations are commonly decisive.

A special case is that of seed-crops, in which freedom from disease in the seed produced is at least as important as size of yield (Neergaard 1977). Sowing dates

Table 2.25. Effect of sowing date on the development of some major diseases of field crops

Crop	Disease	Pathogen	Country	When to sow to minimize risk of disease	References
Cereal Crops					
Winter wheat	Snowmould	*Gerlachia nivalis*, *Typhula idahoensis*	Washington, USA	Sow early to get plants established before snowfall; affected crowns will then recover in spring	Bruehl et al. (1966)
	Root rot	*Rhizoctonia solani*	Indiana, USA	Plants sown early are less susceptible, will produce more tillers from non-invaded crowns	Huber (1979)
	Covered smut	*Tilletia caries*, *T. foetida*	USA	Sow early in autumn, so seedlings grow less susceptible by the time secondary sporidia develop	Nyvall (1979)
Spring wheat	Root rot	*Drechslera sativa*, *Fusarium* spp.	Canada	Sow early May; rising soil temp. in later sowings favours the disease	Greaney (1946)
Wheat, barley	Powdery mildew	*Erysiphe graminis*	temperate areas	See Chapter 2.10.1 (p. 169)	Jenkyn and Bainbridge (1978)
Wheat, barley, rye		*Cephalosporum gramineum*	USA	Sow late, when soil temp. below 13 °C and root growth limited	Nyvall (1979)
Maize	Leaf blight	*Drechslera maydis*	Georgia, USA	Sow early, as rising soil temp. favours disease	Sumner and Littrell (1974)
	Smut	*Ustilago maydis*	USA	Sow early, as rising soil temp. favours disease	Wilcoxson (1975)
	Downy mildew	*Sclerospora sorghi*	Canada	Sow early; rising temp. favours infection from oospores	Cohen and Sherman (1977)
	Stalk rot	*Fusarium* spp.	Yugoslavia	Sow late (May) to avoid drought and high temp. when crop most susceptible	Balaz et al. (1978)

Table 2.25 (continued)

Crop	Disease	Pathogen	Country	When to sow to minimize risk of disease	References
Rice	Wilt	*Cephalosporium gramineum*	USA	Sow late in autumn when soil temp. below 13 °C	Nyvall (1979)
	Nematode	*Pratylenchus* spp.	USA	Sow late in autumn when soil temp. below 13 °C	Nyvall (1979)
Other Field Crops					
Sugar beet	Cyst nematode	*Heterodera schachtii*	California, USA	Sowing in Jan.–Febr. favours the crop, in March–April the nematode	Anon (1968)
Flax	Browning, stem break	*Polyspora lini*	Germany	In early sowings, microclimate at critical crop stage less favorable to disease	Schroedter and Hoffmann (1961)
Sunflower	Downy mildew	*Plasmopara halstedii*	USA	Sow late to escape heavy late rains	Zimmer (1971)
Tobacco	Downy mildew	*Peronospora tabacina*	Greece	Sow in Jan. or later to shorten period of mildew attack	Zachos (1963)
Pea	Powdery mildew	*Erysiphe polygoni*	Wisconsin, USA	Sow early to avoid pod-filling period coinciding with maximum disease	Gritton and Ebert (1975)
	Root rot	*Fusarium solani* f. sp. *pisi*	India	Sow late (Dec.), when low temp. gives the crop an advantage	Majumdar and Raychandhuri (1976)
Rape	Blackleg	*Leptosphaeria maculans*	France	Sow early (August): plants then become resistant before autumn weather favours disease	Delhoval, pers. comm.

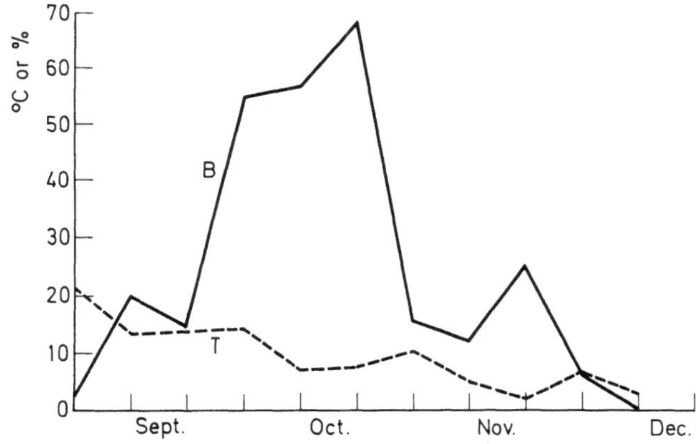

Fig. 2.20. Effect of date of sowing on incidence of covered smut *(Tilletia caries)* on winter wheat. T = 5-day temperature means; B Percentage incidence of smut. (Gaeumann 1951)

for seed-crops have to be chosen as as to minimize attack of systemic seed-borne diseases all through the growing period. Of special importance is the timing of flowering and seed-ripening periods to coincide with weather not conducive to direct attack of pathogens on fruits and seeds. Thus, in climates with dry seasons, most seed-crops are sown to ripen their seeds during the dry months. This precludes infection by almost all seed-borne bacterial pathogens, such as *Pseudomonas phaseolicola* on bean, *X. malvacearum* on cotton, and *X. vesicatoria* or *Pseudomonas tomato* on tomato seed. It also reduces chances of many fungal pathogens attaining the seed, e.g., *Cercospora beticola* on beets, and *Alternaria* and *Macrosporium* diseases on Brassicae and carrots.

2.10.4 Seasonal Massing of Air-Borne Inoculum

The cardinal role played in disease development by the quantity of air-borne inoculum available at any given season has been described in Chap. 1.3.1 in the context of the collective approach to crop disease control. This is an aspect of special importance where moisture is clearly the limiting factor for diseases on leaf and fruit, as massive inoculum can compensate the pathogen and help disease develop under moisture conditions otherwise considered inadequate. Massive loads of inoculum can, however, also, to some extent, overcome other factors restricting disease, such as moderate varietal resistance or a certain amount of age resistance in leaves (Rotem 1978; Aust et al. 1980).

Important objectives of choosing the right date for sowing are therefore not to sow when inoculum can reach peak multiplication rates, and in particular to know the risk of sowing when large amounts of inoculum are already available from earlier crops.

2.10.5 Management of Periods of Bud Burst, Flowering and Fruiting to Minimize Disease

Operations for inducing early or late bud burst, flowering and fruiting of crops can influence disease in much the same way as varying the sowing date. Although the influence of such operations is fairly limited in scope, it can be important where marked changes in weather, e.g., beginning or cessation of rainy periods, can be expected at that time.

In countries with winter rains and summer drought, irrigated apple trees can to some extent be protected from scab (*Venturia inaequalis*) by pruning and dormancy breaking practices aimed at delaying bud burst by 2–4 weeks. This is practicable only with varieties that develop fairly late anyway, but can then be well worth-while.

Operations to delay fruiting of tomatoes, by removal of early flower clusters by hand or hormone spray, can reduce spring infection to tomato fruits by bacterial pathogens (*Xanthomonas vesicatoria*) and probably also by *Stemphylium* sp.

Conversely, induction of early fruiting before the onset of the rainy period can also be beneficial. In Ecuador, supplementing natural pollination of cocoa during the early dry season has been shown to reduce losses from pod rot *(Crinipellis perniciosa, Monilia roreri)*, which becomes severe during the rainy period (Edwards 1978).

The more basic effects of manipulating growing seasons in semi-arid climates by means of irrigation in the dry season has been exemplified in Chap. 2.8.2 by reference to Muller's (1975, 1977) work on coffee.

Other methods of advancing or delaying yield, in order to have the produce ripen with less risk of disease, have been referred to in the chapter on crop age (Chap. 1.6.2.6). These include close planting of orchards and dense sowing of cotton.

2.10.6 Manipulation of Periods when Crop Meets Pathogen

The above examples of the manipulation of sowing, planting, flowering and fruiting, with a view to reducing disease, clearly demonstrate the practical importance of this approach. Although economic considerations often guide the timing of crop growth, especially in crops grown for marketing as fresh produce, much more can be done in this way to reduce disease attack. As said above, this applies in particular to climates with pronounced seasonal changes of weather.

This aspect of plant disease management has not received the attention due to it in phytopathological literature and practice. Some attempts have been made in Israel to draw up tables to indicate the chances of disease development in specific crop/pathogen combinations when the crop is grown in various seasons (Palti and Ausher, unpubl.), but we are not aware of similar work in other countries. Relevant data are stored in card indexes and data banks in the plant protection institutions of various countries. They should be used to crystallize conclusions on disease-free periods, and periods when disease is particularly dangerous for many major crops. This could help the grower to shorten the period during which his crop, at a susceptible stage, has to meet aggressive pathogens.

2.11 Harvesting Dates and Practices

The choice of suitable dates for harvesting, as of those for sowing and planting, often depends primarily on market requirements, availability of storage facilities, and demands of processing plants. However, when the produce matures, and after all investments in the crop have already been made, the danger of disease at or after harvest looms large in the grower's mind and plays an important part in his decision when and how to harvest. It must also be kept in mind that in many, especially perennial, crops the level disease attains at harvest often determines the base line from which it will develop in the following season.

This discussion will be limited to operations the grower can perform before or at harvest to reduce incidence of disease in storage and transport. It will not deal with post-harvest operations, such as threshing, grading, packing, storing, and marketing.

In considering both the time and the mode of harvesting from the angle of disease development, reference must be made to all that has been said in Chap. 1.6 on disease in the senescent plant. Most produce is senescent when harvested, the exceptions being chiefly crops picked or mowed repeatedly to utilize the more tender stages of their growth (tea, mint, some forage legumes). The most important aspect of harvesting mature produce is its pronounced susceptibility to facultative parasites; this is heightened even more by the wounding of stem and stalk, which is often unavoidable, and the wounding of leaf and fruit, which can frequently be minimized with proper care.

2.11.1 Timing of the Harvest

Appropriate timing of harvest, from the point of view of crop disease, means essentially the attempt to escape disease: the intention is to harvest either before environmental conditions (weather, crop climate) become very favourable to the

Table 2.26. Crops helped by proper timing of harvest to escape disease

Crop	Pathogen	Country	Effects of early harvest	References
Potato	*Globodera rostochiensis* *G. pallida*	France England	Harvest precedes new generation of nematodes	Mugniery (1978), Anon (1977a)
Potato	X, Y and leafroll viruses	Germany	Harvest before viruses reach tubers	Klinkowski (1967)
Tomato	*Erwinia carotovora*	USA	Green fruits less affected	Bartz and Crill (1978)
Groundnuts	*Pythium myriotylum*, *Fusarium solani*	Israel	Escape onset of autumn rains	Frank (1972)
Groundnuts	*Rhizoctonia solani*, *Fusarium* sp. *Diplodia* sp.,	Libya		Martin (1960)
Peppermint	*Puccinia menthae*	Tunisia	Younger leaves less affected	Anon (1977b)
Lucerne	*Leveillula taurica*	Israel	Younger leaves less affected	Palti (1945)
	Uromyces striatus var. *medicaginis*	Israel	Younger leaves less affected	Palti (1945)
Lucerne	*Pseudopeziza medicaginis*, *Stemphylium botryosum*, *Rhizoctonia crocorum*	USA	Less infection at pre-bloom stage	Nyvall (1979)
Red clover	*Rhizoctonia leguminicola*, *Pseudoplea trifolii*, *Phoma trifolii*	USA	Harvest before incidence of disease reaches serious proportions	Nyvall (1979)
Sugarcane	*Glomerella tucumanensis*	Australia	Harvest before susceptible cultivars reach peak of maturity	Hughes (1976)

pathogen, or before the crop becomes highly susceptible to age-related pathogens, or both.

Table 2.26 lists a number of crops that are helped by early harvesting to escape pathogens.

2.11.2 The Danger of Inoculum Spreading at Harvest

A distinction must be made between infection at harvest of uninjured produce by its exposure to inoculum, and infection related to wounding which is, at least in part, inherent in the harvesting operation.

Successive harvesting of the same crop in plots in close proximity often constitutes a danger to the plots harvested last. This applies to many crops, and a good example is that of celery in Florida (USA). Here Berger (1973) measured the amount of inoculum of *Cercospora apii* released at harvest in a crop in which percentage of disease was only 3%–5%; harvesting resulted in a high spore count in blight-free young celery 150 m downwind, and the closer the young celery was to the harvested area, the more severe was the resultant infection.

In the harvesting of cereals in France, masses of spores of *Fusarium culmorum* have been found to spread to neighbouring fields sown later, just when these are most highly susceptible to infection by head blight (Cassini, pers. comm.). Where smut of wheat (*Tilletia foetida*) is still a problem, the clouds of spores released by combine harvesting are apt to spread the disease over much larger areas than would be affected if the wheat was harvested by mowing only (Ponchet, pers. comm.). The tiny seeds of broomrape (*Orobanche* sp.) have been found in Morocco to spread widely where broad beans were harvested by combine (Rieuf, pers. comm.). Figure 2.21 shows a carrot field severely infested by broomrape, just before harvest.

What all the above brings out clearly is that avoidance of inoculum spread at harvest means proper planning before the crop is sown. The crop that will ripen first must be sown upwind of any other crop liable to infection by pathogens carried by the first crop.

The danger of potato tubers becoming infected by *Phytophthora infestans* at harvest has been realized for a long time. Haulm destruction will lessen danger of blight, when carried out a few days, and of virus infection of potato seed crops, if carried out a few weeks, before the crop is lifted. Less well-known is the effect of haulm-killing on infection of tubers by *Rhizoctonia solani*. Gudmestad et al. (1978) found that sclerotial development of *R. solani* reached a maximum 3–4 weeks after the haulm was killed, and that seed tubers lifted after than time produced the greatest number of sclerotia in their progeny. The large number of small tubers that may be left in combine-harvested potato fields serve to transmit virus diseases (Spire, pers. comm.). The enlargement of foci of the potato cyst nematode (*Globodera rostochiensis*) by the use of potato spinners during harvest has been described by Jones (1980), as shown in Fig. 2.22.

The risk of produce becoming infected at harvest can sometimes be lessened by harvesting under conditions unfavourable to the pathogen. Thus lifting of potato tubers in dry and warm weather reduces, or even removes, danger of infection by

Fig. 2.21. A carrot field severely infested with broomrape (*Orobanche* sp.), just before harvest. (By courtesy of the Division of Plant Pathology, Agricultural Research Organization, Bet Dagan, Israel)

P. infestans in Israel, even though the foliage may be blighted (Palti and Netzer 1963), and also lessens the rate of infection by the skin spot due to *Oospora pustulans* in England (Anon 1976b). Picking fruit only when dry greatly reduces the danger of pathogen transmission by hand, and this has frequently been observed in relation to bacterial diseases, such as *Pseudomonas lachrymans* on cucumbers.

The danger of spreading *Verticillium albo-atrum* by transporting wilt-infected hop vines to picking machines, is described in Chap. 3.6.

The spread of inoculum by equipment moving through dense crops has been mentioned in Chap. 2.2.2.4, and this obviously applies to harvesting equipment. The spread of sap-borne inoculum at harvest is considered in the following section.

2.11.3 Harvesting Techniques as Related to Crop Injury and Disease

The harvesting operation itself can promote spread of disease by physically carrying inoculum from plant to plant, and by simultaneously wounding plants and facilitating entry of parasites both into the organs harvested and into the plants left in the field. Such spread of inoculum that does not require wounds for penetration

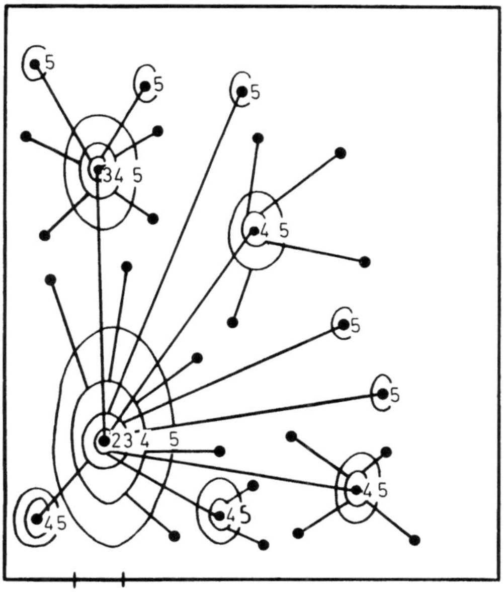

Fig. 2.22. Notional scheme for the spread of potato cyst-nematode in Scottish potato fields, in which successive foci have been enlarged by the use of potato spinners during harvest. (Jones 1980)

into the host is particularly dangerous in the case of sap-borne pathogens, e.g., viruses not dependent on vector transmission, bacteria, and numerous fungi. As pointed out in Chap. 2.2.2.4, the risk of disease transmission by sap can be reduced by refraining from operations in the plot while the crop is wet.

In picking fruit, hand-picking in general results in less injury than machine-picking, provided labour can be trained to handle the fruit properly, e.g., picking apples with stalks attached to lessen incidence of *Sclerotinia fructigena* (Anon 1965 a). However, in larger fruit-growing enterprises, pip- and stone-fruit, as well as olives and other fruits, are increasingly picked by machine, and the risk inherent in such operations must be realized. In the USA, Ogawa et al. (1963) found that machine-harvesting significantly increased the fruit rot of peaches induced by *S. fructicola*, but not that due to *Rhizopus stolonifer*, while on apricots incidence of *R. stolonifer* was actually lower in fruit picked by machine than by hand. Barnes (1964), also in the USA, has reported that harvesting by means of tree-shakers increased incidence of canker *(Ceratocystis fimbriata)* in trees of almond, prune, peach, and apricot, and fruit decay in peach and apricot, and in Israel such harvesting by shakers increases the number of apricot and plum trees affected by the dieback due to *Hendersonula toruloides* (Golan, pers. comm.).

Disease transmission by mowing or cutting is related to the direct cutting injuries as well as, in the case of mechanized equipment, to the above-mentioned crushing effects of wheels. Lucerne wilt *(Verticillium albo-atrum, V. dahliae)* spreads due to both these effects (Heale and Isaac 1963; Haass and Märtin 1977). On bluegrass *(Poa pratensis)*, research in the USA has shown that the amount of disease after mowing depended on mowing height, with less infection in grass mown at 2 cm than at 4 cm height, both in relation to *Fusarium roseum* blight (Turgeon and

Meyer 1974) and to leaf spots caused by *Drechslera dictyoides* and *D. sorokiniana* (Gibbs et al. 1973). Low rates of nitrogen fertilization further reduced the amount of disease caused by the species of *Drechslera* after mowing. However, in Germany, incidence of *D. poae* on the same species of lawn grass was found to be only half as much at mowing heights of 5 cm as compared to 2,5 cm (Teuteberg 1978).

The frequency at which crops are mowed obviously affects the rate of cutting and crushing injury and of inoculum transmission. Thus, root rot of red clover, caused predominantly by *Fusarium oxysporum*, has been found to increase as the number of cuttings between May and August increased from one (in May) to eight (2 each per month) (Siddiqui et al. 1968). The defoliation strain imposed on crops by repeated mowing has also been mentioned in the chapter on stress and strain (Chap. 1.5.4).

2.12 Planning to Minimize Influx of Air- or Vector-Borne Inoculum to Neighbouring Crops

The importance of inoculum management has been referred to in the context of communal approach to disease control (Chap. 1.3) and of sanitation measures (Chap. 2.2). This chapter will consider planning of crops in more or less closely neighbouring fields and orchards as a means of reducing risks of disease transmission by air-borne or vector-borne inoculum.

2.12.1 What Makes a Neighbouring Crop a Dangerous Source of Inoculum?

The *pathogen* characteristics making for dangerous neighbourhood are:

1. The Host Range of the Pathogen. Botrytis cinerea with its hundreds of hosts (see Chap. 1.6.2.8), and many multi-host viruses endangers far more crops than a host-specific rust.

2. Survivability and Amount of Inoculum. Pathogens with spores highly susceptible to rigours of the environment, such as drought, are frequently among those producing inoculum in largest amounts. Neighbouring fields, quickly reached while conditions are favourable, are endangered much more than fields further away. For this reason, downy mildews are among the most dangerous pathogens transmitted between crops growing in close proximity. The added danger of massive inoculum compensating pathogens for environmental deficiencies has been mentioned in Chap. 1.3.1.

3. Period of Inoculum Supply. Virus-infected plants supply vectors with inoculum as long as the plant is alive, but the danger lasts only as long as vectors are active (see below). Fungi that sporulate on dead plant tissue, e.g., *Alternaria* or *Stemphylium* on unharvested crops and on plant remnants after harvest, or *Septoria* and *Rhynchosporium* on cereal stubble, or that disperse from resting bodies (*Sclerotinia sclerotiorum*), continue to do so for a long time. The period over which such pathogens may supply inoculum is therefore much more extended than that of pathogens which cease to sporulate as affected host tissue dries up (powdery and downy mildews). The more rapidly an obligate parasite kills its host, the shorter the period over which it endangers its neighbours.

4. Mode of Inoculum Dispersal. Spores dispersed by splashing are dangerous only to immediately adjacent crops. Inoculum dispersed by wind can carry to vastly different distances, depending on spore characteristics and strength of wind; but it must be borne in mind that, in particular in wind-borne inoculum, not all the spores that reach distant locations remain viable. Inoculum carried by pollen usually only gets as far as the bees that carry the pollen: this includes Prunus Ne-

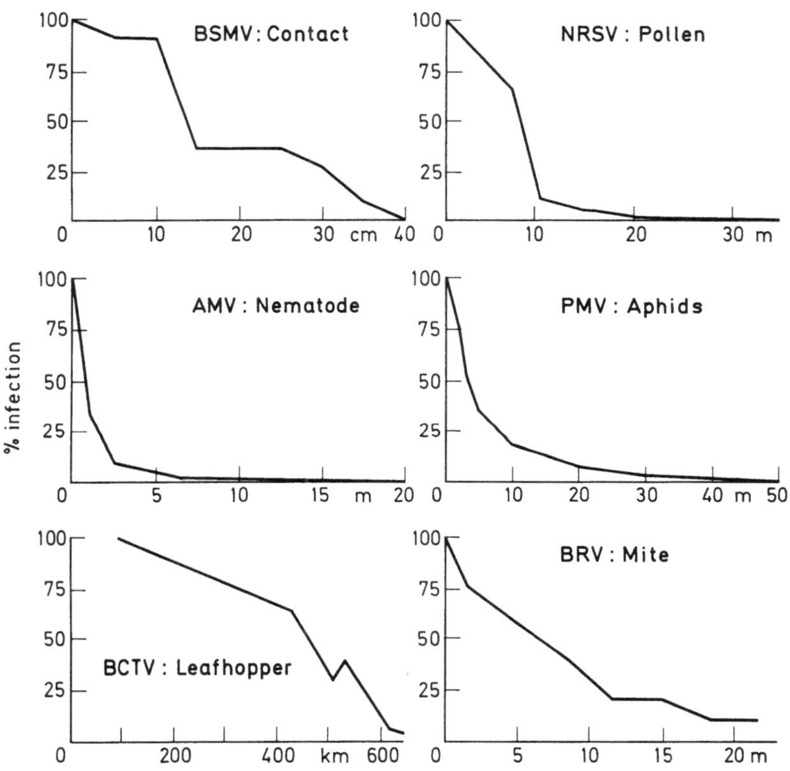

Fig. 2.23. Disease gradients due to the spread of Barley Stripe Mosaic Virus *(BSMV)* by contact (distances in cm), Prunus Necrotic Ringspot Virus *(NRSV)* by pollen (distances in m), Arabis Mosaic Virus, hop strain *(AMV)* by nematodes (distances in m), Peanut Mottle Virus *(PMV)* by aphids (distances in m), Beet Curly Top Virus *(BCTV)* by leafhoppers (distances in km), Black Current Reversion Virus *(BRV)* by eriophyid mites (distances in m). (Tresh 1980b)

crotic Ringspot Virus (Cameron et al. 1973), the viruses causing Peach Rosette and Decline (Smith et al. 1977), the fireblight bacterium *(Erwinia amylovora)* on pears (Anon 1969b), and fungi spread by wind as well as by pollen, such as *Sclerotinia sclerotiorum* on rape (Stelfox et al. 1978). Where danger of pathogen transmission by pollen exists, this must be taken into account when beehives are moved periodically to promote pollination in different crops, especially in seed-crops and orchards. Among insect-carried viruses, those that are stylet-borne will not usually reach very long distances, are chiefly dangerous to close-by crops, and their spread can to some extent be checked by barriers of taller crops (see Chap. 2.3.7.2). Persistent viruses, on the other hand, can disperse over wide distances, again depending to some extent on wind direction and strength. Thresh (1980b) has published a diagram in which the disease gradients due to spread of virus diseases by contact or by various carriers are depicted (Fig. 2.23).

The *crop* characteristics determining the extent to which neighbouring crops may constitute dangerous sources of inoculum are the following:

1. Degree of Susceptibility. Highly susceptible crops may be endangered by amounts of inoculum too low to affect more resistant crops seriously. The massive amount of inoculum building up on susceptible varieties may break the resistance of normally resistant varieties (see Chap. 1.3.1 and 1.3.3), presenting an acute danger in orchards in which fairly resistant varieties are interplanted with, or planted in proximity to, susceptible pollinator varieties.

2. Duration of the Period of Susceptibility. Where only young or intermediate stages of growth, or organs formed over limited periods (flowers, seeds), are susceptible, the period of danger is restricted, but where susceptibility mounts with age and culminates in senescence, the danger is prolonged and often progressive.

3. Seasonal Flexibility of Crops. Since most crops have larger ecological ranges than their parasites, the danger is lessened if one of the neighbouring crops can be grown, at least in part, in the period free from pathogen or vector activity.

2.12.2 Planning to Reduce Infection Risks from Neighbouring Crops

The principal environmental factor to be considered in planning to reduce risks of inoculum transfer from neighbouring crops is, of course, the direction of winds. In regions in which dew is frequent, direction of winds in daytime is more important than of those during the dew period, when air-borne spores of many fungi cannot take off. To minimize blowing of inoculum from infected to healthy crops, the crop likely to be infected first – an older crop, a more highly susceptible variety – must always be sown upwind.

Successive sowings of one crop species, or of several closely allied crops, in neighbouring plots or strips is often convenient for sequential harvesting, e.g., to produce a steady supply of roughage for livestock; but from the point of view of plant disease, this is a cardinal sin. The earlier sowings are most likely to transmit their accumulated pests and diseases to the later crops, and these then become infested at progressively earlier stages.

The economically most harmful infections originating in neighbouring crops are those capable of affecting a young crop and stunting its growth. This applies to numerous virus diseases, such as Tomato Yellow Leaf Curl and various viruses of leguminous crops. Their profuse infection of young growth also makes the downy mildews so dangerous, as pathogens likely to pass from older to younger crops and destroy the latter. The downy mildews of lettuce, onion, sugarbeet, and cucurbits are notorious in this respect.

The ageing crops that present the greatest dangers as sources of inoculum are seed-crops (Neergaard 1977). The pathogens they are most likely to spread are viruses (sugarbeet, lettuce, legumes), downy mildews (sugarbeet, onion), and leaf-spot fungi, such as species of *Alternaria* on Brassicae, carrots, or cotton, and of *Cercospora* on sugarbeet.

A less obvious, but not less important, source of inoculum for neighbouring fields are pathogens borne on stubble and plant debris. Among the fungi persisting on stubble and likely to infect fields in the vicinity are *Septoria nodorum*, *S. tritici*, and *Rhynchosporium secalis* on cereals, *Mycosphaerella pinodes* on peas, and *Phoma lingam* on Brassicae. Air-borne inoculum from plant debris is constituted by ascospores from debris carrying sclerotia of *Sclerotinia sclerotiorum* or by spores from potato debris, infected by *Alternaria solani*.

Protected crops sown early make bad neighbours for crops sown later in the open. Spread of *Peronospora tabacina* from covered nurseries of chilli pepper has been reported from Greece (Zachos 1963) and the disease frequently spreads from covered tobacco seedbeds. Similar spread of *P. brassicae* from nurseries of Brassicae has been observed in many countries.

In planning to minimize risks of inoculum transfer between fields, the grower is faced with a number of decisions, some of which are listed here.

a) If certain crops are to be grown for seeds, especially if the seed-crops are biennial, the same crop should not be grown for the market anywhere near the seed-crop.
b) The prevalent direction of winds must always be taken into account.
c) The shape of the field is important: a crop grown in an elongated field will be able to transmit much more inoculum into a neighbouring field running parallel to its long axis than a field of similar acreage, but more or less quadratic in shape.
d) Nurseries, especially when covered, and early crops should be kept as far away as possible from the main crop, and should be kept free from disease by all available means, including spraying at closer than usual intervals (see Chap. 3.2).
e) Stubble likely to carry inoculum must be incorporated in the soil well before a new, susceptible crop is sown in the vicinity. Burning stubble rarely gives uniformly adequate results.
f) Successive sowings or plantings of the same crop in immediately adjacent fields must be avoided. Even a distance of a few dozen metres between the fields can reduce the risk of disease appreciably.

2.13 Pruning and Grafting

2.13.1 Pruning and Bark Ringing

Pruning as a sanitary measure has been described in Chap. 2.2.3.4 and 2.2.4.1, and the danger of pathogen transmission by pruning in Chap. 2.2.2.4. A number of additional aspects of pruning as related to crop disease remain to be pointed out.

Timing of Pruning. An important rule is that crops should be pruned when least susceptible to pathogen infection. In the case of deciduous fruit and vine, this usually means the dormant season, as indicated for example by Ride et al. (1978) for pruning vine liable to attack by *Xanthomonas ampelina*.

The possibility of slightly manipulating bud burst and flowering of some fruit trees by early or late pruning, to increase their chances of escaping disease, has been mentioned in Chap. 2.10.5.

Shaping Trees or Bushes. Pruning is an important means of shaping the habit of the tree and bush crops, to result in open, airy growth with considerable exposure of inner leaves and fruits to sun, or in closed, shady habits with higher humidity. Since the large majority of fungal and bacterial pathogen prefer humid conditions, the disease control function of pruning is usually to reduce density of tree crown and bush canopies, but in areas with intensive radiation care must be taken not to expose fruit and tree trunks to sun scald.

The amount of pruning necessary to give trees desirable shapes depends to some extent on the vigour of stocks used: trees grafted to very vigorous stocks may have to be pruned more severely, and this in turn may make them more susceptible to pathogens such as *Stereum purpureum*, the cause of the silver leaf disease (Grosclaude, pers. comm.).

Bark Ringing as a means of disease control has mainly been advocated in the tropics. Dadant (1963) reports that ringing barks of coffee trees on the periphery of infected patches will stop the spread of the root rot caused by *Clytocybe tabescens*. As mentioned previously in Chap. 2.2.4, bark ringing can also be effective in combating *Armillariella mellea* on some of its tropical hosts such, as oil palm.

2.13.2 Grafting

Grafting is a universally accepted means of utilizing vigour and disease or pest resistance of stocks to produce high-quality yields of tree and plantation crops. It is much less common as a practice used on vegetable or field crops for the same purpose.

Vegetables. Grafting vegetables, cucurbits and Solanaceae has been developed principally in France and in the Netherlands. The merits of this practice have been summed up by Louvet (1974). On cucurbits, advantage is taken of the resistance of *Benincasa cerifera* to *Fusarium* wilts of melons and cucumbers, and to *Verticillium alto-atrum*, to produce these crops on soils known to be infected by these pathogens. Exact procedures for grafting and growing these crops have been worked out (Louvet and Peyrière 1962).

Tomatoes have been grafted successfully on a variety of *Solanum integrifolium* to resist bacterial wilt *(Pseudomonas solanacearum)*, and on hybrids of *Lycopersicon esculentum* x *L. hirsutum* to resist corky root *(Pyrenochaeta lycopersici)* and *Verticillium* wilt. This is considered a temporary measure, pending the breeding of tomato varieties resistant to these pathogens. On eggplants, however, grafting on tomato varieties resistant to *Verticillium* and *Fusarium* wilts is to be considered as a more permanent means of avoiding losses from these diseases, since no genes for resistance to *Verticillium* have so far been found in eggplants. To resist bacterial wilt, eggplants can be grafted on *S. integrifolium* or *S. torvum*, the latter being resistant also to root-knot nematodes (Beyries 1974, 1979).

Fruit Trees. Detailed information on the resistance of rootstocks for fruit crops to disease is not within the scope of this book. We shall therefore mention only briefly that for most stone fruits stock resistant to root-knot nematodes (*Meloidogyne* spp.), but not against ring nematodes *(Criconemoides xenoplax)*, are available, as are stocks resistant to crown-rot *(Phytophthora cactorum)* for apples and pears (Merrill and Petersen 1979). Avocado rootstocks partly resistant to *Phytophthora cinnamomi* and *Verticillium albo-atrum* are to be found among the Mexican varieties (Zentmyer, Wilhelm in Kranz et al. 1977).

The complexity of producing rootstocks resistant against a variety of diseases may be exemplified by reference to gummosis of citrus trees caused by *Phytophthora citrophthora* and *P. parasitica*. Various stocks with satisfactory resistance are available, but each one has another drawback: Sour orange *(Citrus aurantium)* renders trees highly susceptible to tristeza, *Poncirus trifoliata* to exocortis, and the "Cleopatra" mandarine is less resistant to *P. parasitica* than to *P. citrophthora* (Boccas and Laville 1978).

An important point in grafting on stock resistant to soil-borne pathogens which may also affect branches and fruits is to graft high enough to reduce the danger of such pathogens reaching the graft union and scion. This is of particular importance with regard to *Phytophthora* diseases, such as *P. cactorum* infecting stone fruit.

2.14 Effects of Physical Barriers on Crop Infection and of Optical Means on Virus Vector Control

2.14.1 Effects of Windbreaks and Tall Crops on Infection

Windbreaks influence velocity of the wind that hits the crop, movement of cold air masses down slopes, light, humidity, and temperature conditions in adjacent plots, and deposition of wind-borne particles and organisms. Windbreaks may therefore affect spread of pathogens and virus vectors, susceptibility of hosts and development of crop diseases, in various ways.

Winds of high velocity may directly injure crop tissues, especially leaf tips and margins, and may cause plant organs to rub against each other or knock flowers or fruits off vegetable or fruit crops. They thus facilitate entry of wound-related pathogens. Kuhara (1978) found that incidence of citrus canker (*Xanthomonas citri*) became severe when wind velocity exceeded 8 m/s; windbreaks effectively reduced losses from canker.

High winds are even more destructive when they carry sand or salt particles. In the desert regions of Israel, sandstorms injure leaves and markedly predispose tomatoes and potatoes to infection by *Alternaria solani*. Infection near windbreaks is much reduced.

The effect of trees on the movement of cold air down slopes has been referred to in Chap. 1.1 and their shade effects are noted in Chap. 1.2.

Movement and deposition of inoculum around windbreaks, and subsequent disease development, have been investigated with interesting results. Eversmeyer and Skidmore (1974) found that in the United States development of leaf rust (*Puccinia recondita*) on wheat was more severe leeward of a windbreak, but overwintered in larger amounts on its windward side. Primary foci of stem rust (*Puccinia graminis tritici*) were found near the windbreak, where dew persisted $^1/_2$ h per day longer; urediospore numbers decreased leeward of the windbreaks as far as six times its height, and then increased with distance beyond that point.

A detailed study of the effects of cypress windbreaks on infection of melons by Cucumber Mosaic Virus has been carried out in south-east France, over several years. Quiot et al. (1979) report that in eight out of their ten melon crops infections appeared earliest in a strip 18 m leeward of the 6 m high windbreaks, and this is where aphid vectors were first trapped. Measurement of wind velocity showed that it was lowest at this point. Marrou et al. (1979), observing spread of the virus in relation to 4-m-high windbreaks, found the latter useful in delaying the spread of CMV for some weeks from infected melons to near-by vegetable plots, including tomato, pepper, and squash, in spring and early summer. The duration of this effect depended on the level of virus infection in the melon plots.

On soybeans in the United States, Halbert et al. (1979) reported reduction of the spread of Soybean Mosaic Virus (SMV) by sunflower barriers, for about 17 m downwind. The effects of tall crops intersown, or sown in alternate strips, with low crops to protect the latter from virus diseases, has been described in Chap. 2.3.7.2.

2.14.2 Physical Protection of Crops from Infection by Soil-Borne-Inoculum

The idea of protecting plant parts in direct contact with soil by interposing sheets between them and soil-borne inoculum is not new, but has assumed greater practical interest with the availability of relatively cheap plastic sheeting.

One of the most successful of such uses is the mulching of soil of strawberry beds with polyethylene sheets. This has proved its worth in controlling fruit rot caused by *Botrytis cinerea* and *Rhizopus stolonifer* in Israel, and the rot caused by *Phytophthora cactorum* in France (Viennot-Bourgin 1967).

Physical protection of plant collars has succeeded in various crops. Thus incidence of *Sclerotium rolfsii* on tomatoes in the United States was reduced by a collar dressing of aluminium foil or plastic wrap (Baxter et al. 1977). In Ceylon, such dressings impregnated with PCNB prevented infection by the white root disease *(Fomes lignosus)* of rubber trees adjacent to diseased trees (Peries 1970).

Placing small squares of plastic material under fruit in direct contact with soil has proved effective in Israel in restricting infection of pomelo citrus fruits by brown rot *(Phytophthora citrophthora)* and of melon fruits by *Rhizoctonia solani* and *Sclerotium rolfsii.*

2.14.3 Optical Means for Virus Vector Control

Optical means include coloured material to attract virus vectors, reflectant surfaces to repel them, and netting which repels vectors and makes it more difficult or impossible for them to locate hosts.

Attraction. Success in the attraction of aphids to yellow polyethylene sheets placed vertically along the edges of bell pepper plots, trapping them by application of sticky substances to the sheets, and thus reducing incidence of Cucumber Mosaic Virus and Potato Virus Y, has been reported from Israel by Cohen and Marco (1973). This method, which is effective as long as populations of these vectors are not more than moderate in size, is now used by growers on a limited scale. The limitations are due to the fact that sheets tend to tear in strong winds and that they lose efficacy as their surface is covered by dust or sand carried by winds (S. Cohen, pers. comm.).

Effects of straw mulches in reducing virus vector populations of white fly *(Bemisia tabaci)* were at first ascribed to repellance of the vector. Later studies, however, showed these effects to be due to attraction of white flies to yellow straw

Table 2.27. Field experiments for control of non-persistent viruses with reflective surfaces. (Loebenstein and Racah 1980)

Crop or plant	Virus	Reflective material	Aphids in treated plots (% of control)	% Control
Gladiolus	Cucumber mosaic virus	Aluminum foil	7– 9	73–80
		White polyethylene	12	70
		Aluminum powder	19	55
Squash	Watermelon mosaic virus	Aluminum foil	7–10	72–94
Lettuce	Cucumber mosaic virus	Aluminum foil	25	94
Bell peppers	Potato virus,	Aluminum foil	2–18	86
	+ Cucumber mosaic virus	White on black polytheylene	9–35	88
		Grey plastic	6–18	90

Fig. 2.24. **A** Plots of pepper mulched with grey plastic and making normal growth: **B** Control plots severely stunted by viruses. (Loebenstein and Raccah 1980)

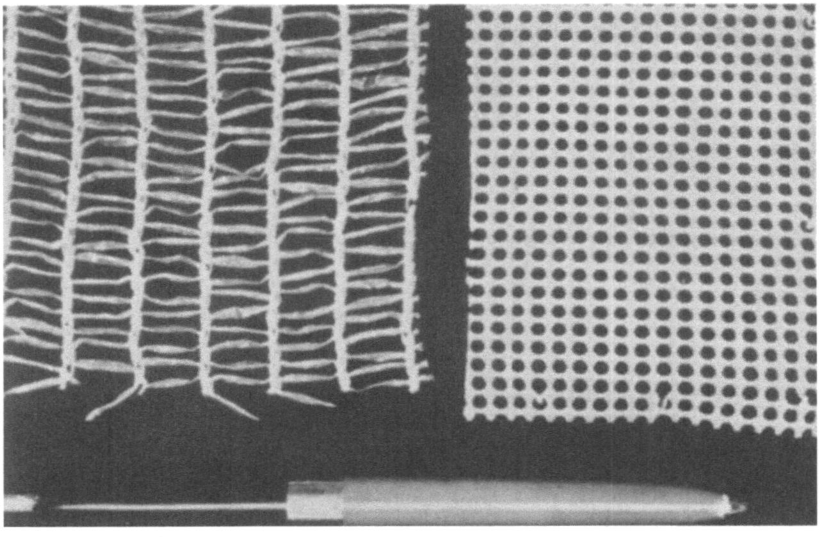

Fig. 2.25. Samples of white nets placed above and around plots to protect them from viruses. (Loebenstein and Raccah 1980)

and their death by high temperatures or alighting on the straw. These mulches have been used successfully in Israel to protect two crops from viruses borne by white flies, viz. cucumbers from Cucumber Vein Yellowing Virus (Nitzany et al. 1964) and tomatoes from Tomato Yellow Curl Virus (Cohen et al. 1974). Infection by the latter virus has also been delayed by the use of yellow polyethylene sheets, which were found to attract the vector more strongly than aluminium- or blue-coloured polyethylene (Cohen and Melamed-Madjar 1978). The above yellow mulches delayed virus infections by about 3 weeks. With the increasing cost of polyethylene, use of straw for mulching may offer economic advantages, and yellowish wood shavings are now also used for this purpose (Cohen, pers. comm.).

Reflectance. The use of reflectant surfaces to control non-persistent viruses transmitted by aphids has been reviewed by Loebenstein and Raccah (1980), as shown in Table 2.27. In experiments with bell pepper, Loebenstein et al. (1975) found that good protection of the crop by grey plastic mulch resulted in normal growth, while control plants were heavily infected by Cucumber Mosaic Virus and Potato Virus Y, and were stunted (Fig. 2.24). Trapping of winged aphids in the mulched plots amounted to only 6%–12%, of that in control plots.

The efficacy of reflectant mulches in protecting crops from viruses is, of course, limited by the time it takes the crop to overgrow most of the mulch, and by other factors interfering with reflectancy, e.g., sandstorms.

Not all aphids, however, are repelled by reflectant surfaces. Smith and Webb (1969) found numbers of *Aphis gossypii* on squash plots mulched with aluminium foil to be five times as high as that of aphids on non-mulched plots.

Bar-Joseph and Frankel (1979) have succeeded in reducing colonization of lime (*Citrus aurantifolia*) by winged aphids by spraying the plants with white clay suspensions containing kaolinite and montmorillonite. The effect of such applications on citrus viruses remains to be assessed.

White Nets. A novel approach to decrease virus dissemination has recently been used with success by Cohen (1981) in Israel. In repeated tests with white, yellow, and light grey polyethylene nets placed 0.5 m above pepper plots, the white nets were most effective in reducing infection by Cucumber Mosaic Virus and Potato Virus Y. A white net with a hole size of 10×3 mm and a thread width of 1.3 mm (see Fig. 2.25), giving a 20% reduction in mean radiation, was cheapest and highly effective: the mean number of aphids trapped on a 30×30 cm square under this net was 6.6, as compared with 46.0 under a yellow net and 55.3 in the control. The white nets have a double effect: they repel aphids and obscure the host from the vectors whose vision is limited to less than 50 cm.

Part 3 Interactions Between Cultural Practices, Resistance Breeding, and Application of Chemicals: Integrated Control

The aim of integrated control of plant diseases, as of that of plant pests, is to prevent parasite damage exceeding economic thresholds, and to achieve this by the means most economical from the point of view not only of a single crop, but of a whole cropping system, and not only of a single farmer, but of an entire farming community.

For this purpose, preventive and curative measures have to be considered, and to be integrated with each other and with essential practices not primarily related to disease control, e.g., fertilization and irrigation, or with the use of pesticides against pests and weeds.

The preventive measures are largely the domain of resistance breeding and of cultural practices, the curative measures (in the sense that they are mostly applied after disease appearance) are predominantly, though not solely, the domain of chemical control. Figure 3.1 depicts an integrated approach to the control of barley powdery mildew *(Erysiphe graminis hordei)* as given by Jones and Clifford

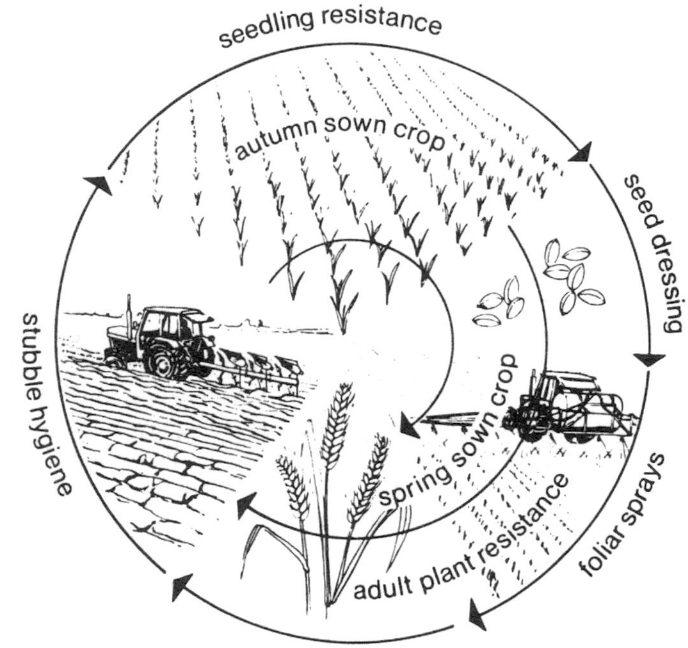

Fig. 3.1. A scheme for the integrated control of barley powdery mildew *(Erysiphe graminis hordei)*. (Jones and Clifford 1978)

(1978). It shows the role of tillage in reducing amounts of overseasoning inoculum, and the role of seed dressings and foliar sprays in protecting the young stages of the crops until adult resistance protects them adequately.

Production of resistant crop varieties is certainly the ideal way of meeting the danger of disease. The successes achieved in major crops are impressive, and a boon to mankind. In fact, efforts to promote crop production free from disease in developing countries rely predominantly on resistance breeding. Additional benefits could be derived by increased breeding of plants not only resistant to specific pathogens, but resistant to the stresses that render plants susceptible to so large a variety of diseases, in particular water deficit and low temperature stresses.

We shall consider the role of cultural practices in the framework of integrated control by discussing combined ways of keeping disease out, minimizing its spread once it has arrived, and timing chemical control optimally. This will be followed by an attempt at generalizing about the economic, human, pathogen, crop, and environmental factors that determine prospects for effective use of cultural practices for disease control, by an assessment of the mutual influence of cultural and chemical methods, and lastly by some thoughts, or perhaps speculations, on the future of integrated control in world crop production.

3.1 Keep Inoculum Out – by Any Available Means

Where there is no inoculum, there is no disease. Thus control always begins with a concentrated effort to keep inoculum out. The significance of this for the farm community as a whole, and the fact that frequently only communal or legal action can be effective, has been set out in Chap. 1.3.1. Exclusion of inoculum as a sanitary measure is treated in Chap. 2.2.

Among the various modes of penetration of inoculum into a region, there is little the farmer can do against wind and water carrying pathogens, but a great deal he can do to prevent their introduction by propagating material, equipment and labour. The watchword here is disinfection.

Disinfection of seeds and of much vegetative propagating material is nowadays often carried out routinely by the supplier. For seeds, the treatment applied should be stated on the label. However, this is not always the case, and on some seeds stored for extended periods, the action of protectant seed-dressings may have weakened. This enables pathogens borne within the seed, or soil-borne pathogens, to attack when the seed germinates. Seed treatment by chemicals is usually not difficult to perform, and the fungicides recommended are safe, but buying disinfected seed from reliable sources is generally preferable. Disinfection by heat treatment is a more delicate matter, as it attempts to take advantage of the often limited difference in the level of heat the crop can stand but the pathogen cannot. Heat treatment is the most practical way of controlling some diseases caused by viruses, fungi, bacteria, and nematodes in propagating material. Its use against virus diseases has been reviewed by Nyland and Goheen (1969). A list of some crops, the vegetative propagating material or seed of which can be heat-disinfected to control disease, is presented in Table 3.1. Such treatments, if directed against fungi, can be

Table 3.1. Some crops and pathogens amenable to heat treatment vegetative propagating material

Crop	Disease	Pathogen	Parts treated	References
Sugarcane	Chlorotic streak virus		Cane cuttings (setts)	Benda (1973)
	Ratoon stunting	Bacterium	Cane cuttings (setts)	Benda (1973)
	Red rot	*Colletotrichum falcatum*	Cane cuttings (setts)	Singh (1973)
	Streak, mosaic viruses		Bud tissue	Roth (1973)
Sweet potato	Black rot	*Ceratocystis fimbriata*	Tubers	Nielsen (1977)
	Scurf	*Monilochaetes infuscans*		
	Nematode	*Meloidogyne incognita*		
Yam	Nematode	*Scutellonema bradys*	Tubers	Bridge (1972)
Onion	Downy mildew	*Peronospora destructor*	Bulbs	Yarwood (1943)
Peppermint	Rust	*Puccinia menthae*	Rhizomes	Fletcher (1962)
Anemone	Rust	*Tranzschelia pruni-spinosae* var. *discolor*	Corms	Afek and Sztejnberg (1980)
Carnation	Mottle Virus		Plants	Hakkaart and Jordanova (1968)
Gladiolus	Wilt	*Fusarium oxysporum* f. sp. *gladioli*	Cormels	Magie (1971)
Narcissus	Leaf scorch	*Stagonospora curtisii*	Bulbs	Anon (1974)
	Nematode	*Ditylenchus dipsaci*	Bulbs	Anon (1974)
Strawberry	Nematode	*Aphelenchoides fragariae*, *A. ritzemabosi*	Sets	Klinger (1969)
Peach	Yellows Virus		Buds	Gaeumann (1951)
Pear	Vein Yellows Virus, Red Mottle Virus		Tips of seedlings	Minoiu (1974)

combined with treatment by systemic fungicides, e.g., combined heat-benomyl treatment for control of *Fusarium oxysporum* f. sp. *gladioli* on gladiolus corms.

Disinfection also has to be applied to equipment and labour. If soil-borne diseases threaten, everything that moves on wheels, tracks, or feet should be disinfected before it gets into still healthy fields. Cheap and safe disinfectants are available, and if the grower is sufficiently aware of the dangers of disease transmission by soil, such disinfection is mostly a practical proposition. This also applies to disinfection of shoes or feet of labourers as they pass from one field to another, and where viruses are concerned the clothing may also have to be disinfected. Material and equipment used for staking, irrigation, and packing must also be disinfected, as it is moved from field to field.

3.2 Prevent Multiplication and Spread of Inoculum

In preventing spread of inoculum in whole regions or individual fields, the use of resistant varieties, of cultural practices, and of pesticides, all fulfil functions that can be combined.

3.2.1 Use of Resistant Varieties to Delay Inoculum Build-up

Varieties are mostly bred for resistance against one or very few major pathogens, but most resistant varieties have some drawbacks (otherwise all other varieties would disappear). Thus, the varieties resistant to an important disease may be more susceptible to stress, to pests, or indeed to other diseases, or may give quantitatively or qualitatively inferior yields. The most publicized cases of varieties resistant to some diseases failing to yield as expected because they were attacked by other diseases are those of Mexican wheat bred to resist rust, but highly susceptible to *Septoria* blotch, and of rice unexpectedly attacked by bacterial diseases.

This illustrates the dangers of replacing too rapidly and too widely varieties grown over extended periods by imported varieties with unknown susceptibility to local diseases or pathogen strains. Other examples, relating to tomato varieties introduced to Israel from the USA and succumbing to typically Mediterranean pathogens, are the introduction of the "Moneymaker" variety that suffered extensive loss from attack by Tomato Yellow Leaf Curl Virus, and of VF processing tomatoes that proved highly susceptible to *Leveillula* mildew.

The proper use of resistant varieties therefore requires an overall view of the crop's problems. In the present context of keeping inoculum down, their use is fairly simple: where a pathogen, such as *Phytophthora infestans* on potatoes, is likely to mount increasingly severe epidemics by passing through successive cycles of multiplication during the whole of the season in which the crop can be grown, then its period of multiplication, the number of its cycles, must be reduced; this can be done either by not growing the crop at all in the early part of the season, or by growing then only resistant varieties. If this is not done, inoculum in mid-season, or certainly late in the season, may mount to such levels that the later crops do not stand a chance. It may then be impossible to protect these crops by sprays applied in schedules that were perfectly adequate at lower levels of inoculum. The resistant variety must then be used for the early crops, in the interest of later crops, even if its yield is somewhat inferior.

Resistant varieties and cultural practices, especially choice of sowing dates, irrigation and fertilization, have basic influence on the rate of inoculum production; but in all crops that can stand the expense, depression of the rate of infection, and prevention of disease exceeding thresholds of economic loss is the major task of pesticides and has led to their wide, and sometimes excessive, use.

3.2.2 Cultural Practices and the Timing of Pesticide Use

Disease/loss relationships have been worked out for a few important crop/pathogen pairs, e.g., leaf rust of wheat caused by *Puccinia recondita tritici* (Burleigh et al. 1972) and potato blight caused by *Phytophthora infestans* (James et al. 1972). Forecasts whether and when loss thresholds will be reached provide a rational guide when to begin chemical control, under conditions approaching those under which the studies were made, in temperate climates. However, neither forecasts of epidemics nor determinations of disease/loss relationships have so far succeeded in other climates.

The problems of when to initiate chemical control in field or orchard, how often to repeat applications, and when to discontinue them, is of course closely linked to cultural aspects. If the sources of inoculum are near, or even adjacent, prophylactic applications may well be needed where fungicides would normally be applied only after appearance of symptoms. In densely planted row crops, or in crops that have responded to fertilization with dense growth, it is often better to make a first application before the rows close, even if no disease is apparent at that time. Early applications may also be needed in crops that develop a prostrate or a very bushy habit, as the older leaves cannot easily be covered at a later stage.

Irrigation affects the timing of applications of protectant fungicides decisively, and the reason are various:

a) the rate of growth is largely determined by irrigation, i.e., the amount of growth made after the previous application;
b) the wetting of the soil by irrigation makes spraying by heavy machinery impossible, and by knapsack sprayer difficult, on the days immediately following irrigation. One of the advantages of aircraft application, is that it is independent of ground conditions;
c) overhead irrigation may promote disease, as described in Chap. 2.8, and also washes off at least part of most spray deposits. The dilemma then arises: should the application be made before sprinkling, perhaps with addition of stickers, but yet incurring the risk of losing some deposit, or should application be delayed for a few days, until the ground has dried, risking disease development on as yet unprotected foliage (and this is when flushes of new growth are apt to form), during the high moisture period just after irrigation.

In practice, the intervals between pesticide applications in crops irrigated regularly must be adapted to irrigation intervals, if tractor-mounted or tractor-drawn equipment is to be used.

3.3 The Proper Place for Cultural Practices in Integrated Disease Control

Rotem and Palti (1980) have recently summarized what they called the prospects of disease control by cultural practices. These summaries will here serve as a basis for the assessment of the proper position of such practices in the ensemble of integrated control. They refer to farm practices in developed agricultural economies, not to subsistence farmers.

3.3.1 Economic and Human Factors

Table 3.2 lists economic considerations and human factors that affect the probability of cultural measures being used for crop disease control. The principal economic factor, crop value and therefore level of potential loss due to disease, may be so overriding in its importance, that it makes any further considerations superfluous. A case in point is that of flower crops, the high value of which demands

Table 3.2. Economic and human factors affecting the prospects for use of cultural practices in the integrated control of crop diseases. (Modified after Rotem and Palti 1980)

	Prospects for control by cultural practices	
	Improve when	Diminish when
Crop value and level of potential crop loss is:	Low	High
Cost of chemical control, relative to overall growing expense, is:	High (e.g., cereals)	Low
Choices for pre-sowing practices are:	Numerous	Few
Chances for manipulation of field conditions are:	Many (e.g., irrigated crops)	Limited
Need for specialized and costly machinery is:	Essential	Not essential
Chances for communal planning to minimize inoculum build-up are:	Good	Bad
Experience and educational level of farmer are:	Good	Bad
Advisory services are:	Available	Unavailable
Availability of labour when needed is:	No problem	Problematic

the quickest and safest measures available, and this means generally control by chemicals.

Where the cost of chemical control, relative to total growing expenses, is high and the margin of profit left by such control is small, growers will tend to prefer resistant varieties or cultural practices. Cereals are the obvious example, because varieties resistant to major diseases are available, and cultural measures, such as variation of sowing dates and rates, and tillage, can have marked effects. However, as pointed out in Chap. 2.5 and 2.1, availability of labour and equipment at certain seasons may outweigh all other considerations.

The role of the educational level of the farmer, a factor of supreme importance in his choice of means to control diseases, will be referred to in Chap. 3.6. Whatever that level, farmers are sure to require much more advisory assistance where pest and disease control is performed by integrated management techniques, preferably based on economic thresholds, than where control is chiefly by pesticides (Furtick 1976).

3.3.2 Pathogen, Crop, and Environmental Factors

Table 3.3 presents a list of pathogen and crop factors that affect the likelihood of cultural practices being used to advantage to minimize disease. The pathogen factors mentioned in the table do not fit any particular group of pathogens. The most important of these factors are the pathogen's host range and mode of dispersal, closely followed by rate of inoculum build-up.

Among the crop factors, the question whether the economically valuable (marketable) produce is affected directly or not, is the crucial question. On perennial crops, the slow but steady influence of cultural practices obviously has more scope than on annual crops. Irrigated crops are always more amenable to cultural

Table 3.3. Pathogen and crop factors affecting the prospects for use of cultural practices in the integrated control of crop diseases. (Modified after Rotem and Palti 1980)

	Prospects for control by cultural practices	
	Improve when	Diminish when
Pathogen factors		
Host range is:	Narrow	Wide
Dispersal of inoculum is:	Short-range (soil, splashing, neighbouring fields)	Long-range (wind, vectors)
Rate of inoculum build-up is:	Rapid	Slow
Wetting period required for infection is:	Long	Short
Temperature ranges for development are:	Narrow	Wide
Sensitivity of inoculum to heat and drought is:	High	Low
Crop factors		
Marketable produce is affected:	Indirectly	Directly
Growth period is:	Extended (e.g., perennials)	Short
Water supply by:	Irrigation	Rain
Amount of susceptible tissue at any one time is:	Limited	Plentiful
Sensitivity to stress is:	High	Low

manipulation than rain-fed crops, especially where the relief of moisture-deficit stress is concerned.

Since most crops are liable to attack by more than one pathogen or pest, the nature of the latter often dictates the way in which control measures are integrated. Rabb et al. (1976) have given a good example of management systems suggested for tobacco crops in North Carolina (USA), when affected to various degrees by two diseases, black shank (*Phytophthora parasitica* var. *nicotianae*) and bacterial wilt (*Pseudomonas solanacearum*), and a root-knot nematode (*Meloidogyne incognita*). Their scheme is reproduced in Table 3.4 and shows the flexibility demanded of the grower to meet the conditions created by the attack of each of these pathogens at various levels. In addition, the grower was expected to take into account two further diseases, mosaic, and brown spot (*Alternaria longipes*), as well as three insect pests.

As for the environmental factors listed in Table 3.5, widely divergent daily or seasonal levels of temperature and humidity frequently make it necessary to alleviate stress situations they create, by suitable cultural treatments.

3.4 Profit in Fungicide Applications, as Related to Cultural Factors

The profit derived from fungicide applications is obvious when they control diseases directly affecting the quality of the produce, e.g., control of apple and pear scab, *Septoria* or *Cercospora* blights on celery, downy mildew on lettuce or tobacco, or powdery mildew of roses (*Sphaerotheca pannosa*). In these cases the extent

Table 3.4. Summary of systems suggested for growers to use in combinating tobacco disease problems of different intensities in North Carolina, USA. (After Rabb et al. 1976)

A	B	C	D	E	F
Relative infestation levels of three major tobacco diseases					
Black shank, wilt – high	Black shank, wilt – low to moderate	Black shank, wilt – low to moderate	Black shank, wilt – low	Black shank, wilt – very low	Black shank, wilt – very low
Root-knot – low to high	Root-knot – low to high	Root-knot – low to high	Root-knot – low to high	Root-knot – low to high	Root-knot – very low
Suggested management systems[a] for each infestation level					
Cropping system – 2 or 3 year rotation	Cropping system – no rotation	Cropping system – 2 year rotation	Cropping system – 2 year rotation	Cropping system – 2 year rotation	Cropping system – 2 or 3 year rotation
Variety[b] – high black shank, wilt, root-knot	Variety – high black shank, wilt, root-knot	Variety – high black shank	Variety – low to moderate	Variety – low to moderate	Variety – low to moderate
Multipurpose chemical	Multipurpose chemical	Nematicide	Multipurpose chemical	Nematicide	No chemical

[a] Four important control practices should be included in each system: 1. Stalk destruction and ploughing out tobacco stubble. 2. Plant bed disease control. 3. Prevention of mosaic. 4. Control of brown spot *(Alternaria longipes)*

[b] Varietal resistance level to be chosen

Table 3.5. Environmental factors affecting the prospects for use of cultural practices in the integrated control of crop diseases. (Modified after Rotem and Palti 1980)

	Prospects for control by cultural practices	
	Improve when	Diminish when
Climatic conditions tend to produce stress situations	Frequently	Rarely
e.g.,		
Distribution of rain is:	Not uniform	Uniform
Occurrence of dry periods is:	Frequent	Infrequent
Differences between seasonal or daily maximum and minimum temperatures are:	Wide	Limited
Radiation is:	High	Low

of disease development may also be influenced by cultural practices, which can render chemical control more effective and reduce the number of applications required; but their influence is not decisive, and the profitability of the use of fungicides needs no further proof.

Not so with diseases that affect the yield indirectly, e.g., by attacking the leaves of cereals, sugarbeet, potato, or olives, but not the part for which such crops are chiefly grown. To determine the benefits of fungicide applications in such crops, and to relate them to the varieties sown and the cultural practices employed requires large-scale study in a great number of field tests.

This has been undertaken in France in 1973 and 1974 by Lescar et al. (1975) and has been evaluated by Cassini et al. (1975) in the wider context of the integrated control of crop diseases. The study comprised no less than 100 tests on winter wheat grown after wheat or potatoes in two regions, and 341 tests on wheat grown after maize or sugarbeet in three regions. The results relate to five varieties of wheat, two types of tillage (by rotovator to a depth of 3–5 cm, and by ploughing to a depth of 18–20 cm); in addition, in the wheat sown after maize or sugarbeet, three rates of sowing were compared.

The results of this work are presented in Fig. 3.2, and may be summarized as follows:

a) In wheat sown after wheat or potatoes, 54% of the tests showed profitable response of the crops to two applications of benomyl + mancozeb, while in wheat sown after maize or sugarbeet only 29% responded likewise.
b) With the wheat varieties Champlein and Top, the response was consistently less than with the varieties Joss, Hardi, or Capitol.
c) Fungicide application was profitable in fewer of the wheat fields sown at a low, than in those sown at a normal or high rate.
d) Treatments in region 1 tended to give higher returns for the fungicide than in the other regions.

The conclusions the authors draw from these results are that in the regions concerned wheat grown after maize did not need fungicides if the Champlein or Top varieties were sown at low rates after superficial tillage. This is also true for wheat sown after ploughing in region 1, if the rate of sowing is low, but higher rates may, in that region, necessitate spraying in about one third of the number of fields. On the other hand, wheat following wheat or potatoes will respond profitably to fungicides in three cases out of four, especially when tillage has been superficial and varieties such as Joss, Hardi, or Capitol are sown.

The above study is rare in its scope and complexity, but effects of individual cultural practices on economic results of fungicidal treatments have, of course, been variously investigated. Certain practices are essential to obtain high yield potentials and indeed to secure a crop that is worth protecting. An example is the late application of nitrogenous topdressings to cereals. This increases yield as well as incidence of powdery mildew caused by *Erysiphe graminis* (Jenkyn and Bainbridge 1978), and *Septoria nodorum* (Oppitz and Hoeser 1978), but properly timed fungicide applications enable the crop to realize the potential given by the nitrogen.

The efficacy of fungicides is frequently related to the amount of inoculum in field and orchard. Only if inoculum is kept within certain limits can fungicides give

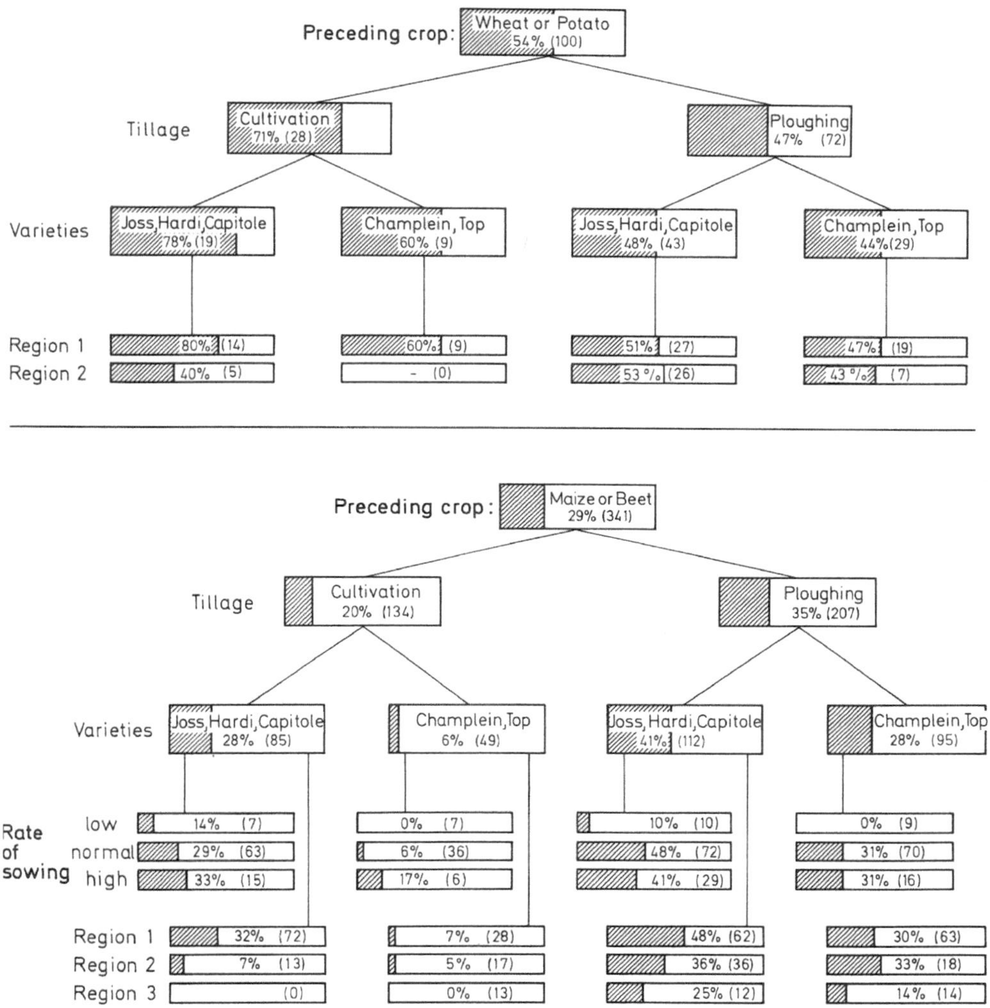

Fig. 3.2. Evaluation of the profitability of fungicidal treatments applied to five wheat varieties in three regions of France in 1973 and 1974; according to crop sequence, mode of tillage and rate of sowing. (After Lescar et al. 1975, as cited by Cassini et al. 1975). The numbers indicate the percentage of tests in which the loss by diseases in untreated crops caused the sprayed plots to yield at least 500 kg/ha more. This percentage conforms to the shaded area in each rectangle. Fungicidal treatments consisted of two applications of benomyl + mancozeb. The *numbers in brackets* represent the number of field tests evaluated in each case. *Top* Wheat grown after wheat or potatoes. *Bottom* Wheat grown after maize or sugarbeet

economic control of many diseases. We need here to refer only to a few examples where cultural practices can do this. Sanitation, such as pruning and removal of dead growth and mummies, are essential for chemical control of crops as diverse as cocoa, to control *Phytophthora palmivora* pod rot (Muller 1974), and grapevine, to control *Guignardia* black rot (Stevens and Stevens 1952); sowing at a distance from suspected sources of inoculum (neighbouring fields) is of obvious importance, and so is management of debris, if this is where the inoculum comes from.

There are, conversely, cultural practices applicable only when backed up by fungicides. Overhead sprinkling in California so much increased incidence of *Cercospora beticola* on sugarbeet, that only regular spraying permitted farmers to avail themselves of this mode of irrigation (Paulus et al. 1971). High rates of sowing are necessary to gain maximum revenue from many crops, but since they promote various diseases, such dense sowing presents an acceptable risk only where fungicides can check disease outbreak. Sowing some vegetables under plastic cover, in Israel, proved so conducive to development of *Sclerotinia sclerotiorum* rot, that these crops sometimes became profitable only after the advent of systemic fungicides which controlled the rot.

3.5 Cultural Practices and the Use of Herbicides and Physiologically Active Chemicals

3.5.1 Herbicides and Desiccants

One of the most remarkable successes of chemical industry, and one of the most significant developments in post-World War II crop production, is the widespread use of herbicides. It has already been mentioned that herbicides have in many cases done away with the need for crop rotation. In fact, crop rotation has in many cropping systems been replaced by "herbicide rotations" (Horowitz, pers. comm.): crops are grown in sequences that permit the selective herbicides applied to each crop to keep down all the potentially harmful weeds.

One may likewise speak of "herbicide monoculture" where residual effects of herbicides highly effective in a given crop present hazards for other following crops. The classical example is the use of simazine in maize: this persists in the soil well beyond the growth span of a single maize crop, and is an important factor in growers' decisions whether or not to grow maize in monoculture (Messiaen, pers. comm.).

Interactions between herbicides and plant pathogens have been reviewed by Katan and Eshel (1973) and Altman and Campbell (1977). Examples of the diverse ways in which such interactions take place are given in Table 3.6.

However, of greater practical importance than these biological effects of herbicides are their agronomic and environmental effects relating to crop disease. These effects are exerted by:

a) removing additional or alternate hosts of pathogens with wide host ranges, prominent among them many viruses (see Chap. 2.2.3.1), and including symptomless carriers of destructive pathogens;
b) affecting the crop climate, which is bound to be more humid in most crops overgrown with weeds;
c) facilitating manipulation of sowing dates: where previously sowing often had to wait until soil conditions permitted weeds to be turned in by tillage, killing

Table 3.6. Examples of interactions between herbicides and soil-borne crop pathogens. (Data from Katan and Eshel 1973)

Herbicide effect	Crop and disease	Pathogen	Herbicide applied
Direct Effect on Pathogen			
Growth stimulation	Stem rots on various crops	*Sclerotium rolfsii*	Atrazine, trifluralin
	Cotton wilt	*Fusarium oxysporum* f. sp. *vasinfectum*	Atrazine, prometryne
Growth retardation	Collar rot on various crops	*Phytophthora cactorum*	2–4, D
	Potato scab	*Streptomyces scabies*	2–4, D ester
Effect on Host Susceptibility			
Increased susceptibility	Damping-off of cotton	*Rhizoctonia solani*	Trifluralin
	Tomato wilt	*Fusarium oxysporum* f. sp. *lycopersici*	Maleic hydrazide
Decreased susceptibility	Tomato wilt	*Fusarium oxysporum* f. sp. *lycopersici*	Propham, T.C.A.
	Wheat foot rot	*Pseudocercosporella herpotrichioides*	Diuron
Effect on Relationship Between Pathogen and Soil Microflora			
Depression of pathogen by increased microflora activity		*Sclerotium rolfsii*	Simazine
		Fusarium spp.	Linuron, diuron
Depression of microflora, increased pathogen activity	Damping-off of cotton and pepper	*Rhizoctonia solani*	Trifluralin, diphenamid
	Cotton wilt	*Fusarium oxysporum* f. sp. *vasinfectum*	Prometryne

the weeds in time by means of herbicides gives the farmer a much wider choice of sowing dates;

d) affording a way of killing weeds without injuring the roots of crops, and thus lessening chances of root infection, as pointed out by Renard (1979) in relation to the infection of oil palms by *Fusarium oxysporum* f. sp. *elaeidis*.

Although on the whole clearly beneficial to disease control, use of herbicides has in some cases had a negative influence, especially where they caused certain weeds not susceptible to their selective action to replace the weeds they controlled. A case in point is that of the ergot disease caused by *Claviceps purpurea*. In France, Rapilly (1966) found this disease to increase in wheat grown after another cereal crop, in which hormone-type weed-killers had been used to kill broad-leafed weeds, but did not affect graminaceous weeds such as species of *Alopecurus*, *Lolium*, *Agrostis*, and *Poa*. These weeds flowered early in the following season's wheat field, just when the pathogen released its spores, and were infected; subsequently the wheat, which flowers later, was infected through the *Sphacelia* stage of the disease that had developed on the weeds. Another example is wild beet, which is hard to control in sugarbeet; the wild beet is highly susceptible to downy mildew (*Peronospora farinosa*), much more so than many varieties of sugarbeet, and its profuse growth in fields in which other weeds have been suppressed by selective herbicides tends to increase infection of the sugarbeet crop.

The flexibility that herbicides, often those of the desiccant type such as paraquat, afford the grower in choosing sowing dates is of great practical importance. From the aspect of disease management, it may make it possible to sow in disease-free parts of the season, or, in regions with arid seasons, crops can be sown at the end of the rainy period to ripen, free from disease, in the rainless season. The desiccant treatments have been instrumental in making direct-drilling of crops a widespread practice; the disease problems this involves in growing cereals have been referred to in Chap. 2.5.

3.5.2 Chemicals Applied to Affect Crop Physiology

Among the chemicals applied to change host physiology, the widest use has been made of chlormequat (C.C.C.) which prevents lodging of cereals. It has been found to reduce incidence of eye-spot (*Pseudocercosporella herpotrichioides*), but not that of *Septoria tritici* or *Fusarium culmorum* on wheat (Rapilly 1968). C.C.C. and two other growth retardants have also been reported to mitigate cotton wilt (*Verticillium dahliae*) and thereby increase yield of the variety SJ-2 in California (Erwin et al. 1979).

The hormone sprays, such as dilute 2,4-D, used to prevent preharvest drop of deciduous and citrus fruit, decrease the amount of rotting fruit on the ground, which can only be beneficial from the point of view of sanitation.

Oil or oil-sprays amended with chemicals, are employed to break the dormancy of deciduous fruit trees in regions with mild winters. Uniform bud-burst and flowering are induced in this way at a time which, within certain limits, can be preselected. This may reduce likelihood of hail and frost damage and minimize subsequent entry by pathogens (see Chap. 2.9.4).

Mention has been made in the chapter on stress and strain (Chap. 1.5.3) of substances that alleviate the strain of water-deficit.

A rewarding field for agrochemical research lies in the search for chemicals that mitigate major plant stresses, e.g., chilling, with all that implies for crop disease control.

3.6 Hop Wilt in England: Success of Integrated Control

The wilt of hops is a soil-borne disease caused by several strains of *Verticillium albo-atrum*. It spreads systemically in the plant by transport of conidia in the xylem sap, pervades all vegetative organs of the shoot, and is readily dispersed by debris, even from hops showing no external symptoms. In addition many weeds carry the disease, some with, others without symptoms (Fig. 3.3).

Successful control of this destructive disease in England by combined resistance breeding, sanitation, weed control, improved tillage and fertilizer practices, crop sequence management and communal (legislative) action will here be described

Fig. 3.3. A hop garden devastated by wilt, with weed hosts of *Verticillium*, photographed in 1941 when the disease was rapidly spreading throughout south-east England. (Courtesy Dr. G.F.W. Sewell, East Malling Research Station, Kent, UK)

briefly, mainly according to publications by Sewell and Wilson (1961, 1966a, b, 1967, 1974a, 1974b).

The first fundamental step in combating hop wilt was the breeding of wilt-resistant varieties: the resistant "Target" variety is now grown on more than two thirds of the acreage of hops in south-east England. However, fluctuating outbreaks of disease even in these varieties condition their successful growth on various cultural practices.

Two factors have been found to increase susceptibility of hops to wilt: low soil temperatures (especially below 11 °C) and high doses of nitrogen, and in particular the combination of both. This is illustrated by Fig. 3.4. Growers are accordingly advised to reduce applications of nitrogen in infected gardens to the lowest level commensurate with acceptable yields (Sewell and Wilson 1974b).

Persistence of the hop wilt pathogen in the field, and its spread from isolated foci to more generalized infection has been demonstrated to be associated with spread of infected hop debris and weeds by tillage. Weed control by over-all herbicides and introduction of non-tillage practices have greatly contributed to wilt control: roots are not cut or injured by tillage, their pattern of spread over the whole area is more uniform, use of triazine herbicides reduces root growth in the topmost

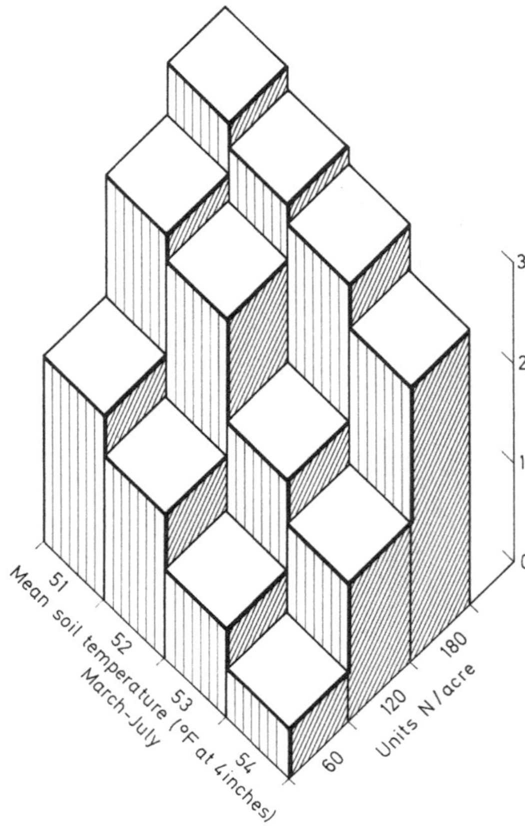

Fig. 3.4. Effects of three rates of inorganic nitrogen application and of soil temperatures during four years of wilt incidence in a hop garden. (From Sewell and Wilson 1974b)

soil layer, and thus diminishes their contact with surface-borne inoculum – all this reduces chances of infection. Moreover, elimination of weed carriers of the pathogen restricts sources of infection (Sewell and Wilson 1974a).

It has been found possible to eradicate *V. albo-atrum* from infected soil by keeping it under grass for 3–5 years, provided that weeds are strictly controlled (Sewell and Wilson 1966b).

Hops in England are now largely picked by machine. As this involves transporting all the aerial parts of the crop to the machine, it presents serious hazards of the spread of wilt by infected debris along the transport route and around the picking machine. The risk of spread by transport is reduced by burning all aerial parts of bines showing disease symptoms before they are sent for picking; this is a practical proposition because the use of resistant varieties has so greatly diminished the number of infected bines (Sewell and Wilson 1961). Sanitation at the site of the picking machine consists in carefully composting the hop waste and, of course, in applying the compost to crops other than hops (Sewell et al. 1962).

None of the above sanitation measures could have been fully effective without communal action. By request of the hop growers, legislation has been passed to oblige growers to notify the authorities of outbreaks of wilt, to prohibit the transport of hops from one district to another (to prevent spread of specific strains of the wilt fungus and of virus diseases), and to ensure elimination of diseased plants by grubbing and destruction of debris.

3.7 Some Thoughts on the Future of Integrated Disease Control and its Components in World Crop Production

The principal tendencies in world crop production to serve rapidly expanding populations may be expected to be:

a) extension of agricultural production to areas not so far exploited;
b) intensification of production on areas already under cultivation;
c) growing new crops with improved nutritional and other properties;
d) education of farmers to make better use of available resources.

How will the diverse approaches and techniques of disease control serve these ends?

3.7.1 Bringing New Areas into Production

The large areas not so far cultivated are the deserts or semi-deserts, mountainous areas, and tropical jungles.

Soil in many jungles seems to be such that, as shown lately in Brazil, once the jungle is cleared, there are no extraordinary cultural problems. However, high humidity and high temperatures are so favourable to development of diseases that, perhaps after an initial disease-free period, introduction of pathogens to susceptible crops could be catastrophic. The emphasis in crop production in these areas will therefore be on growing disease-resistant crops spaced widely enough to permit application of fungicides when needed (see also Chap. 2.6.2).

Deserts and semi-deserts of the world are an enormous land reservoir that awaits irrigation. The breakthrough here should come if and when seawater can be desalinified cheaply enough to permit its use in irrigation, but even before that time, the many opportunities of adapting irrigation so as to minimize disease development will make the irrigation technique one of the outstanding tools of producing healthy crops. Sanitation in crops to be grown all the year round will also be of supreme importance. Resistant varieties will be used to the full, but use of fungicides may perhaps be less essential in these very dry, than in more humid, regions.

Mountainous areas can be brought into production where a sufficiently deep layer of soil is available and erosion can be controlled. New techniques of range management should succeed in making better use of mountain areas for livestock. For crop production, use of mountainous areas in semi-arid countries seems possible only with fairly sophisticated techniques of applying water and nutrients through trickle-irrigation systems. Experience in Israel has shown that high-value crops, such as bananas, can thus be grown on shallow mountain soil, with the added advantage that the site is free from frost. The difficulties of using mechanical equipment on steep slopes can probably be solved in time. From the crop disease aspect, slopes with marked air movement present no special problem, and where trickle irrigation systems are employed they can also be used for application of fungicides to the soil.

3.7.2 Intensification of Production on Areas Already Cultivated

Wider and better use of water and fertilizers are the basis for raising agricultural production in developing areas. This can be expected to increase the danger of disease in crops which then reach fuller development, but also raises crop value so as to repay application of fungicides.

As mentioned in the chapter on crop sequence, in Chap. 2.3.9, monoculture, as well as multiple cropping and intercropping, can be expected to be practiced much more widely. Monoculture is always suspect of raising problems of disease of the "irreversible" type (see Chap. 2.3.6.1), and these can be solved chiefly by resistant varieties. It must, however, be noted that in growing cereals in the temperate zone, some of the more important pathogens have proved to be of the "reversible" type, and monoculture is now widely accepted as profitable.

Multiple cropping and intercropping are tried and proved ways of raising production where land and/or water are scarce, but labour is available. These practices can certainly be expected to be adopted by farmers in additional areas, e.g., in fields newly irrigated, and this presents numerous disease problems. As long as labour is cheap enough to keep up a running battle of sanitation, i.e., remove infected material constantly as the crop grows, this can be effective in delaying epidemics; this is in fact being done by farmers in the tropics, and probably also in intensive vegetable plots in China. Without cheap labour, disease control in intensified cropping will have to rely greatly on the use of resistant varieties and of fungicides.

3.7.3 Widening the Geographic Range of Valuable Crops

The classic example of a crop spreading to regions it had not been expected to reach is that of maize. Once considered a warm-climate crop, maize is now grown in Europe and America in distinctly temperate climates. This has been achieved by the breeding of varieties adapted to cooler conditions. It has also raised certain disease problems, especially where both maize and wheat are grown in the same area, since both crops are susceptible to diseases such as *Fusarium culmorum* and certain viruses. What has been done for maize, and to a certain extent for potatoes, can probably also be done for other crops. Other directions of breeding can also be envisaged. For example, it could be attempted to improve shallow-rooted crops, which are highly sensitive to moisture fluctuations, to develop deeper-going root systems, that may make it possible to grow them in areas with irregular rainfall, or to breed crops for tolerance rather than for resistance to pathogens (Mussell and Malone 1979), where new pathogens or pathogen races, adaptable to change, may be expected to attack a crop in a new environment.

3.7.4 Education of Farmers to Employ Cultural Practices for Disease Control

The education of farmers is crucial for the use of cultural practices for disease control. Of all the means of crop protection, proper use of cultural measures demands most understanding and experience. As outlined in the chapter on economic aspects (2.1), decision-making is not very complicated for the farmer where resistant varieties are concerned, as these do not generally require important changes in growing techniques. Use of fungicides demands a little more discernment, because they may be expensive, but their application is a simple process that requires little training.

To be capable of using cultural practices in controlling crop disease, the farmer needs experience and judgement. The details of their application depend very much on the specific conditions of the fields or orchards, and they cannot be applied by rule of thumb. In developing countries farmers have the experience to grow traditional crops, sometimes even with sophisticated techniques, such as those required to build and maintain underground irrigation systems in Iran and elsewhere. However, intensified crop protection in the future cannot rely on traditional crops alone, cannot be content with the levels of yield now obtained, and will certainly involve new techniques of fertilization, irrigation, crop protection etc. Adapting tradition-bound farmers to new techniques is a process of considerable complexity and social significance.

The best, and sometimes the only, way to educate farmers to this end – including large populations of illiterate or semi-literate men and women – is the institution of an efficient advisory service. This is true for all agricultural development, and applies fully to the use of cultural measures for crop protection. It involves formation of a crops of farm advisors who can disseminate information by all the media, and demonstrate new techniques in the field.

The future use of cultural practices in world food production, including the protection of crops in the field and store, lies to a large extent in the success of such educational efforts.

French, German and Spanish Translation of Some of the English Terms Used in This Book

English	French	German	Spanish
Ball of earth	Motte	Wurzelballen	Cepellón
Booting (of cereals)	Gonflement	Schwellen der Ähre	Hinchamiento
Calyx	Calice	Blütenkelch	Cáliz
Carnation	Oeillet	Nelke	Clavel
Combine harvester	Moissoneuse-batteuse	Mähdrescher	Cosechadora
Cutting	Bouture	Steckling	Estaca
Damping-off (of seedlings)	Fonte des semis	Keimlingskrankheit	Enfermedad de los semilleros
Deciduous	À feuilles caduques	Laubabwerfend	Caducifolio
Decoy plant	Plante piège	Fangpflanze	Planta-trampa
Deficiency of nutrients	Carence en substances nutritives	Nährstoffmangel	Deficiencia de elementos nutritivos
Downy mildew	Mildiou	Falscher Mehltau	Mildiú falso (falso oidio)
Drip irrigation	– see trickling		
Ear emergence	Épiaison	Ährenschieben	Espigazón
Earth up	Butter	Anhäufeln	Aporcar
Eggplant	Aubergine	Eierfrucht	Berenjena
Flag leaf (of cereals)	Dernière feuille	Fahnenblatt	Hoja superior
Flax	Lin	Flachs, Lein	Lino
Flooding, irrigation by	Irrigation par déversement	Bewässerung durch Überschwemmung	Riego por inundación
Furrow	Sillon	Furche	Surco
Garlic	Ail	Knoblauch	Ajo
Gooseberry	Groseiller	Stachelbeere	Grosella espinosa
Graft, to	Greffer	Pfropfen	Injertar
Grapefruit	Pamplemousse	Pampelmuse	Pampelmusa
Harrow	Herse	Egge	Rastra
Haulm-killing (potatoes)	Dêfanage	Kartoffelkraut-abtötung	Defoliamiento
Inarch (graft)	Greffer par approche	Ablaktieren	Injerto de aproximación
Incorporation (in soil)	Enfouissage	Unterpflügen	Incorporar
Isolate (of fungus, bacterium or virus)	Souche	Stamm	
Knapsack sprayer	Pulvérisateur à dos	Rückenspritze	Pulverizador de mochila
Lawn	Gazon, pelouse	Rasen	Césped
Minor element	Oligoélément	Spurenelement	Microelemento
Misting	Brumisation	Nebeln	Nebulización
Mow	Faucher	Mähen	Segar
Nozzle	Buse	Düse	Boquilla
Overhead irrigation – see sprinkling			
Pineapple	Ananas	Ananas	Piña
Ploughing	Labour	Pflügen	Arar

English	French	German	Spanish
Pod (bot.)	Gousse	Hülse	Vaina
Prune, to	Tailler	Beschneiden	Podar
Raspberry	Framboise	Himbeere	Frambuesa
Rape	Colza	Raps	Colza
Roguing	Arrachage	Entfernung kranker Pflanzen	Extirpar
Root collar	Collet de la racine	Wurzelhals	Cuello de la raíz
Root-stock	Porte-greffe	Unterlage	Portainjertos
Scion	Greffon	Edelreis	Injerto
Shooting (cereals)	Montaison	Schossen	Crecimiento de los tallos
Slope	Pente	Abhang	Pendiente
Soil sickness	Fatigue du sol	Bodenmüdigkeit	Fatiga del suelo
Soil suppressiveness	Résistance du sol		Resistencia del suelo
Sprinkling	Aspersion	Beregnung	Riego por aspersión
Splashing	Éclaboussure	Spritzen	Salpicar
Sepal	Sépale	Kelchblatt	Sépalo
Stone fruit	Drupe	Steinfrucht	Fruto de hueso
Strawberry	Fraise	Erdbeere	Fresa
Stunting	Rabougrissement	Verkümmern	Depresión vegetativa
Sunflower	Tournesol	Sonnenblume	Girasol
Threshold	Seuil	Schwelle	Umbral
Thin, to	Éclaircir	Lichten	Ralear
Tiller (cereals)	Taller	Bestocken	Hacer pies
Top-dressing	Engrais en couverture	Kopfdüngung	Abono de cubertura
Trickle irrigation	Irrigation goutte à goutte	Tropfenbewässerung	Riego por goteo
Tuber	Tubercule	Knolle	Tubérculo
Volunteer plant	Repousse	Auswuchs, Ausfallsgetreide	Rebrote
Windbreak	Brise-vent	Windschutz -(pflanzung)	Rompevientos

References

Abia JA, Hess WM, Smith BN (1977) Increased susceptibility of Mg- and Mn-deficient pumpkin seedlings to fungal attack. Naturwissenschaften 64:437–438

A'Brook J (1964) The effect of planting date and spacing on the incidence of groundnut rosette disease and of the vector *Aphis craccivora* Koch, at Mokwa, Northern Nigeria. Ann Appl Biol 54:199–208

Abu Salih HS, Ishag HM, Siddig SA (1973) Effect of sowing date on incidence of Sudanese broad bean mosaic virus in, and yield of, *Vicia faba*. Ann Appl Biol 74:371–378

Adams PB (1975) Factors affecting survival of *Sclerotinia sclerotiorum* in soil. Plant Dis Rep 59:599–603

Adeniji MO, Obigbesan GO (1976) The effect of potassium nutrition on the bacterial wilt of cassava. Nigerian J Plant Protect 2:1–3

Aderungboye FO (1973) The blast disease of oilpalm seedlings in Nigeria. 2nd Int Congr Plant Pathol, Minneapolis, Abstr no 0737

Adlerz WC (1972) *Momordica charantia* as a source of watermelon mosaic virus 1 for cucurbit crops in Palm Beach County, Florida. Plant Dis Rept 56:563–564

Afek U, Sztejnberg A (1980) Perennial infections of *Tranzschelia pruni-spinosae* var. *discolor* on *Anemone coronaria* and its control. Hassadeh 60: (in press) (Hebrew, English summary)

Ahenkorah Y, Akrofi GS, Adri AK (1974) The end of the first cocoa shade and manurial experiment at the Cocoa Research Institute of Ghana. J Hortic Sci 49:43–51

Alabouvette C, Rouxel F, Louvet J (1979) Characteristics of *Fusarium* wilt-suppressive soils and prospects for their utilization in biological control. In: Schippers B, Gams W (eds) Soil borne plant pathogens. Academic Press, London New York San Francisco, pp 165–182

All JN, Kuhn CW, Gallaher RN, Jellum MD, Hussey RS (1977) Influence of no-tillage cropping, carbofuran, and hybrid resistance on dynamics of maize chlorotic dwarf and maize dwarf mosaic diseases of corn. J Econ Entomol 70:221–225

Altman J, Campbell CL (1977) Effect of herbicides on plant diseases. Annu Rev Phytopathol 15:361–385

Anon (1948) Gooseberry cluster-cup rust. Adv Leafl 198:Min Agric UK, 3 pp

Anon (1961) *Armillaria* root rot. Adv Leafl 500:Min Agric UK, 5 pp

Anon (1965a) Brown rot and blossom wilt of apples. Adv Leafl 155:Min Agric UK, 7 pp

Anon (1965b) Virus Yellows of sugar beet. Adv Leafl 323:Min Agric UK, 6 pp

Anon (1968) Reduction of nematode populations through land management and cultural practices. In: Control of plant-parasitic nematodes, vol IV of principles of plant and animal pest control. Natl Acad Sci Publ 1969, Washington, pp 94–105

Anon (1969a) Chrysanthemum eelworm. Adv Leafl 339:Min Agric UK, 6 pp

Anon (1969b) Fireblight of apples and pears. Adv Leafl 571:Min Agric UK, 10 pp

Anon (1970a) Pest control on rice. PANS Manual no 3:270 pp

Anon (1970b) Les maladies des plantes. ACTA Journ Fr Etudes Inf 1969, Paris, 745 pp

Anon (1972) Stem eelworm on tulip. Adv Leafl 461:Min Agric UK, 4 pp

Anon (1973a) Eelworms on strawberry. Adv Leafl 414:Min Agric UK, 7 pp

Anon (1973b) Lettuce downy mildew. Adv Leafl 577:Min Agric UK, 3 pp

Anon (1973c) Rapport Annuel 1973, IRAT, Paris, 328 pp

Anon (1973d) A compendium of corn diseases. Am Phytopathol Soc St Paul, 64 pp

Anon (1974) Stem eelworms on narcissus. Adv Leafl 460:Min Agric UK, 7 pp

Anon (1975) Stem eelworm on vegetables. Adv Leafl 440:Min Agric UK, 7 pp

Anon (1976a) Stem eelworm on arable crops. Adv Leafl 178:Min Agric UK, 7 pp

Anon (1976b) Potato and tomato blight. Adv Leafl 271:Min Agric UK, 6 pp

Anon (1977a) Cyst eelworms on potato. Adv Leafl 284:Min Agric UK, 8 pp

Anon (1977b) Parfuempflanzenbau in Tunesien. Publ GTZ, Eschborn, Germany
Anon (1978) Cereal aphids. Adv Leafl 586:Min Agric UK, 6 pp
Anon (1979) Anleitung zur Schadenerregerüberwachung im Rahmen des Pflanzenschutzwarndienstes. Bayer Landesanst Bodenkul
Armstrong GM, Armstrong JK (1948) Non-susceptible hosts as carriers of wilt Fusaria. Phytopathology 38:808–826
Arnon I (1972) Crop production in dry regions, vol I. Background and principles. Barnes and Noble Books
Ashworth LJ (1974) Crop rotation for control of *Verticillium albo-atrum* in central California. Abstr Proc Am Phytopathol Soc 1:101
Aust HJ, Bashi E, Rotem J (1980) Flexibility of plant pathogens in exploiting ecological and biotic conditions in the development of epidemics. In: Palti J, Kranz J (eds) Comparative epidemiology. PUDOC, Wageningen, pp 46–56
Avizohar-Hershenzon Zehara (1962) The influence of some summer crops on the sclerotial population of *Sclerotium rolfsii* in the soil. Tech Rep 384; Agric Res Org Israel, 5 pp (Hebrew, English summary)
Avizohar-Hershenzon Zehara, Palti J (1962) *Sclerotium rolfsii* in Israel. Plant Dis Rep 46:418–423
Avizohar-Hershenzon Zehara, Lachover D, Ruderig H (1969) Survival of sclerotia of *Sclerotium rolfsii* Sacc. in cattle manure. Phytopathol Mediterranea 8:171–174
Aycock R (1966) Stem rot and other diseases caused by *Sclerotium rolfsii.* N C Agric Exp Stn Tech Bull 174:202 pp
Ayres PG (1977) Effect of powdery mildew *Erysiphe pisi* and water stress upon the water relations of pea. Physiol Plant Pathol 10:139–145
Backman PA, Rodriguez-Kabana R (1975) A system for the growth and delivery of biological control agents to the soil. Phytopathology 65:819–821
Bains SS, Jhooty JS, Sokhi SS, Rental HS (1978) Rôle of *Digitaria sanguinalis* in outbreaks of brown stripe downy mildew of maize. Plant Dis Rep 62:143
Baker KF (1949) Observations on some *Botrytis* diseases in California. Plant Dis Rep 30:145–155
Baker KF, Cook RJ (1974) Biological control of plant pathogens. WH Freeman and Co, San Francisco, 433 pp
Balaz F, Maric A, Markovicz Z (1978) Influence of some agrotechnical measures on incidence and harmfulness of *Fusarium* stalk rot of maize. Zast Bilja 28:63–73 (Yugoslav, English summary)
Bar-Joseph M, Frankel H (1979) Spraying lime plants with clay suspensions prevents their colonization by aphids. Hassadeh 60:65–67 (Hebrew, with English summary)
Barnes EH (1964) Changing plant disease losses in a changing agriculture. Phytopathology 54:1314–1319
Bartz JA, Crill JP (1978) The postharvest susceptibility of green, pink, and red tomato fruit to *Erwinia carotovora* var. *carotovora.* Plant Dis Rep 62:582–585
Baxter LW, Witcher W, Fagan SG, Owen MG (1977) Control of southern stem blight of tomatoes by physical means. Plant Dis Rept 61:341–342
Bayon F, Ayrault J-P (1977) Importance de la jaunisse nanissante de l'orge (JNO ou BYDV) en Poitou-Charentes: incidence économique, nuisibilité et approche des moyens pratiques de lutte. Ann Phytopathol 9:235–239
Belcher JV, Hussey RS (1977) Influence of *Tagetes patula* and *Arachis hypogaea* on *Meloidogyne incognita.* Plant Dis Rep 61:525–528
Bell TS (1980) Stunt disorders in arable crops. Min Agric UK, A.D.A.S, Aerial Photography Unit; Progress Report on Investigations in Eastern and East-Midland Regions in 1979. Multilithed 6 pp and 6 plates
Benda GTA (1973) Heat adaptation of sugarcane and heat cure of its virus diseases. 2nd Int Congr Plant Pathol Minneapolis, Abstr 0900
Benson DM (1974) Fate of *Verticillium albo-atrum* in the rhizosphere and rhizoplane of alfalfa, barley, sorghum, and corn. Proc Am Phytopathol Soc 1:102
Béreau M, Messiaen CM (1975) Réceptivité comparée des sols à l'infestation par *Pseudomonas solanacearum.* Ann Phytopathol 7:191–193
Berger RD (1973) Early blight of celery: analysis of disease spread in Florida. Phytopathology 63:1161–1165
Bernaux P (1976) La Piriculariose en 1975. Rizic Fr 163:15–17

Bertrand PF, English H, Uriu K, Schick FJ (1976) Late season water deficits and development of *Cytospora* canker in French Prune. Phytopathology 66:1318–1320

Beugnon M, Brun J, Melin Ph (1970) La lutte contre *Trachysphaera fructigena* (Tabor et Bunting), parasite des bananes au Cameroun. Fruits 25:187–197

Beyries A (1974) Le greffage des Solanacees maraîchères. Pépinieristes Hortic Maraichèrs 152:27–32

Beyries A (1979) Le greffage, moyen du lutte contre les parasites telluriques des Solanées cultivées pour leur fruits. Thèse, Univ du Langedoc, Acad Montpellier, 166 pp

Bhargava B, Bhargava KS, Joshi RD (1975) Perpetuation of watermelon mosaic virus in eastern Uttar Pradesh, India. Plant Dis Rep 59:634–636

Bird GW, Hogger Ch (1973) Nutsedges as hosts of plant-parasitic nematodes in Georgia cotton fields. Plant Dis Rep 57:402

Bird GW, Brooks DL, Perry CE, Futral JG, Canerday TD, Boswell FC (1974) Influence of subsoiling and soil fumigation on the cotton stunt disease complex, *Hoplolaimus columbus* and *Meloidogyne incognita*. Plant Dis Rep 58:541–544

Black CA (1968) Soil-plant relationships, 2nd edn. John Wiley and Sons Inc, New York London Sydney, 792 pp

Blanc D, Tramier R, Montarone M, Bettachini A (1979) Influence de la nutrition calciques sur les fusarioses vasculaires de l'oeillet, du melon et de la Reine-Marguerite. Ann Phytopathol 11:132–133

Bliss DE (1946) The relation of soil temperature to the development of *Armillaria mellea* root rot. Phytopathology 36:302–318

Boccas B, Laville E (1978) Les maladies à *Phytophthora* des agrumes. Inst Rech Fruits Agrumes (IRFA), 162 pp

Bochow H (1975) Phytosanitäre Auswirkungen der Beregnung in der Gemüseproduktion. Gartenbau 22:155–156

Booker RH (1963) The effect of sowing date and spacing on rosette disease of groundnut in Northern Nigeria, with observations on the vector *Aphis craccivora*. Ann Appl Biol 52:125–131

Boosalis MG, Colville WC, Sumner DR (1969) Effect of intensive cultural practice on soil-borne and related corn diseases. In: Browning JA (ed) Disease consequences of intensive and extensive culture of field crops. Iowa State Exp Stn Spec Rep 64:12–20

Boughey AS (1947) The effect of rainfall on plant disease distribution in the Anglo-Egyptian Sudan. Imp Mycol Inst Mycol Pap no 19; 13 pp, Kew Surrey

Bouhot D (1979) Un test biologique à deux niveaux pour l'étude des fatigues du sol. Application à l'étude des nécroses des racines de céleri-rave. Ann Phytopathol 11:95–109

Bouhot D, Perrin R (1980) Mise en évidence de resistance biologique aux *Pythium* en sol forestier. Eur J Forest Pathol 10:77–90

Bowen GD (1978) Dysfunction and shortfalls in symbiotic responses. In: Horsfall JG, Cowling EB (eds) Plant disease, vol III. Academic Press, London New York San Francisco, pp 231–256

Brathwaite CWD (1973) Effect of crop sequence and fallow on populations of *Rotylenchulus reniformis* in fumigated and untreated soil. Plant Dis Rep 58:259–261

Bridge J (1972) Nematode problems with yams (*Dioscorea* spp.) in Nigeria. PANS 18:89–91

Broadbent L (1969) Disease control through vector control. In: Maramorosch K (ed) Viruses, vectors, and vegetation. Interscience Publishers, New York, pp 593–630

Broadbent P, Baker KF (1975) Soil suppressive to *Phytophthora* root rot in Eastern Australia. In: Bruehl GW (ed) Biology and control of soil-borne pathogens. Am Phytopathol Soc St Paul Minn, pp 152–157

Brodie BB, Murphy WS (1975) Population dynamics of plant nematodes as affected by combinations of fallow and cropping sequences. J Nematol 7:91–92

Brown JG (1937) Relation of livestock to the control of sclerotiniosis of lettuce. Phytopathology 27:1045–1050

Browning JA (1974) Relevance of knowledge about natural ecosystems to development of pest management programs for agro-ecosystems. Proc Am Phytopathol Soc 1:191–199

Browning JA, Simons MD, Frey KJ, Murphy HC (1969) Regional deployment of oat crown-rust resistant genes. In: Browning JA (ed) Disease consequences of intensive and extensive culture of field crops. Iowa Agr Exp Sta Spec Rept 64:49–56

Bruehl GW (1975) Systems and mechanisms of residue possession by pioneer fungal colonists. In: Bruehl GW (ed) Biology and control of soil-borne plant pathogens. Am Phytopathol Soc, St Paul Minn, pp 77–83

Bruehl GW, Sprague R, Fisher WR, Nagamitsu M, Nelson WL, Vogel OA (1966) Snow molds of winter wheat in Washington. Wash Agric Exp Stn Bull 677:1–21
Buchholtz WF (1944) Crop rotation and soil drainage effects on sugar beet tip rot and susceptibility of other crops to *Aphanomyces cochlioides*. Phytopathology 34:805–812
Burke DW (1965a) Plant spacing and *Fusarium* root rot of bean. Phytopathology 55:757–759
Burke DW (1965b) *Fusarium* root rot of beans and behavior of the pathogen in different soils. Phytopathology 55:1122–1126
Burke DW, Miller DE, Holmes LD, Barker AW (1972) Counteracting bean root rot by loosening the soil. Phytopathology 62:306–309
Burleigh JR, Roelfs AP, Eversmeyer MG (1972) Estimating damage to wheat caused by *Puccinia recondita tritici*. Phytopathology 62:944–946
Busch LV, Smith EA, Njoh-Elango F (1978) The effect of weeds on the value of rotation as a practical control for *Verticillium* wilt of potato. Can Plant Dis Surv 58:61–64
Butt DJ (1978) Epidemiology of powdery mildews. In: Spencer DM (ed) The powdery mildews. Academic Press, London New York San Francisco, pp 51–81
Butterfield, EJ, DeVay JE, Pullman J, McCutcheon OD, Garber RH (1976) Soil flooding and short-term crop rotations for control and management of *Verticillium* wilt in cotton. 2nd Int Verticillium Symp California, pp 11–12
Butterfield EJ, DeVay JE, Garber RH (1978) The influence of several crop sequences on the incidence of *Verticillium* wilt of cotton and on the population of *Verticillium dahliae* in field soil. Phytopathology 68: 1217–1220
Byford WJ, Hull R (1967) Some observations on the economic importance of sugar beet downy mildew in England. Ann Appl Biol 60:281–296
Cameron HR, Milbrath JA, Tate LA (1973) Pollen transmission of Prunus Ringspot Virus in prune and sour cherry orchards. Plant Dis Rep 57:241–243
Cardin M-C (1979) Influence of temperature on the relationships of *Meloidogyne hapla* with cucumber roots. Rev Nematol 2:169–175
Carlson GA (1979) Pest control risks in agriculture. In: Roumasset JA, Boussard JM, Singh I (eds) Risk uncertainty and economic development. Agric Dev Council, New York, pp 199–200
Carlson GA, Main CE (1976) Economics of disease-loss management. Annu Rev Phytopathol 14:381–403
Carter WW, Nieto S Jr (1975) Population development of *Meloidogyne incognita* as influenced by crop rotation and fallow. Plant Dis Rep 59:402–403
Cassini R, Cassini R, Pauvert P (1975) La lutte intégrée dans les rotations des céréalières. Bull Tech Inf 297:201–210
Chamberlain DW, Gerdemann JW (1966) Heat-induced susceptibility of soybeans to *Phytophthora megasperma* var. *sojae*, *Phytophthora cactorum* and *Helminthosporium sativum*. Phytopathology 56:70–73
Cheo PC, Pound GS, Weathers LG (1952) The relation of host nutrition to the concentration of cucumber virus I in spinach. Phytopathology 42:377–381
Chet I, Hadar Y, Elad Y, Katan J, Henis Y (1979) Biological control of soil-borne plant pathogens by *Trichoderma harzianum*. In: Schippers B, Gams W (eds) Soil-borne plant pathogens. Academic Press, London New York San Francisco, pp 585–591
Cohen S (1981) Reducing the spread of aphid-transmitted viruses in peppers by coarse-net cover. Phytoparasitica 9:69–76
Cohen S, Marco S (1973) Reducing the spread of aphis-transmitted viruses in peppers by trapping the aphids on sticky yellow polyethylene sheets. Phytopathology 63:1207–1209
Cohen S, Melamed-Madjar Venezia (1978) Prevention by soil mulching of the spread of tomato yellow leaf curl virus transmitted by *Bemisia tabaci* (Gennadius) (Hemiptera: Aleyrodidae) in Israel. Bull Entomol Res 68:465–470
Cohen S, Melamed-Madjar Venezia, Hameiri J (1974) Prevention of spread of tomato yellow leaf curl virus transmitted by *Bemisia tabaci* (Gennadius) (Homoptera, Aleyrodidae) in Israel. Bull Ent Res 64:193–197
Cohen Y, Sherman Y (1977) The role of airborne conidia in epiphytotics of *Sclerospora sorghi* on sweet corn. Phytopathology 67:515–521
Cohen Y, Perl M, Rotem J, Eyal Helena, Cohen J (1974) Ultrastructure and physiological changes in sporangia of *Pseudoperonospora cubensis* and *Phytophthora infestans* to water stress. Can J Bot 52:447–450

Cohn E (1976) Report on investigations on nematodes of citrus and subtropical fruit crops in South Africa 1975/76. Citrus and Subtropical Fruit Research Inst, Nelspruit, S Africa, multilithed 43 pp

Colbran RC (1979) Problems in tree replacement. III. The effects of cultural practices and soil fumigation on root lesion nematodes and on the growth of apple replants. Aust J Agric Res 30:113–123

Colhoun J (1961) Spore load, light intensity and plant nutrition as factors influencing the incidence of club-root of Brassicae. Trans Br Mycol Soc 44:593–600

Colhoun J (1973) Effects of environmental factors on plant disease. Annu Rev Phytopathol 11:343–364

Conroy JJ, Green RJ (1974) Interactions of the root-knot nematode *Meloidogyne incognita* and the stubby root nematode *Trichodorus christiei* with *Verticillium albo-atrum* on tomato at controlled inoculum densities. Phytopathology 64:1118–1121

Cook RJ (1973) Influence of low plant and soil water potentials on disease caused by soil-borne fungi. Phytopathology 63:451–458

Cook RJ (1977) Management of the associated microbiota. In: Horsfall JG, Cowling EB (eds) Plant disease, an advanced treatise, vol I. Academic Press, London New York San Francisco, pp 146–166

Cook RJ (1979) Antagonism and biological control: concluding remarks. In: Schippers B, Gams W (eds) Soil-borne plant pathogens. Academic Press, London, New York, San Francisco, pp 653–657

Cook RJ, Papendick RI (1970) Effect of soil water on microbial growth, antagonism and nutrient availability in relation to soil-borne fungal diseases of plants. In: Toussoun TA, Bega RV, Nelson PE (eds) Root diseases and soil-borne pathogens. Univ Calif Press, Berkeley Los Angeles London, pp 81–89

Cook RJ, Papendick RI (1972) Influence of water potential of soils and plants on root disease. Annu Rev Phytopathol 10:349–374

Cook RJ, Boosalis MG, Doupnik B (1978) Influence of crop residues on plant diseases. In: Crop residue management systems. Am Soc Agron Spec Publ 31, Madison Wis, pp 147–163

Costa AS (1975) Increase in the populational density of *Bemisia tabaci*, a threat of widespread virus infection of legume crops in Brazil. In: Bird J, Maramorosch K (eds) Tropical diseases of legumes. Academic Press, London New York San Francisco, pp 27–45

Couch HB, Bloom JR (1960) Influence of environment on diseases of turfgrasses. II. Effect of nutrition, pH and soil moisture on *Sclerotinia* dollar spot. Phytopathology 50:761–763

Cowling EB (1978) Agricultural and forest practices that favor epidemics. In: Horsfall JG, Cowling EB (eds) Plant disease, an advanced treatise, vol II. Academic Press, London New York San Francisco, pp 361–381

Crowther FM, Cochran WG (1942) Rotation experiments with cotton in the Sudan Gezira. J Agric Sci 32:390–405

Croxall HE, Smith LP (1976) The epidemiology of potato blight in the East Midlands 1923–1974. Ann Appl Biol 82:451–466

Cunfer BM, Touchton J (1979) Effect of phosphorus and potassium fertilization on glume blotch of wheat in a minimum tillage system. Ninth Int Congr Plant Protect, Washington, Abstr 562

Cunnungham PC (1975) Some consequences of cereal monoculture on *Gaeumannomyces graminis* (Sacc.) Arx and Olivier and take-all disease. EPPO Bull 3:297–317

Curl EA (1963) Control of plant diseases by crop rotation. Bot Rev 29:413–479

Dadant R (1963) Contribution à l'etude de *Clitocybe tabescens* et de ses relations avec *Trichoderma viride*. Agron Trop 18:265–329

Dakwa JT, Danquah O-A (1978) A *Colletotrichum* leaf blight of cocoa in Ghana. Plant Dis Rep 62:369–373

Damsteegt VD (1976) A naturally occurring corn virus epiphytotic. Plant Dis Rep 60:858–861

Dange SRS, Jain KL, Siradhana BS, Rathore RS (1974) Perpetuation of sorghum downy mildew (*Sclerospora sorghi*) of maize on *Heteropogon contortus* in Rajasthan, India. Plant Dis Rep 58:285–286

Dasgupta MK, Chattopadhyay SB (1977) Effect of different doses of N and P on the susceptibility of rice to brown spot caused by *Helminthosporium oryzae*. Z Pflanzenkrankh Pflanzenschutz 84:276–285

Davis JR, Dole RE, Callihan RH (1976) Fertilizer effects on common scab of potato and the relation of calcium and phosphate-phosphorus. Phytopathology 66:1236–1241

De Boodt M (1979) Soil conditioning for better soil management. Outlook Agric 10:63–70

Dengel H-J (1978) Untersuchung über das Maniokmosaik im Süden Togos. Ph D Diss, Univ Gießen, Germany, 140 pp

Dhavantari BN (1978) Cold predisposition of dormant peach twigs to nodal cankers caused by *Leucostoma* spp. Phytopathology 68:1779–1783

Dhingra OD, Sinclair JB (1977) An annotated bibliography of *Macrophomina phaseolina* 1905–1975. Univ Fed Viscosa, Brazil, and Univ Illinois, Urbana-Champaing USA, 244 pp

Dhingra OD, Sinclair JB (1978) Biology and pathology of *Macrophomina phaseolina*. Imp Univers Univ Fed Vicosa, Brazil, 166 pp

Dickson JG (1923) Influence of soil temperature and moisture on the development of the seedling blight of wheat and corn caused by *Gibberella saubinetti*. J Agric Res 23:837–870

Diener UL, Morgan-Jones G, Hagler WM, Davis ND (1976) Mycoflora of activated sewage sludge. Mycopathology 58:115–116

Diercks R (1975) Cultural measures. Crop protection and economics in cereal monocultures in Bavaria. EPPO Bull 5:153–168

Dinoor A (1980) Control of *Septoria* leaf blotch in wheat by means other than disease resistance. Proc 5th Congr Mediterranean Phytopath Patras p 174

Dinoor A (1981) Epidemics caused by fungal pathogens in wild and crop plants. In: Thresh JM (ed) Pests, pathogens and vegetation. Pitman Publications, London, pp 143–158

Domsch KH, Gams W, Andersen T-H (1980) Compendium of soil fungi, vols I, II. Academic Press, New York

Douine L, Quiot JB, Marchoux G, Archange P (1979) Recensement des espèces végétales sensibles au Virus de la Mosaique de Concombre (CMV). Étude bibliographique. Ann Phytopathol 11:439–475

Doupnik B, Boosalis MG (1980) Ecofallow – a reduced tillage system – and plant diseases. Plant Dis 64:31–35

Duckham AN, Masefield GB (1970) Farming systems of the world. Chatto and Windhus Ltd, London, 542 pp

Duniway JM (1973) Pathogen-induced changes in host-water relations. Phytopathology 63:458–466

Duniway JM (1977) Predisposing effects of water stress on the severity of *Phytophthora* root rot in safflower. Phytopathology 67:884–889

Dunn RA, Mai WF (1973) Reproduction of *Pratylenchus penetrans* in roots of seven cover crop species in greenhouse experiments. Plant Dis Rep 57:728–730

Duvdevani S, Reichert I, Palti J (1946) The development of downy and powdery mildew of cucumbers as related to dew and other environmental factors. Palest J Bot Rehovot Ser 5:127–151

Easton G (1976) Late blight of potatoes under center pivot sprinkler irrigation in arid South Central Washington. Am Potato J 53:400–401

Easton DG, Nagle ME (1977) Tuber-rot – the result of ponding by irrigation water. Plant Dis Rep 61:1064–1066

Eden-Green SJ, Billing Eve (1974) Fireblight. Rev Plant Pathol 53:353–366

Edmunds LK (1964) Combined relation of plant maturity, temperature, and soil moisture to charcoal stalk rot development in grain sorghum. Phytopathology 54:514–517

Edmunds LK, Voigt RL (1966) Role of seed production in predisposition of sorghum to charcoal rot. Phytopathology 56:876

Edwards DF (1978) Studies on the manipulation of the timing of crop maturity of cocoa in Ecuador in relation to losses from pod diseases. J Hortic Sci 53:243–254

Ehrenpfordt V (1976) Ergebnisse zur Wirkung von Fruchtfolge- und Intensivierungsmaßnahmen auf die Erträge im konzentrierten Kartoffelanbau auf leichtem Boden. Tagungsber Akad Landwirtsch Wiss DDR Berlin 148:229–240

El-Behadli AH, Al-Hassan SA (1979) Effect of seeding dates on severity of damping-off and root rot of sugar beet. 9th Int Congr Plant Protect, Washington, Abstr 553

Ellis MB, Waller JM (1974) *Sclerotinia fuckeliana* (conidial state: *Botrytis cinerea*). C.M.I. descriptions of pathogenic fungi and bacteria no 431:2 pp

Endo RM (1972) The turfgrass community as an environment for the development of facultative fungal parasites. In: Younger VB, McKell CM (eds) The biology and utilization of grasses. Academic Press, London New York San Francisco, pp 171–202

Endo RM, Amacher RH (1964) Influence of guttation fluid on infection structures of *Helminthosporium sorokinianum*. Phytopathology 54:1327–1334

Erinle ID, Edmunds LK (1970) Effect of nitrogen on development of seedling disease caused by *Fusarium moniliforme* in sorghum. Phytopathology 60:1291

Erwin DC, Tsai SD, Khan RA (1979) Growth retardants mitigate *Verticillium* wilt and influence yield of cotton. Phytopathology 69:283–287
Eshed N, Wahl I (1975) Role of wild grasses in epidemics of powdery mildew on small grains in Israel. Phytopathology 65:57–63
Eversmeyer MG, Skidmore EL (1974) Wheat leaf and stem rust development near a wind barrier. Plant Dis Rep 58:459–463
Farkas GL (1978) Senescence and plant disease. In: Horsfall JG, Cowling EB (eds) Plant disease, an advanced treatise, vol III, pp 391–412
Farrell JAK (1976) Effects of intersowing with beans in the spread of groundnut rosette virus by *Aphis craccivora* Koch (Hemiptera, Aphididae) in Malawi. Bull Entomol Res 66:331–333
Feakin SD (ed) (1971) Pest control in bananas. PANS manual no 1, new edn, London, 128 pp
Felton MW, Walker JC (1946) Environal factors affecting downy mildew of cabbage. J Agric Res 72:69–81
Fenne SB (1941) Tobacco black shank is spreading in Virginia. Plant Dis Rep 25:534–535
Fett WF (1978) Volunteer soybeans: survival sites for soybean pathogens between seasons in Southern Brazil. Plant Dis Rep 62:1013–1016
Filho ES, Dhingra OD (1980) Survival of *Macrophomina phaseolina* sclerotia in nitrogen amended soils. Phytopathol Z 97:136–143
Fletcher JT (1962) Experiments in the control of mint rust. Plant Pathol 11:115–120
Foster RE, Walker JC (1947) Predisposition of tomato to *Fusarium* wilt. J Agric Res 74:165–185
Fox RA (1970) A comparison of methods of dispersal, survival and parasitism in some fungi causing root diseases of tropical plantation crops. In: Toussoun TA, Bega RV, Nelson PE (eds) Root diseases and soil-borne pathogens. Univ Calif Press, Berkeley Los Angeles London, pp 179–187
Frank JA, Leach SS (1976) *Rhizoctonia* damage on potatoes as influenced by alternate cropping with buckwheat. Am Potato J 53:401
Frank ZR (1972) Notes on soil management in relation to *Pythium* rot of peanut pods. Plant Dis Rep 56:600–601
Fuchs WH, Grossmann F (1972) Ernährung und Resistenz von Kulturpflanzen gegenüber Krankheitserregern und Schädlingen. In: Linser H (ed) Handbuch der Pflanzenernährung und Düngung. Springer, Berlin Heidelberg New York, pp 1006–1107
Furtick WR (1976) Implementing pest management programs: an international perspective. In: Apple JL, Smith RF (eds) Integrated pest management. Plenum Press, New York London
Furuya H, Owada M, Ui T (1979) A suppressive soil of common bean rot in Kitami District, Hokkaido. Ann Phytopathol Soc Jpn 45:608–617 (Japanese, English abstract)
Gaeumann E (1951) Pflanzliche Infektionslehre. Verlag Burkhaeuser, Basel, 681 pp
Garber RH, Carter L, DeVay JE, Miller J (1976) The effects of certain cultural practices on the incidence of *Verticillium* wilt of cotton. 2nd Int Verticillium Symp Calif, p 18
Garren KH (1961) Control of *Sclerotium rolfsii* through cultural practices. Phytopathology 51:120–124
Garrett SD (1970) Pathogenic root-infecting fungi. Cambridge Univ Press, Cambridge UK, 294 pp
Gassert WL (1977) Die Ausbreitung der Kaffeekirschenkrankheit (*Colletotrichum coffeanum* Noack sensu Hindorf) in Äthiopien. In: Agrarwissenschaftliche Forschung in den humiden Tropen, Gießener Beiträge zur Entwicklungsforschung, Reihe 1, Bd 3, Tropeninstitut, Gießen, pp 103–110
Gassert WL (1978) Rinde und mumifizierte Kirschen als Quelle für das Primärinokulum der Kaffeekirschenkrankheit in Äthiopien. Z Pflanzenkrankh Pflanzenschutz 85:30–40
Georgiev D, Dzumalieva D, Konstantinova K (1976) Anbau von Weizen in Monokultur mit und ohne Bewässerung. Tagungsber Akad Landwirtsch Wiss DDR Berlin 148:193–199
Gerechter-Amitai ZK (1973) Stem rust *Puccinia graminis* Pers. on cultivated and wild grasses in Israel. Ph D Thesis, Hebrew Univ Jerusalem, 269 pp
Ghaffar A, Erwin DC (1969) Effect of soil water stress on root rot of cotton caused by *Macrophomina phaseoli*. Phytopathology 59:795–797
Gibbs AF, Wilcoxson RD, Thomas HL (1973) The effect of nitrogen fertilization and mowing on *Helminthosporium* leaf spot and sugar content of bluegrass leaves. Plant Dis Rep 57:544–548
Gilman DF, Jones JE, Williams C, Birchfield W (1978) Cotton-soybean rotation for control of reniform nematodes. Lousiana Agric 21:10–11
Glaser HJ, Ogbogu FC (1974) Problems and experiences of plant protection extension in the control of cassava bacterial blight disease in East Central State of Nigeria. Proc 1 st Workshop on Cassava Bacterial Blight in Nigeria, pp 44–51

Golden JK, Van Gundy SD (1975) A disease complex of okra and tomato involving the nematode *Meloidogyne incognita* and the soil-inhabiting fungus *Rhizoctonia solani*. Phytopathology 65:265–273

Goldweber S (1975) The use of chelated iron to restore normal vigor to *Verticillium* wilt infected mango trees. Proc Fla State Hortic Soc 88:499–500

Gray PM, Guthrie JW (1977) The influence of sprinkler irrigation and post-harvest residue removal practices on the seed-borne population of *Drechslera poae* on *Poa pratensis* "Merion". Plant Dis Rep 61:90–93

Gray TRG (1976) Survival of vegetative microbes in soil. In: Gray TRG, Postgate JR (eds) The survival of vegetative microbes. Cambridge Univ Press, Cambridge London New York Melbourne, pp 327–364

Greaney FJ (1946) Influence of time, rate, and depth of seeding on the incidence of root rot in wheat. Phytopathology 36:252–263

Greb BW (1979) Reducing drought effects on croplands in the west-central Great Plains. USDA Inf Bull 420:31

Grinstein A, Orion D, Greenberger A, Katan J (1979) Solar heating of the soil for the control of *Verticillium dahliae* and *Pratylenchus thornei* in potatoes. In: Schippers B, Gams W (eds) Soil-borne pathogens. Academic Press, London New York San Francisco, pp 431–438

Gritton ET, Ebert RD (1975) Interaction of planting date and powdery mildew on pea plant performance. J Am Soc Hortic Sci 100:137–142

Grosclaude C, Dupouy J (1963) Rôle des „Cytospora“ dans certains dépérissements des rameaux du pêcher. Rév Zool Agricol Appl 4–6:1–10

Grossmann F (1967) Gründüngung als Pflanzenschutzmaßnahme. Z Pflanzenkrankh Pflanzenschutz 74:143–149

Grossmann F (1970) Einfluß der Ernährung der Pflanze auf den Befall durch Krankheitserreger und Schädlinge. Landwirtsch Forsch 23:Sonderh 25/1, 79–91

Gudmestad NC, Huguelet JE, Zink RT (1978) The effect of cultural practices and straw incorporation into the soil on *Rhizoctonia* disease of potato. Plant Dis Rep 62:985–989

Guerout R (1975) Nematodes of pineapple: a review. PANS 21:123–140

Guntz M (1959) Epidémiologie du mildiou de la pomme de terre. Eléments de prévision. Phytiatrie-Phytopharm 8:55–63

Gupta RN (1976) Impact of nitrogen, phosphorus and manganese on the stem gall disease of coriander. Acta Bot Indica 4:30–35

Haass J, Märtin B (1977) Untersuchungen zur Übertragung des Luzernewelkeerregers *Verticillium albo-atrum* Rke. et Berth. innerhalb des Pflanzenbestandes durch Befahren der Luzerne mit Ernteaggregaten und Transportfahrzeugen. Nachrichtenbl Pflanzenschutz DDR 31:240–243

Habicht WA (1975) The nematicidal effects of varied rates of raw and composted sewage sludge as soil organic amendments on a root-knot nematode. Plant Dis Rep 59:631–634

Hackney RW, Dickerson OW (1975) Marigold, castor bean, and chrysanthemum as controls of *Meloidogyne incognita* and *Pratylenchus alleni*. J Nematol 7:84–90

Hakkaart FA, Jordanova J (1968) Heat treatment experiments with carnations for the elimination of carnation mottle and etched ring viruses. Neth J Plant Pathol 74:146–149

Halbert SE, Irwin ME, Goodman RM (1979) Altering aphis behavior through barriers and row spacing variation: some implications for controlling soybean mosaic virus. 9th Int Congr Plant Protect, Washington, Abstr 125

Hanan JJ, Holley Wd, Goldsberry KL (1978) Greenhouse management. Springer, Berlin Heidelberg New York, 530 pp

Hancock JG (1979) Occurrence of soils suppressive to *Pythium ultimum*. In: Schippers B, Gams W (eds) Soil-borne plant pathogens. Academic Press, London New York San Francisco, pp 183–189

Hardison JR (1976) Fire and flame for plant disease control. Annu Rev Phytopathol 14:355–379

Harlan JR (1976) Diseases as a factor in plant evolution. Annu Rev Phytopathol 14:31–51

Harpaz I (1972) Maize rough dwarf. Isr Univ Press, Jerusalem, 251 pp

Harrison MD (1976) The effect of barley straw on the survival of *Verticillium albo-atrum* in naturally infested field soil. Am Potato J 53:385–394

Hassebrauk K, Roebbelen G (1975) Der Gelbrost, *Puccinia striiformis* West. IV. Epidemiologie. – Bekämpfungsmaßnahmen. Mitt Biol Bundesanst 164:1–183

Hawkins JE (1976) A cauterisation method for the control of cankers caused by *Pseudomonas syringae* in stone fruit trees. Plant Dis Rep 60:60–61

Heald CM, Burton GW (1968) Effect of organic and inorganic nitrogen on nematode populations on turf. Plant Dis Rep 52:46–48

Heale JB, Isaac I (1963) Wilt of lucerne caused by species of *Verticillium*. IV. Pathogenicity of *V. albo-atrum* and *V. dahliae* to lucerne and other crops; spread and survival of *V. albo-atrum* in soil and in weeds; effect upon lucerne production. Ann Appl Biol 52:439–451

Heale JB, Isaac I, Milton JM (1980) The administrative control of *Verticillium* wilt. In: Ebbels DL, King JE (eds) Plant health: the scientific basis for administrative control of plant diseases and pests. Blackwell Sci Publ, London, pp 71–79

Heathcote GD (1968) Protection of sugar-beet stecklings against aphids and viruses by cover crops and aluminum foil. Plant Pathol 17:158–161

Hendrix FF, Nielsen LW (1958) Invasion and infection of crops other than the forma suscept by *Fusarium oxysporum* f. *batata* and other formae. Phytopathology 48:224–228

Henis Y (1976) Effect of mineral nutrients on soil-borne pathogens and host resistance. In: Fertilizer use and plant health. Proc 12th Colloq Int Potash Inst, Bern

Henis Y, Chet I (1967) Mode of action of ammonia on *Sclerotium rolfsii*. Phytopathology 57:425–427

Henis Y, Katan J (1975) Effect of inorganic amendments and soil reaction on soil-borne plant diseases. In: Bruehl GW (ed) Biology and control of soil-borne plant diseases. Am Phytopathol Soc, St Paul Minn, pp 100–106

Henneberry TJ, George BW (1969) Economic entomology – a challenge to our ingenuity. Proc North Central Branch Entomol Soc Am 24:74–81

Herrera L, Seidel D (1978) Einfluß der Bewässerung und Bodenplanierung auf den Befall von Reispflanzen durch *Cochliobolus miyabeanus* (Ito et Kuribayashi) Drechsler ex Dastur. Arch Phytopathol Pflanzenschutz 14:263–265

Hijink MT, Winoto Suatmadji R (1967) Influence of different Compositae on population density of *Pratylenchus penetrans* and some other root-infesting nematodes. Neth J Plant Pathol 73:71–82

Hirata K (1966) Host range and geographical distribution of the powdery mildews. Niigata Univ, Niigata Japan, 472 p

Hirling W (1977) Ölrettich *(Raphanus oleiferus)* eine Feindpflanze für *Pratylenchus neglectus*. Z Pflanzenkrankh Pflanzenschutz 84:410–429

Hoeflich G, Steinbrenner K, Roth R (1977) Wirkung verschiedener Maßnahmen der organischen und mineralischen Düngung, der Bodenbearbeitung und der Beregnung auf den Befall des Getreides mit Fußkrankheiten. Arch Acker-Pflanzenbau Bodenkd Berlin 21:733–747

Hoitink HAJ (1980) Composted bark, a lightweight growth medium with fungicidal properties. Plant Dis 64:142–147

Hoppe PE (1949) Difference in *Pythium* injury to corn seedlings at high and low temperatures. Phytopathology 39:77–84

Horner CE (1965) Control of mint rust by propane gas flaming and contact herbicides. Plant Dis Rep 49:393–395

Horowitz M (1981) Chemical weed control, principles, and application. Am Oved, Tel Aviv (Hebrew)

Hough A, Mulder NJ, La Grange JM (1979) Heat treatment for the control of *Phytophthora* gummosis in citrus. Plant Dis Rep 63:40–43

Howell WE, Mink GL (1977) The rôle of weed hosts, volunteer carrots, and overlapping growing seasons in the epidemiology of carrot thin leaf and carrot motley dwarf viruses in central Washington. Plant Dis Rep 61:217–222

Huber DM (1979) Multi-adversity relationships of seeding date as a cultural pest control tactic. 9th Int Congr Plant Protect Washington, Abstr 77

Huber DM, Anderson GR (1976) Effect of organic residues on snowmold of winter wheat. Phytopathology 66:1028–1032

Huber DM, Watson RD (1970) Effect of organic amendment on soil-borne plant pathogens. Phytopathology 60:22–26

Huber DM, Watson RD (1974) Nitrogen form and plant disease. Annu Rev Phytopathol 12:139–165

Hughes CG (1976) Diseases of sugarcane – a review. PANS 24:143–159

Hussey RS, Roncadori RW (1977) Vertical distribution of soil microorganisms following subsoiling in a cotton management system. Phytopathology 67:783–786

Ioannou N, Schneider RW, Grogan RG (1977) Effect of flooding on the soil gas composition and the production of microsclerotia by *Verticillium dahliae* in the field. Phytopathology 67:651–656

Ismunadji M, Sisimyati S, Yazawa F (1973) The effect of fertilization on growth, nitrogen nutrition, and the occurrence of *Helminthosporium* leaf spot in lowland rice. Contr Res Inst Agric Bogor 5:12 (Abstr Rev Plant Pathol 57:2515, 1978)

James WC, Shih CS, Hodgson WA, Callbeck LC (1972) The quantitative relationship between late blight of potato and loss of tuber yield. Phytopathology 62:92–96

Jenkyn JF, Bainbridge A (1978) Biology and pathology of cereal powdery mildews. In: Spencer DM (ed) The powdery mildews. Academic Press, London New York San Francisco, pp 283–321

Johnson LF, Chambers AY, Reed HE (1967) Reduction of root-knot of tomatoes with crop residue amendments in field experiments. Plant Dis Rep 51:219–222

Johnson SR, Berger RD (1972) Nematode and soil fungi control in celery seedbeds on muck soil. Plant Dis Rep 56:661–664

Johnstone GR, Rapley PEL (1979) The effect of time of sowing on the incidence of subterranean clover red leaf virus infection in broad bean *(Vicia faba)*. Ann Appl Biol 91:345–351

Jones DG, Clifford BC (1978) Cereal diseases, their pathology and control. BASF UK Ltd, Ipswich Suffolk UK, 279 pp

Jones FWG (1980) Some aspects of the epidemiology of plant parasitic nematodes. In: Palti J, Kranz J (eds) Comparative epidemiology. PUDOC, Wageningen, pp 71–92

Jones FWG, Parrot DM, Ross GJS (1967) The population genetics of the potato cyst nematode, *Heterodera rostochiensis*: mathematical models to simulate the effects of growing eelworm-resistant potatoes bred from *Solanum tuberosum* var. *andigena*. Ann Appl Biol 60:151–171

Jones GH, Abd-el-Ghani S (1940) The influence of sowing depth and moisture on smut disease and the prospects of a new method of control. Ann Appl Biol 27:35–57

Jones JP, Woltz SS (1972) Effect of soil pH and micronutrient amendments on *Verticillium* and *Fusarium* wilt of tomato. Plant Dis Reptr 56:151–153

Kasasian L (1971) Orobanche spp. PANS 17:35–41

Katan J (1971) Symptomless carriers of the tomato Fusarium wilt pathogen. Phytopathology 61:1213–1217

Katan J (1980) Solar pasteurization of soils for disease control: status and prospects. Plant Dis 64:450–454

Katan J, Eshel Y (1973) Interactions between herbicides and plant pathogens. Residue Rev 45:145–177

Katan J, Greenberger A, Alon H, Grinstein A (1976) Solar heating by polyethylene mulching for the control of diseases caused by soil-borne pathogens. Phytopathology 66:683–688

Katan J, Rotem I, Finkel Y, Daniel J (1980) Solar heating of the soil for the control of pink root and other soil-borne diseases in onion. Phytoparasitica 8:39–50

Kelman A, Cook RJ (1977) Plant pathology in the people's Republic of China. Annu Rev Phytopathol 15:409–429

Kevan DKM (1965) The soil fauna – its nature and biology. In: Baker KF, Snyder WC (eds) Ecology of plant pathogens. Univ Calif Press, Berkeley Los Angeles, pp 33–50

Kincaid RR, Martin FG (1972) Regression of tobacco black shank index on soil calcium. Phytopathology 62:302

King SB, Webster OJ (1970) Downy mildew of sorghum in Nigeria. Indian Phytopathol 23:342–349

Király Z (1976) Plant disease resistance as influenced by biochemical effects of nutrients in fertilizers. In: Fertilizer use and plant health. Int Potash Inst, Bern, pp 33–46

Klaus H (1976) Ergebnisse und Lösungsvorschläge zur Konzentration der Zuckerrübenproduktion in der Fruchtfolge. Tagungsber Akad Landwirtsch Wiss DDR Berlin 148:241–250

Klingler J (1969) Die Wirkung der Warmwasserbehandlung gegen Erdbeerblattälchen und Erdbeermilben. Schweiz Z Obst-Weinbau 105:605–609

Klinkowski M (1967) Bekämpfung pflanzenpathogener Viren. In: Klinkowski M (ed) Pflanzliche Virologie, Bd I. Akademie Verlag, Berlin, pp 266–285

Klisiewicz JM (1977) Effects of flooding and temperature on incidence and severity of safflower seedling rust and viability of *Puccinia carthami* teliospores. Phytopathology 67:787–790

Klisiewicz JM, Beard BH (1976) Diseases of sunflowers in California. Plant Dis Rep 60:298–301

Kontaxis DG (1973) Common sugar beet virus diseases in Imperial Valley. Agric Ext Circ 103; Univ California, Imperial County, 4 p

Kraft JM, Giles RA (1979) Increasing green pea yields with root-rot resistance and subsoiling. In: Schippers B, Gams W (eds) Soil-borne plant pathogens. Academic Press, London New York San Francisco, pp 407–413

Kraft JM, Muehlbauer FJ, Cook RJ, Entemann FM (1974) The reappearance of common wilt of peas in eastern Washington. Plant Dis Rep 58:62–64
Kranz J (1966a) Eine Analyse von annuellen Epidemien pflanzlicher Parasiten. I. Die Befallskurven und ihre Abhängigkeit von einigen Umweltsfaktoren. Phytopathol Z 61:59–86
Kranz J (1966b) Vergleich der parasitischen Mycofloren Guineas und der Cyrenaica. Angew Bot 40:30–37
Kranz J, Aust HJ (1979) Schatten als epidemiologische Einflußgröße beim Gerstenmehltau. Z Pflanzenkrankh Pflanzenschutz 86:533–545
Kranz J, Schmutterer H, Koch W (1977) Diseases, pests and weeds in tropical crops. Paul Parey, Berlin Hamburg, 666 pp
Krigsvold DT, Garren KH, Griffin GJ (1977) Importance of peanut field cultivation and soybean cropping in the spread of *Cylindrocladium crotalariae* within and among peanut fields. Plant Dis Rep 61:495–497
Krikun J (1976) *Verticillium* and wheat. 2nd Int Verticillium Symp, Calif, p 28
Krikun J, Frank ZR (1975) Effect of sequestrene on the reaction to *Verticillium dahliae* of peanuts growing in calcareous loess soil. Phytoparasitica 3:77–78
Krikun J, Orion D (1979) *Verticillium* wilt of potato: importance and control. Phytoparasitica 7:107–116
Krishnamurty GVG, Lai R, Nagarajan K (1975) Further studies on the effect of various crops on the germination of *Orobanche* seed. PANS 23:206–208
Kuhara S (1978) Present epidemic status and control of the citrus canker disease (*Xanthomonas citri* (Hasse) Dowson) in Japan. Rev Plant Prot Res 11:132–142
Kupers LJP (1976) Is the loss of cereals in the field as a consequence of a narrow rotation always caused by pathogens? EPPO Bull 6:119–128
Lapwood DH, Hering TF (1970) Soil moisture and the infection of young potato tubers by *Streptomyces scabies* (common scab). Potato Res 13:296–304
Lapwood DH, Wellings LW, Hawkins JH (1973) Irrigation as a practical means to control potato common scab *(Streptomyces scabies)*: Final experiment and conclusions. Pl Pathol 22:35–41
Last FT (1954) The effect of time of application of nitrogenous fertilizer on powdery mildew of wheat. Ann Appl Biol 41:381–392
Laville E (1980) La Fusariose de l'ananas au Brésil. I. Synthèse des connaissances actuelles. Fruits 35:101–113
Leach LD (1947) Growth rates of host and pathogen as factors determining the severity of pre-emergence damping-off. J Agric Res 75:161–179
Leach LD, Garber RH (1970) Control of *Rhizoctonia*. In: Parmeter JR (ed) *Rhizoctonia solani*, biology and pathology. Univ Calif Pres, Berkeley Los Angeles London, pp 181–198
Leach LD, Mead SW (1936) Viability of sclerotia of *Sclerotium rolfsii* after passage through digestive tracts of cattle and sheep. J Agric Res 53:519–526
Lecoq H (1979) Réceptivité des sols aux agents phytopathogènes: cas des virus transmis par champignons. Ann Phytopathol 11:119
Leiteritz R (1977) Der Einfluß verschiedener Intensivierungsmaßnahmen (N, CCC, Beregnung) auf den Befall des Winterweizens mit Ährenkrankheiten. Tagungsber Akad Landwirtsch Wiss DDR 149:165–169
Lemaire JM, Carpentier F, Dalle JF, Doussinault G, Perraton B (1979) Lutte biologique contre le piétin-echaudage des céréales, modifications physiologiques chez le blé inoculé par une souche atténuée d' *Ophiobolus graminis.* 1) Précocité accrue aux prémiers stades de blè. Acad Agric Fr Extr Séance 16/5/1979, pp 766–772
Lescar L, Gouet JP, Maumene J, Poussard C (1975) Interêt des traitements fongicides en cours de végétation du blé tendre d'hiver. Technique nouvelles de production du blé. ICTF, Paris, 179–195
Levitt J (1972) Responses of plants to environmental stress. Acad Press, London New York San Francisco
Lewis JA, Papavizas GC (1975) Survival and multiplication of soil-borne plant pathogens as affected by plant tissue amendments. In: Bruehl G (ed) Biology and control of soil-borne plant pathogens. Am Phytopathol Soc, pp 84–89
Lima JAA, Nelson MR (1977) Etiology and epidemiology of mosaic of cowpea in Ceara, Brazil. Plant Dis Rep 61:864–867
Lin YS, Cook RJ (1979) Suppression of *Fusarium roseum* "Avenaceum" by soil microorganisms. Phytopathology 69:384–388

Line RF (1976) Factors contributing to an epidemic of stripe rust on wheat in the Sacramento Valley of California in 1974. Plant Dis Rep 60:312–316

Ling KC (1972) Rice virus diseases. Int Rice Res Inst, Los Banos Laguna Philippines, 142 pp

Littauer F (1944) *Sclerotium bataticola* Taubenhaus on potatoes in Palestine. Palestine J Bot Rehovot Ser 4:142–147

Liyanage D de S, Royle DJ (1976) Overwintering of *Sphaerotheca humuli*, the cause of hop powdery mildew. Ann Appl Biol 83:381–394

Llano A (1977) The orange coffee rust in Nicaragua. Plant Dis Rep 61:999–1002

Lockhart DAS, Heppel VAF, Holmes JC (1975) Take-all (*Gaeumannomyces graminis* (Sacc.) Arx and Olivier) incidence in continuous barley growing and effect of tillage method. EPPO Bull 5:375–383

Loebenstein G, Alper M, Levy S, Palevitch D, Menagem E (1975) Protecting peppers from aphid-borne viruses with aluminum foil or plastic mulch. Phytoparasitica 3:43–53

Loebenstein G, Raccah B (1980) Control of non-persistently transmitted aphid-borne viruses. Phytoparasitica 8:221–235

Lot H, Jacquemond M, Quiot J-B (1979) La nécrose de la tomate. Phytoma 309:21–23

Louvet J (ed) (1972) Les nématodes dans leurs relations avec les maladies des plantes. Ann Phytopathol 4:405–408

Louvet J (1974) L'utilisation du greffage en culture maraîchère. Pépiniéristes Hortic Maraîchèrs 152:13–16

Louvet J, Peyrière J (1962) Interêt du greffage du melon sur *Benincasa cerifera* Savi. Proc 16th Int Hortic Congr Brussels, pp 167–171

Louvet J, Rouxel F, Alabouvette C (1975) Sols résistants aux maladies et perspectives de lutte biologique. Ann Phytopathol 7:204–207

Lucas GB, DeLoach LR (1964) Tobacco blue mold in Mexico. Plant Dis Rep 48:524

Lukens RJ, Mullamy R (1972) The influence of shade and wet soil on southern corn leaf blight. Plant Dis Rep 56:203–206

Magie RO (1971) Effectiveness of treatments with hot water plus benzimidazoles and ethephon in controlling *Fusarium* disease of gladiolus. Plant Dis Rep 55:82–85

Majumdar M, Raychaudhuri SP (1976) Effect of temperature on *Fusarium* root rot of pea. Acta Phytopathol Hung Acad Sci 11:45–52

Mankau R, Minteer RJ (1962) Reduction of soil population of the citrus nematode by addition of organic materials. Plant Dis Rep 46:375–378

Marrou J, Quiot JB, Duteil M, Labonne G, Leclant F, Renoust M (1979) Ecologie et épidémiologie du Virus de la Mosaique du Concombre. VIII. Influence des brise-vents et des cultures environnantes sur la dissémination du Virus Mosaique du Concombre. Ann Phytopath 11:375–392

Martin H (1960) Maladies et insectes nuisibles aux arachides en Libye. Report FAO Expert, multilithed, 24 pp

Marx DH (1973) Mycorrhzae and feeder root diseases. In: Marks GC, Kozlowski TT (eds) Ectomycorrhizae – their ecology and physiology. Academic Press, London New York San Francisco, pp 351–382

Mattingly GEG, Slope DB (1977) Phosphate fertilizer and "take-all" disease of wheat and barley. J Sci Food Agric 28:658–659

Mayee CD, Rattan BK, Shahi HN (1976) Agro-practices in the incidence of stem-rot disease of rice in the Punjab. Indian J Mycol Plant Pathol 6:166–170

McBeth CW, Taylor AL (1944) Immune and resistant cover crops valuable in root-knot infested peach orchards. Proc Am Soc Hortic Sci 45:158–166

McCain AH, Pyeatt LE, Byrne TG, Farnham DS (1980) Suppressive soil reduces carnation disease. Calif Agric 34:9

McIntyre JL, Horner CE (1973) Inactivation of *Verticillium dahliae* in peppermint stems by propane gas flaming. Phytopathology 63:172–175

Meléndez PL, Powell NT (1970) The *Pythium*-root knot nematode complex in flue-cured tobacco. Phytopathology 60:1303

Menzies JD (1959) Occurrence and transfer of a biological factor in soil that suppresses potato scab. Phytopathology 49:648–652

Menzies JD, Griebel GE (1967) Survival and saprophytic growth of *Verticillium dahliae* in uncropped soil. Phytopathology 57:703–709

Merrill W, Petersen DM (1979) Diseases of economic plants: tree and small fruits. Dep Plant Pathol Penn State Univ, 2nd edn, unnumbered

Mew TW, Howard FL (1969) Root rot of soybean *(Glycine max)* in relation to antagonism of *Rhizobium japonicum* and *Fusarium oxysporum*. Phytopathology 59:401

Milbrath DG, Ryan HJ (1938) A method for control of western celery mosaic. Calif Dep Agric Bull 27:290–295

Miller DE, Burke DW (1975) Temporary excessive soil moisture predisposes bean to *Fusarium* root rot. Bean Improvement Coop Annu Rep 18:48–49

Miller PM (1977) Reducing field populations of several plant parasitic nematodes by leaf mold composts and some other additives. Plant Dis Rep 61:328–331

Miller PM, Taylor GS, Wihrheim SE (1968) Effect of cellulosic soil amendments and fertilizers on *Heterodera tabacum*. Plant Dis Rep 52:441–445

Miller PM, Sands DC, Rich S (1973) Effect of industrial mycelial residues, wood fiber wastes, and chitin on plant parasitic nematodes and some soil-borne diseases. Plant Dis Rep 57:438–442

Minoiu N (1974) Elimination of virus infestation from pear and cherry stock by heat treatments. Anal Inst Cer C Plant Protect X for 1972, pp 11–19

Mittelstaedt H (1979) Pflanzenbauliche und technische Maßnahmen zur Minderung der Schäden durch Spätfröste in der Obstproduktion. Gartenbau 26:51–52

Mizukami T, Wakimoto S (1969) Epidemiology and control of bacterial leaf blight of rice. Annu Rev Phytopathol 7:51–52

Mojitahedi H, Lownsbery BF, Moody EH (1975) Ring nematodes increase development of bacterial cankers in plums. Phytopathology 65:556–559

Moody AR, Guidrat D (1977) Biological control of black root rot by *Gliocladium roseum*. Phytopathology 67:1159–1162

Moore LW (1979) Practical use and success of *Agrobacterium radiobacter* strain 84 for crown gall control. In: Schippers B, Gams W (eds) Soil-borne plant pathogens. Academic Press, London New York San Francisco, pp 553–568

Moore LW, Lagerstedt HB, Hartmann N (1974) Stress predisposes young filbert trees to bacterial blight. Phytopathology 64:1537–1540

Moore WD (1949) Flooding as a means of destroying the sclerotia of *Sclerotinia sclerotiorum*. Phytopathology 39:920–927

Mudich A (1967) Effect of trace elements bound to superphosphate on the resistance of potato tubers to *Phytophthora infestans* (Mont.) de Bary. Acta Phytopathol Akad Sci Hung 2:295–302

Mugniery D (1978) Lutte culturale contre les nématodes à kystes de la pomme de terre *Globodera rostochiensis* (Wollenweber) et *G.pallida* (Stone) et perspectives de lutte integrée. Ann Zool Ecol Anim 10:187–203

Mukibi J (1976) Possible relationship between intercropping and plant disease problems in Uganda. In: Mongo JH, Ker ADR, Campbell M (eds) Intercropping in semi-arid areas. IDRC Publ 076e, Ottawa, Canada, p 45

Muller RA (1974) Integrated control methods. In: Gregory PH (ed) *Phytophthora* disease of cocoa, pp 259–268

Muller RA (1975) L'irrigation précoce, assurance pour une production régulière de haut niveau du caféier Arabica. Café Cacao Thé 19:95–122

Muller RA (1977) L'irrigation précoce du caféier arabica au Cameroun. Methode moderne de contrôle de l'anthracnose des baies (*Colletotrichum coffeanum* Noack sensu Hindorf) integrée aux pratiques culturales. Assurance pour une production régulière de haut niveau. ASIC, 8e Colloq Abidjan, pp 489–506

Mussell HW, Malone MJ (1979) Disease tolerance: reducing the impact of disease-induced stress on crop yields. In: Mussel HW, Staples RC (eds) Stress physiology in crop plants. J Wiley and Sons, New York, pp 15–23

Nash-Smith S (1970) The significance of populations of pathogenic Fusaria in soil. In: Toussoun TA, Bega RV, Nelson PE (eds) Root diseases and soil-borne pathogens. Univ Calif Press, Berkeley Los Angeles, pp 28–30

Naumann K, Lange-de-la Camp M (1977) Wirkung von Pflanzenrückständen auf die parasitische Aktivität von bodenbürtigen Krankheitserregern und die saprophytische Mikroflora des Bodens. II. Einfluß eines Zwischenfruchtanbaus auf den Befall mit *Rhizoctonia solani* Kuehn. Z F Bakter, Abt. II, 132:573–592

Neergaard P (1977) Seed pathology. Mac Millan Press, London

Nelson R (1950) *Verticillium* wilt of peppermint. Mich Agric Exp Stn Tech Bull 221:1–259

Nesbitt HJ, Malajczuk N, Glenn AR (1979) Effect of organic matter on the survival of *Phytophthora cinnamomi* Rands in soil. Soil Biol Biochem 11:133–136

Newhall AG (1955) Disinfestation of soil by heat, flooding and fumigation. Bot Rev 21:189–250
Nielsen LW (1977) Thermotherapy to control sweetpotato sprout-borne root-knot, black rot, and scurf. Plant Dis Rep 61:882–887
Nitzany FE, Geisenberg H, Koch B (1964) Tests for the protection of cucumbers from a white-fly borne virus. Phytopathology 54:1059–1061
Norgaard RB (1976) Integrating economics and pest management. In: Apple JL, Smith RF (eds) Integrated pest management. Plenum Press, New York London, pp 17–27
Norton DC (1978) Ecology of plant-parasitic nematodes. J Wiley and Sons, New York Chichester Brisbane Toronto, 268 pp
Norton DC (1979) Relationship of physical and chemical factors to populations of plant parasitic nematodes. Annu Rev Phytopathol 17:279–299
Norton GA (1976) Analysis of decision making in crop protection. Agro-Ecosystems 3:27–44
Nusbaum CJ, Ferris H (1973) The role of cropping systems in nematode population management. Annu Rev Phytopath 11:423–440
Nyland G, Goheen AC (1969) Heat therapy of virus diseases of perennial plants. Annu Rev Phytopath 7:331–354
Nyvall RF (1979) Field crop diseases handbook. Avi Publ Co, Westport Ct, 436 pp
O'Bannon JH (1968) The influence of an organic soil amendment on infectivity and reproduction of *Tylenchulus semipenetrans* on two citrus rootstocks. Phytopathology 58:597–601
Obst A (1975) Biologie, Epidemiologie und chemische Bekämpfung der wichtigsten Blatt- und Ährenkrankheiten des Getreides. Bayer Landwirtsch Jahrb 52:702–708
Obst A, Graf R, Rueppold H (1977) Beobachtungen in einem Monokultur-Weizensortiment zur Vikarianz der Fußkrankheitserreger. Nachrichtenbl Dtsch Pflanzenschutzdienst (Braunschweig) 29:113–114
Ogawa JM, Sandeno JL, Mathre JH (1963) Comparisons in development and chemical control of decay-causing organisms on mechanical- and hand-harvested stone fruits. Plant Dis Rep 47:129–133
O'Neill NR, Rush MC, Horn NL, Carver RB (1977) Aerial blight of soybeans caused by *Rhizoctonia solani*. Plant Dis Rep 61:713–717
Onesirosan PT (1975) Seedborne and weedborne inoculum in web blight of cowpea. Plant Dis Rep 59:338–339
Oppitz K, Hoeser K (1978) Ertrags- und Qualitätsbeeinflussung durch N-Spätdüngung und Fungizidbehandlung bei Winterweizen. Bayer Landwirtsch Jahrb 55:1005–1014
Orsenigo JR (1979) Integrated sugarcane crop management. 9th Int Congr Plant Protect Washington, Abstr 750
Ou SH (1972) Rice diseases. Commonw Mycol Inst, Kew Surrey UK, 368 pp
Palti J (1945) Diseases of forage crops and their control. Hassadeh 45:242–249 (Hebrew)
Palti J, Netzer D (1963) Development and control of *Phytophthora infestans* (Mont.) De By. under semi-arid conditions in Israel. Phytopathol Medit 2:265–274
Palti J, Rotem J (1981) Control of downy mildews by cultural practices. In: Spencer DM (ed) The downy mildews. Academic Press, London New York San Francisco
Palti J, Shoham H (1979) Plant protection benefits from trickle irrigation. 9th Int Congr Plant Protect Washington, Abstr
Palti J, Chorin M, Nitzany FE (1962) Vegetable diseases in Israel. Hassadeh Publ Co, Tel Aviv (Hebrew, English summary)
Palti J, Nitzany FE, Shoham H, Cohen R (1975) Diseases of field and vegetable crops in a dynamic agriculture: lessons of the past and future outlook. Hassadeh 55:815–823 (Hebrew)
Panagopoulos CG, Crosse JE (1964) Frost injury as a predisposing factor in blossom blight of pear caused by *Pseudomonas syringae* van Hall. Nature (London) 202:no 4939, p 1352
Panagopoulos CG, Psallidas PG, Alivizatos AS (1979) Evidence of a breakdown in the effectiveness of biological control of crown gall. In: Schippers B, Gams W (eds) Soil-borne plant pathogens. Academic Press, London New York San Francisco, pp 569–578
Papavizas GC (1966) Suppression of *Aphanomyces* root rot of peas by cruciferous soil amendments. Phytopathology 56:1071–1075
Papavizas GC, Lewis JA (1979) Integrated control of *Rhizoctonia solani*. In: Schippers B, Gams W (eds) Soil-borne plant pathogens. Academic Press, London New York San Francisco, pp 415–424
Papendick RI, Campbell GS (1975) Water potential in the rhizosphere and plant, and methods of measurement and experimental control. In: Biology and control of soil-borne pathogens. Am Phytopathol Soc, St Paul Minn, pp 39–49

Pappo S (1965) Apple bitter rot caused by *Gloeosporium fructigenum* Berk. PhD Thesis, Hebrew Univ Jerusalem, 134 pp (Hebrew, English summary)

Pappo S, Arenstein Z (1962) The problem of spongy rot in citrus groves and overhead sprinkling. Hassadeh 43:194–195 (Hebrew)

Parker MB, Minton NA, Brooks OL, Perry CE (1973) Influence of subsoiling, fertility and nematicide treatments on soybeans grown on a Marlboro soil infested with lance nematodes. 2nd Int Congr Plant Pathol Abstr 0217

Paulus AO, Harvey OA, Nelson J, Shibuya F, Holland AH (1971) Control of *Cercospora* leafspot of sugarbeet under sprinkler irrigation. Plant Dis Rep 55:449–457

Peries OS (1970) Economics of control of the white root disease *(Fomes lignosus)* of *Hevea brasiliensis* in Ceylon. In: Toussoun TA, Bega RV, Nelson PE (eds) Root diseases and soil-borne pathogens. Univ California Press, Berkeley Los Angeles London, pp 191–193

Perrenoud S (1977) Potassium and plant health. IPI Research Topics. Int Potash Inst, Bern, 218 pp

Perrin RM (1977) Pest management in multiple cropping systems. Agro-Ecosystems 3:93–118

Phillips DH (1959) The destruction of *Didymella lycopersici* Kleb. in tomato haulm composts. Ann Appl Biol 47:240–253

Pieczarka DJ, Lorbeer JW (1974) Control of bottom rot of lettuce by ridging and fungicide application. Plant Dis Rep 58:837–840

Plaut Z, Halevy AH (1966) Increasing the regeneration of wheat plants following wilting by treatment with two growth-retarding chemicals. Naturwissenschaften 53:509–510

Polley RW (1971) Barley leaf blotch epidemics in relation to weather conditions, with observations on the overwintering of the disease on barley debris. Plant Pathol 20:184–190

Populer C (1972) Les épidémies de l'öidium de l'hévéa et la phénologie de son hôte dans le monde. Publ Inst Nat Etude Agron Congo (INEAC) Sér Sci 115:368 pp.

Populer C (1978) Changes in host susceptibility in time. In: Horsfall JG, Cowling EB (eds) Plant disease, an advanced treatise, vol II. How disease develops in populations. Academic Press, London New York San Francisco, pp 239–262

Prasad JS, Rao YS (1978) Influence of crop rotation on the population densities of the rod lesion nematode *Pratylenchus indicus* in rice and rice soils. Ann Zool Ecol Anim 10:627–633

Price DW (1976) Passage of *Verticillium albo-atrum* propagules through the alimentary canal of the bulb mite. Phytopathology 66:46–50

Pullman GS, DeVay JE, Garber RH, Weinhold AR (1979) Control of soil-borne fungal pathogens by plastic tarping of soil. In: Schippers B, Gams W (eds) Soil-borne plant pathogens. Academic Press, London New York San Francisco, pp 439–446

Purdy LH (1979) *Sclerotinia sclerotiorum*: history, diseases and symptomatology, host range, geographic distribution and impact. Phytopathology 69:875–880

Putter CAJ (1979) Quantifying crop losses under subsistence farming conditions. 9th Int Congr Plant Protect Washington, Abstr

Putter CAJ (1980) The management of epidemic levels of endemic diseases under tropical subsistence farming conditions. In: Palti J, Kranz J (eds) Comparative epidemiology. PUDOC, Wageningen, pp 93–103

Quiot JB, Verbrugghe M, Labonne G, Leclant F, Marrou J (1979) Ecologie et épidémiologie du virus du concombre dans le Sud-Est de la France. IV. Influence des brise-vent sur la répartition des contaminations virales dans une culture protégée. Ann Phytopathol 11:307–324

Rabb RL, Todd FA, Ellis HC (1976) Tobacco pest management. In: Apple JL, Smith RF (eds) Integrated pest management. Plenum Press, New York London, pp 71–106

Rangaswami F (1975) Diseases of crop plants in India. Prentice-Hall of India, New Delhi, 520 pp

Raniere LC, Crossan DF (1959) The influence of overhead irrigation and microclimate on *Colletotrichum phomoides*. Phytopathology 49:72–74

Rapilly F (1966) L'ergot du blé et les graminées adventices. Phytiatrie-Phytopharm 4:251–256

Rapilly F (1968) Les conséquences dues a l'emploi du C.C.C. sur le développement de certaines maladies des céréales. Phytiatrie-Phytopharm 4:307–310

Reinking OA, Manns MM (1933) Parasitic and other fusaria counted in tropical soils. Z Parasitenk 6:23–75

Renard JL (1979) Vascular wilt disease *(Fusarium)* in the oil palm. Oléagineux 34:61–62

Richardson MJ (1979) An annotated list of seed-borne diseases. Third Edition. Phytopathol Pap 23, Commonw Mycol Inst, Kew, Surrey UK, 320 pp

Ridé M, Ridé S, Novoa D (1978) Biology and epidemiology of *Xanthomonas ampelina* Panagopoulos on grapevine in France. Proc 4th Int Conf Plant Pathol Bacteriol, Angers, pp 969–971
Roane CW, Harrison RL, Genter CF (1974) Observations on gray leaf spot of maize in Virginia. Plant Dis Rep 58:456–459
Rode H (1979) Zier- und Wildpflanzen als Wirte von Krankheitserregern im Obstbau. Gartenbau 26:31–32
Roelfs AP (1977) Foliar fungal diseases of wheat in the People's Republic of China. Plant Dis Rep 61:836–841
Roelfs AP, McVey DV, Long DL, Rowell JB (1972) Natural rust epidemics in wheat nurseries as affected by inoculum density. Plant Dis Rep 56:410–414
Roessner J (1979) Einfluß langjähriger Monokulturen und Unkrautbekämpfungsmaßnahmen auf die Populationsdichte wandernder Wurzelnematoden. Z Pflanzenkrankh Pflanzenschutz 86:257–265
Rogers CH (1939) The relation of moisture and temperature to growth of the cotton root rot fungus. J Agric Res 58:701–709
Romanko RR (1965) Factors affecting the spread of hop downy mildew in an arid climate. Plant Dis Rep 49:247
Rondomanski W (1966) The possibilities of limiting the chemical control of late blight in tomato and of onion downy mildew. Zeszyty Problem. Postepow Nauk Rolniczych 60:271–276
Ross H (1965) Vorbedingungen des Befall. In: Klinkowski M, Muehle E, Reinmuth E (eds) Grundlagen und allgemeine Probleme der Phytopathologie und des Pflanzenschutzes. Akademie Verlag, Berlin, pp 35–44
Rotem J (1968) Thermoxerophytic properties of *Alternaria porri* f. sp. *solani*. Phytopathology 58:1284–1287
Rotem J (1978) Climatic and weather influences on epidemics. In: Horsfall JG, Cowling EB (eds) Plant disease, an advanced treatise, vol II. Academic Press, London New York San Francisco, pp 317–337
Rotem J, Cohen Y (1966) The relationship between mode of irrigation and severity of tomato foliage diseases in Israel. Plant Dis Rep 50:635–639
Rotem J, Palti J (1969) Irrigation and plant diseases. Annu Rev Phytopathol 7:267–288
Rotem J, Palti J (1980) Epidemiological factors as related to plant disease control by cultural practices. In: Palti J, Kranz J (eds) Comparative epidemiology. PUDOC, Wageningen, pp 104–116
Rotem J, Reichert I (1964) Dew – a principal moisture factor enabling early blight epidemics in a semi-arid region in Israel. Plant Dis Rep 48:211–215
Rotem J, Cohen Y, Spiegel S (1968) Effect of soil moisture on the predisposition of tobacco to *Peronospora tabacina*. Plant Dis Rep 52:310–313
Rotem J, Palti J, Lomas J (1970) Effects of sprinkler irrigation at various times of the day on development of potato late blight. Phytopathology 60:839–843
Roth G (1973) The elimination of the virus diseases "streak" and "mosaic" by thermotherapy and tissue culturing. 2nd Int Congr Plant Pathol Minn, Abstr 901
Rouxel F, Alabouvette C, Louvet J (1979) Recherches sur la résistance des sols aux maladies. IV. Mise en évidence du rôle des *Fusarium* autochtones dans la résistance d'un sol à la Fusariose vasculaire du melon. Ann Phytopathol 11:199–207
Royle DJ (1975) Host, fungal, and cultural factors affecting control of hop powdery mildew. Ann Appl Biol 81:437–444
Russell GE (1979) New approaches to sugar beet virus yellow control. 9th Int Congr Plant Pathol Washington, Abstr 757
Sahni VM, Gera SD (1967) Early sowings and fungicides help arrest the angular leaf-spot of cotton. Indian Farming 16, 10:p 27
Sastry KSM, Singh SJ, Sastry KS (1976) Density of tomato plant population and application of insecticides in relation to incidence of tomato leaf curl virus. Indian J Mycol Plant Pathol 6:8–13
Saxena PK, Chhabra HK, Jasial K (1977) Effect of certain soil amendments and nematicides on the population of nematodes infesting grape-vines. Z Angew Zool 64:325–330
Sayre RM, Habicht WA, Walker JM (1973) Effect of sewage sludge as soil organic amendments on the incidence of root-knot. 2nd Int Congr Plant Pathol Minneapolis, Abstr 0557
Schenck NC, Kinloch RA (1974) Pathogenic fungi, parasitic nematodes and endomycorrhizal fungi associated with soybean roots in Florida. Plant Dis Rep 58:169–173
Schieber E (1969) Enfermedad del cacao „Mal de Machete" provocada par *Ceratocystis fimbriata* en la Republica Dominicana. Turrialba 19:340–344

Schieber E (1975) Present status of coffee rust in South America. Annu Rev Phytopath 13:375–382
Schiffmann-Nadel M, Waks J, Chalutz E (1975) Frost injury predisposes grapefruit to storage rots. Phytopathology 65:630
Schmid G, Bucher R, Weigelt H (1973) Einfluß von Klärschlamm auf Pflanze und Umwelt. Landwirtsch Forsch Sonderh 28:1, 363–373
Schmidle A (1979) Zur Einsparung von Fungiziden im Obstbau im Rahmen des integrierten Pflanzenschutzes. Ber Landwirtsch 57:513–520
Schoenbeck F (1979) Endomycorrhiza in relation to plant diseases. In: Schippers B, Gams W (eds) Soil-borne plant pathogens. Academic Press, London New York San Francisco, pp 271–280
Schoeneweiss DF (1975) Predisposition, stress, and plant disease. Annu Rev Phytopathol 13:193–211
Schoeneweiss DF (1978) Water stress as a predisposing factor in plant disease. In: Kozlowski TT (ed) Water deficits and plant growth. Academic Press, London New York San Francisco, pp 61–99
Schoenrock-Fischer R, Schwaebe L (1976) Wirkung von Intensivierungsmaßnahmen zur Minderung nematodenbedingter Schadeffekte im konzentrierten Getreideanbau. Tagungsber Akad Landwirtsch Wiss DDR 148:327–335
Schroedter H, Hoffmann GM (1961) Der Einfluß der Witterung und des Mikroklimas auf den Befall des Leins durch *Polyspora lini* Laff. Zentralbl Bakteriol II Abt 114:15–44
Schroth MN, Hendrix FF (1962) Influence of non-susceptible plants on the survival of *Fusarium solani* f. *phaseoli* in soil. Phytopathology 52:906–909
Schulz FA (1968) Untersuchungen über den Einfluß der Gründüngung auf die Halmbruchkrankheit des Weizens und ihren Erreger *Cercosporella herpotrichioides* Fron. I. Beeinflussung einzelner Entwicklungsphasen des Pilzes im Boden. II. Beeinflussung des Krankheitsbefalls. Phytopathol Z 62:1–20, 127–141
Schwartz HF, Steadman JR (1978) Factors affecting sclerotium populations of, and apothecium production by, *Sclerotinia sclerotiorum*. Phytopathology 68:383–388
Seemann J, Chirkov YI, Lomas J, Primault B (1979) Agrometeorology. Springer, Berlin Heidelberg New York, 324 pp
Sequeira L (1958) Bacterial wilt of bananas: dissemination of the pathogen and control of the disease. Phytopathology 48:64–69
Sequeira L (1962) Influence of organic amendments on survival of *Fusarium oxysporum* f. *cubense* in the soil. Phytopathology 52:976–983
Sewell GWF (1981) Soil-borne fungal pathogens in natural vegetation and weeds of cultivation. In: Thresh JM (ed) Pests, Pathogens, and vegetation. Pitman, London, pp 175–190
Sewell GFW, Wilson JF (1961) Machine-picking in relation to progressive *Verticillium* wilt of the hop. I. A study of the infectivity of machine-picked hop waste. Annu Rep East Malling Res Stan 1960, pp 100–104
Sewell GFW, Wilson JF (1966a) *Verticillium* wilt of the hop: fluctuating outbreaks in wilt-resistant varieties. Ann Appl Biol 58:251–258
Sewell GFW, Wilson JF (1966b) *Verticillium* wilt of the hop: the survival of *Verticillium albo-atrum* in the soil. Ann Appl Biol 58:241–249
Sewell GWF, Wilson JF (1967) *Verticillium* wilt of the hop: field studies on wilt in a resistant cultivar in relation to nitrogen fertilizer applications. Ann Appl Biol 59:265–273
Sewell GWF, Wilson JF (1974a) The influence of normal tillage and of non-cultivation on *Verticillium* wilt of hop. Ann Appl Biol 76:37–47
Sewell GWF, Wilson JF (1974b) Hop wilt, soil temperature and nitrogen. Ann Rep East Malling Res Stn 1973, pp 203–204
Sewell GWF, Wilson JF, Martin DG (1962) Machine-picking in relation to progressive *Verticillium* wilt of the hop. II. The effect of composting on the infectivity of machine-picked hop waste. Ann Rep East Malling Res Stn 1961, pp 102–106
Sharma SL, Chowfla SC, Sohi HS (1976) Control of buck-eye rot of tomato by cultural practices. Indian J Mycol Plant Pathol 6:51–54
Shipton PJ (1977) Monoculture and soil-borne plant pathogens. Annu Rev Phytopathol 15:387–407
Shipton PJ (1979) Experimental evidence for the effect on soil-borne diseases of changes in techniques of crop and soil cultivation. In: Schippers B, Gams W (eds) Soil-borne plant pathogens. Academic Press, London New York San Francisco, pp 385–397
Shipton PJ, Cook RJ, Sitton JW (1973) Occurrence and transfer of a biological factor in soil that suppresses take-all of wheat in Eastern Washington. Phytopathology 63:511–517

Shokes FM, McCarter SM (1976) Occurrence of plant pathogens in irrigation ponds in Southern Georgia. Am Phytopathol Soc Proc 3:342
Shokes FM, Lyda SD, Jordan WR (1977) Effect of water potential on the growth and survival of *Macrophomina phaseolina*. Phytopathology 67:239–241
Siddiqui WM, Halisky PM, Lund S (1968) Relationship of clipping frequency to root and crown deterioration in red clover. Phytopathology 58:486–488
Singh K (1973) Hot air therapy against red rot of sugarcane. Plant Dis Rep 57:220–222
Singh RS, Singh B, Beniwal SPS (1967) Observations on the effect of sawdust on incidence of root knot and on yield of okra and tomatoes in nematode-infested soil. Plant Dis Rep 51:861–863
Skadow K (1966) Unkräuter als Wirtspflanzen für *Verticillium albo-atrum* Rke. et Berth. einschließlich *Verticillium dahliae* Kleb. Zentralb Bakteriol II Abt 120:49–59
Skadow K (1969) Untersuchungen über die Welkeerreger *Verticillium albo-atrum* Rke. et Berth. and *V. dahliae* Kleb. II. Unkräuter als Infektionsreservoire. Zentralbl Bakteriol II Abt 123:736–765
Slykhuis JT (1976) Virus and virus-like diseases of cereal crops. Annu Rev Phytopathol 14:189–210
Smiley RW (1975) Forms of nitrogen and the pH in the root zone and their importance to root infections. In: Bruehl GW (ed) Biology and control of soil-borne pathogens. Am Phytopathol Soc, St Paul Minn, pp 52–62
Smiley RW, Cook RJ (1973) Relationship between take-all of wheat and rhizosphere pH in soils fertilized with ammonium vs. nitrate nitrogen. Phytopathology 63:882–889
Smith AM (1972a) Drying and wetting sclerotia promotes biological control of *Sclerotium rolfsii* Sacc. Soil Biol Biochem 4:119–123
Smith AM (1972b) Biological control of sclerotia in soil. Soil Biol Biochem 4:131–134
Smith FF, Webb RE (1969) Repelling aphids by reflective surfaces, a new approach to the control of insect-transmitted viruses. In: Maramorosch K (ed) Viruses, vectors, and vegetation. Wiley Interscience Publishers, New York, pp 631–639
Smith MN, Olien CR (1978) Pathological factors affecting survival of winter barley following controlled freeze tests. Phytopathology 68:773–777
Smith PR, Stubbs LL, Challen DI (1977) Studies on the epidemiology of Peach Rosette and Decline disease in Victoria. Aust J Agric Res 28:103–113
Smith SN, Snyder WC (1972) Germination of *Fusarium oxysporum* chlamydospores in soils favorable and unfavorable to wilt establishment. Phytopathology 62:273–277
Snyder WC, Schroth MN, Christou T (1959) Effect of plant residue on root rot of bean. Phytopathology 49:755–756
Solel Z (1967) Dry collar rot of sorghum. Ktavim 17:175–176
Spotts RA, Ferree DC (1979) Effect of overtree misting for bloom delay on *Venturia inaequalis* ascospore maturity and release. Plant Dis Rep 63:108–122
Spotts RA, Stang EJ, Ferree DC (1976) Effect of overtree misting for bloom delay on incidence of fire blight in apple. Plant Dis Rep 60:329–330
Steadman JR, Maier CR, Schwartz HF, Kerr ED (1975) Pollution of surface irrigation waters by plant pathogenic organisms. Water Resourc Bull 11:796–804
Steiner KG (1973) Der Einfluß der Fungizidbehandlung auf den Verlauf der Kaffeekirschen-Krankheit (*Colletotrichum coffeanum* Noack). Z Pflanzenkrankh Pflanzenschutz 80:671–681
Stelfox D, William JR, Soehngen U, Topping RC (1978) Transport of Sclerotinia ascospores by rapeseed pollen in Alberta. Plant Dis Rep 62:576–579
Sterne RE, McCarver TH, Courtney ML (1979) Survival of plant pathogens in composted gin trash. Arkansas Farm Res 28:9
Stetter S, Leroul N (1979) Continuous barley growing. II. Influence on the fungal population of the roots. Tidskrift Plantearb 83:50–72 (Swedish, English summary)
Stevens NE, Stevens RB (1952) Disease in plants. Chronica Botanica, Waltham Mass, 219 pp
Stevens RB (1960) Cultural practices in disease control. In: Horsfall JG, Dimond AE (eds) Plant pathology, vol III. Academic Press, pp 357–429
Stevenson WR, Hagedorn DJ (1973) Overwintering of pea seedborne mosaic virus in hairy vetch, *Vicia villosa*. Plant Dis Rep 57:349–352
Stone OW (1962) Alternate hosts of cucumber powdery mildew. Ann Appl Biol 50:203–210
Stover RH (1954) Flood-fallowing for eradication of *Fusarium oxysporum* f. *cubense*. II. Some factors involved in fungus survival. Soil Sci 77:401–414
Stover RH (1955) Flood fallowing for eradication of *Fusarium oxysporum* f. *cubense*. III. Effect of oxygen on fungus survival. Soil Sci 80:397–412

Stover RH (1956) Studies on *Fusarium* wilt of bananas. I. The behavior of *F. oxysporum* f. *cubense* in different soils. Can J Bot 34:927–942
Stover RH (1959) Growth and survival of root-disease fungi in soil. In: Holton CS et al (eds) Plant pathology, problems and progress 1908–1958. Univ Wis Press, Madison Wis, pp 339–355
Stover RH, Ostmark HE (1979) Integrated plant protection in bananas and plantains. 9th Int Congr Plant Protect Washington, Abstr 752
Stubbs LL (1956) Motley dwarf virus disease of carrot in California. Plant Dis Rep 40:763–764
Sumner DR (1968) The effect of soil moisture on corn stalk rot. Phytopathology 58:761–765
Sumner DR, Littrell RH (1974) Influence of tillage, planting date, inoculum survival, and mixed populations on epidemiology of southern corn leaf blight. Phytopathology 64:168–173
Sumner DR, Johnson AW, Jaworski CA, Chalfant RB (1978) Influence of film mulches and soil pesticides on root diseases and populations of soil-borne fungi in vegetables. Plant Soil 49:267–283
Sutic D, Milovanovic M (1980) Some factors affecting epidemiology of sugar beet rizomania-like disease. Proc 5th Congr Medit Phytopathol Union, Patras Greece, pp 29–30
Swenson KG (1968) Role of aphids in the ecology of plant viruses. Annu Rev Phytopathol 6:351–374
Sztejnberg A, Golan J (1980) Changes in the occurrence of fruit crop diseases as a result of changes in irrigation measures. Proc 5th Congr Medit Phytopathol Union, Patras Greece, pp 127–128
Tagami Y (1968) Epidemiology of the bacterial leaf blight of rice. Rev Plant Protect Res 1:1–14
Tarjan AC, O'Bannon JH (1977) Non-pesticidal approaches to nematode control. Proc Int Soc Citric 3:848–853
Taubenhaus JJ, Ezekiel WN (1936) A rating of plants with reference to their relative resistance or susceptibility to *Phymatotrichum omnivorum*. Texas Agric Exp Stn Bull 527:52 pp.
Taylor CE, Cadman CH (1969) Nematode vectors. In: Maramorosch K (ed) Viruses, vectors, and vegetation. Interscience Publ, New York London Sydney Toronto, pp 55–94
Teakle DS (1969) Fungi as vectors and hosts of viruses. In: Maramorosch K (ed) Viruses, vectors and vegetation. Interscience Publ, New York London Sydney Toronto, pp 23–54
Teuteberg A (1978) *Drechslera poae* (Baudys) Shoem. als Schaderreger an *Poa pratensis* L. Rasen Grünflächen Begrünungen 9:36–38
Thanassoulopoulos CC, Biris DA, Tjamos EC (1980) Dissemination of *Verticillium* propagules in olive orchards by irrigation water. Proc 5th Congr Medit Phytopathol Union, Patras Greece, pp 52–53
Thomason IJ, Faulkner LR (1975) Distribution of plant pathogens to agricultural crop lands in irrigation water. Nat Pest Contr Operation News 35:13–15
Thomson SV, Allen RM (1974) Occurrence of *Phytophthora* species and other potential plant pathogens in recycled irrigation water. Plant Dis Rep 58:945–949
Thomson SV, Allen RM (1976) Mechanism of survival of zoospores of *Phytophthora parasitica* in irrigation water. Phytopathology 66:1198–1202
Thorne DW, Thorne MD (1979) Soil, water and crop production. Avi Publ Co Inc, Westport Conn, 353 pp
Thorne G (1942) Distribution of the root-knot nematode in high ridge plantings of potatoes and tomatoes. Phytopathology 32:650
Thresh JM (1974) Vector relationships and the development of epidemics: The epidemiology of plant viruses. Phytopathology 64:1050–1056
Thresh JM (1980a) The origins and epidemiology of some important plant virus diseases. Appl Biol 5:1–65
Thresh JM (1980b) An ecological approach to the epidemiology of plant virus diseases. In: Palti J, Kranz J (eds) Comparative epidemiology. PUDOC, Wageningen, pp 57–70
Thresh JM (1980c) Patterns of virus spread. Proc 5th Congr Medit Phytopathol Union, Patras Greece, pp 19–22
Thresh JM (1981) The rôle of weeds and wild plants in the epidemiology of plant virus diseases. In: Thresh JM (ed) Pests, pathogens and vegetation. Pitman Publication, London, pp 53–70
Thurston HD (1980) International potato disease research for developing countries. Plant Dis 64:252–257
Trocmé O (1972) Contribution à l'étude d'une maladie du cacaoyer en Ouganda: le désséchement écofongique des branches. Café Cacao Thé 16:219–235
Tu C, Cheng YH, Chen M (1975) Flax *Fusarium*-wilt suppressive soil in Taiwan. Plant Protect Bull Taiwan 17:390–399
Tu JC (1978) Protection of soybean from severe *Phytophthora* root rot by *Rhizobium*. Physiol Plant Pathol 12:233–240

Tu JC (1980) Incidence of root rot and overwintering of alfalfa as influenced by rhizobia. Phytopathol Z 97:97–108
Tulang MC, Bedish JW (1975) The soil conservation service and drip irrigation. Proc 33 rd Hawai Sugar Technol Conf, pp 35–40
Turgeon AJ, Meyer WA (1974) Effects of mowing height and fertilization level on disease incidence in five Kentucky bluegrasses. Plant Dis Rep 58:514–516
Turner PD (1970) Some factors in the control of root diseases of oil palm. In: Toussoun TA, Bega RV, Nelson PE (eds) Root diseases and soil-borne pathogens. Univ Calif Press, Berkeley Los Angeles, pp 194–200
Ullstrup AJ (1976) Highlights in corn pathology during the past half-century. Proc 9 th Meet Braz Soc Phytopathol, Camnas, pp 113–122
Vanderplank JE (1947) The relation between the size of the plant and the spread of systemic diseases. I. A discussion of ideal cases and a new approach to problems of control. Ann Appl Biol 34:376–387
Vanderplank JE (1963) Plant diseases: epidemics and control. Academic Press, London New York San Francisco, 349 pp
Vanderplank JE (1975) Principles of plant infection. Academic Press, London New York San Francisco, 216 pp
Van Gundy SP, Tsao PH (1963) Growth reduction of citrus seedlings by *Fusarium solani* as influenced by the citrus nematode and other soil factors. Phytopathology 53:488–489
Van Schreven DA (1948) Investigations on certain pests and diseases of Vorstenlanden tobacco. Tijdschr Plantenziekten 54:149–174 (Dutch)
Varghese G, Chew PS (1973) Studies into the biology and control of *Ganoderma* root disease of tea in Malaysia. 2nd Int Congr Plant Pathol, Minneapolis, Abstr 1183
Varma MK, Sharma HC, Pathak VN (1978) Efficacy of *Tagetes patula* and *Sesamum orientale* against root knot of eggplant. Plant Dis Rep 62:274–275
Vesely D (1979) Use of *Pythium oligandrum* to protect emerging sugar beet. In: Schippers B, Gams W (eds) Soil-borne plant pathogens. Academic Press, London New York San Francisco, pp 593–595
Viennot-Bourgin G (ed) Les champignons parasites des arbres fruitiers à noyau. Ed M Ponsot, Paris, 165 pp
Vigouroux A (1973) La verticilliose de l'amandier. Bull Tech Inf 279:343–346
Vigouroux A (1975) *Verticillium dahliae*, agent d'un dépérissement de l'olivier en France. Ann Phytopathol 7:37–44
Vigouroux A (1979) Incidence des basses températures sur la sensibilité du pêcher au dépérissement bactérien (*Pseudomonas mors-prunorum* f. sp. *persicae*). Ann Phytopathol 11:231–239
Vigouroux A, Castelain C (1969) La Verticilliose de l'abricotier. Ann Phytopathol 1:427–448
Vuittenez A (1980) Une nouvelle virose de la betterave en France: la rhizomanie. CR Acad Agric Fr 66:376–390
Walker JC (1969) Plant pathology. 3 rd edn. McGraw Hill, New York
Walker JC, Hooker WJ (1945) Plant nutrition in relation to disease development. Am J Bot 32:314–320
Walker JC, Patel PN (1964) Splash dispersal and wind as factors in epidemiology of halo blight of beans. Phytopathology 54:140–141
Waller JM (1967) Varietal resistance to sugar cane smut in Kenya. East Afr Agric Forest J 32:399–403
Wastie RL (1975) Diseases of rubber and their control. PANS 21:268–288
Weaver DJ, Wehunt EJ (1975) Effect of soil pH on susceptibility of peach to *Pseudomonas syringae*. Phytopathology 65:984–989
Webster RK, Bolstad J, Wick CM, Hall DH (1976) Varietal distribution and survival of *Sclerotium oryzae* under various tillage methods. Phytopathology 66:97–101
Weinhold AR, Oswald JW, Bowman T, Bishop J, Wright D (1964) Influence of green manures and crop rotation on common scab of potatoes. Am Potato J 41:265–273
Weischer B (1977) Stand der Bekämpfung pflanzenschädigender Nematoden. Mitt Biol Bundesanst 178:35–52
Welch ER, May RD, Riggs RD (1977) An economic analysis of controlling soybean cyst nematodes on soybeans. Arkansas Farm Res 26 (6):5
Wellman FL (1972) Tropical American plant disease. The Scarecrow Press, Metuchen New Yersey
Westerlund FV, Campbell RN, Grogan RG, Duniway JM (1978) Soil factors affecting the reproduction and survival of *Olpidium brassicae* and its transmission of Big Vein Agent to lettuce. Phytopathology 68:927–935
Wiese MV (ed) (1977) Compendium of wheat diseases. Am Phytopathol Soc, St Paul Minn, 106 pp

Wilcoxson RD (1975) The relationship between corn plant populations and smut infection. Plant Dis Rep 59:678–680
Wilhelm S (1965) Analysis of biological balance in natural soil. In: Baker KF, Snyder WC (eds) Ecology of soil-borne plant pathogens. Univ Calif Press, Berkeley Los Angeles, pp 509–517
Wilhelm S, Evans G, Snyder WC, George A, Mathre D, Garber RH, Hall D (1966) Cultural control of *Verticillium* in cotton ... a three-point approach. Calif Agric April 1966; 1–4
Wilhelm S, Sagan JE, Tietz H (1976) Optimizing *Verticillium* wilt resistance in upland cotton with short-branching, arching phenotypes and in-row crowding. 2nd Int Verticillium Symp Univ Calif, Berkeley, pp 41–42
Wilhelm S, Sagan JE, Tietz H (1978) *Verticillium* control in cotton through resistance breeding and dense plant cultivation. Proc Am Phytopathol Soc 5
Williams PM (1979) Vegetable crop protection in the People's Republic of China. Annu Rev Phytopathol 17:311–324
Wilson VE (1971) Downy mildew of peas in northern Idaho and eastern Washington in 1970. Plant Dis Rep 55:730
Woltz SS, Jones JP (1972) Control of *Fusarium* wilt of tomato by varying the nutrient regimes in soils. Phytopathology 62:799
Wu JH, Wildman SG (1963) Heat activation of different strains of tobacco mosaic virus infection on varieties of beans *(Phaseolus vulgaris)*. Phytopathology 53:1144
Wutscher HK, Shull AV (1975) Machine-hedging of citrus trees and transmission of exocortis and xyloporosis viruses. Plant Dis Rep 59:368–369
Yarham DJ (1973) Take-all decline and intensive cereal growing. ADAS Sci Arm Annu Rep 1973:200–205
Yarham DJ (1979a) The effect on soil-borne diseases of changes in crop and soil management. In: Schippers B, Gams W (eds) Soil-borne plant pathogens. Academic Press, London New York San Francisco, pp 371–383
Yarham DJ (1979b) Cultural methods for the control of plant pathogens. Br Assoc Adv Sci Annu Meet 1978, Bath Univ, pp 128–135
Yarham DJ, Hirst JM (1975) Diseases in reduced cultivation and direct drilling systems. EPPO Bull 5:287–296
Yarwood CE (1939) Control of powdery mildews with a water spray. Phytopathology 29:288–290
Yarwood CE (1943) Onion downy mildew. Hilgardia 14:595–691
Yarwood CE (1956) Heat-induced susceptibility of beans to some viruses and fungi. Phytopathology 46:323–325
Yarwood CE (1959) Microclimate and infection. In: Plant pathology problems and progress 1908–1958. Univ Wisconsin Press, Madison, pp 548–556
Yoshii K, Agudelo V de F (1977) Los efectos de *Tagetes minuta* y *Crotalaria spectabilis* en la poblàcion de nematodos y en el rendimiento posterior de la soya. Fitopatologica 12:15–19
Zachos DG (1963) Le nombre de germes en tant que facteur épidémiologique à l'évolution de l'épidémie du mildiou du tabac en Grèce pendant l'année 1963. Annal Inst Phytopathol Benaki NS 5:380–390
Zakaria MJ, Lockwood JL (1980) Reduction in *Fusarium* populations in soil by oilseed meal amendments. Phytopathology 70:240–243
Zentmyer GA, Bald JG (1977) Management of the environment. In: Horsfall JG, Cowling EB (eds) Plant disease, an advanced treatise, vol I. Academic Press, London New York San Francisco, pp 122–144
Ziegler G, Meister B, Kopp R (1976) Möglichkeiten zur Minderung der Schwarzbeinigkeit und der Knollenfäule mittels acker- und pflanzenbaulicher Maßnahmen. Tagungsber Landwirtsch Wiss DDR 140:261–273
Zimmer DE (1971) A serious outbreak of downy mildew in the principal sunflower production area of the United States. Plant Dis Rep 55:11–12
Zitter TA, Simons JN (1980) Management of viruses by alteration of vector efficiency and by cultural practices. Annu Rev Phytopathol 18:289–310
Zogg H (1948) Untersuchungen über die Epidemiologie des Maisrostes *Puccinia sorghi* Schw. Phytopathol Z 15:143–192
Zogg H (1963) Studien über die biologische Bodenentseuchung. IV. Eliminierung von *Ophiobolus graminis* im Boden unter Anwesenheit verschiedenartiger monokotyler und dikotyler Futterpflanzen. Phytopathol Z 48:272–286
Zogg H, Amiet JA (1980) Laboratory studies on decline with different foot and root rot pathogens of wheat. Phytopathol Z 97:193–213

Pathogen Index

Abutilon Mosaic Virus 81
Acyrthosiphon sesbaniae Dav. 171
Agrobacterium radiobacter (Beijerinck & van Delden) Conn 36
A. tumefaciens (E.F. Sm. & Town.) Conn 36, 107–109
Agrobacterium sp. 38
Alternaria citri Ell. & Pierce apud Pierce 51
A. longipes (Ell. & Ev.) Mason 141, 196, 197
A. solani Sor. 64, 65, 104, 141, 145, 151–154, 156, 157, 183, 186
Alternaria sp. 38, 50, 61, 68, 82, 85, 136, 158, 174, 181, 183
Anguillulina tumefaciens (Cobb) Goodey 94
Anguina agrostis (Steinbuch) Goodey 94
Anguina sp. 37, 38, 94
Aphanomyces cochlioides Drechs. 143
A. euteiches Drechs. 29, 120, 137, 141
Aphanomyces sp. 37, 38
Aphelenchoides fragariae (Ritzema-Bos) Christie 89, 192
A. oryzae Yokoo 95
A. ritzema-bosi (Schwartz) Steiner & Buhrer 165, 192
Aphelenchoides sp. 37, 38
Aphis citricola 167
A. craccivora Koch 167, 170, 171
A. gossypii Glov. 167, 189
Apple Mosaic Virus 81
Arabis Mosaic Virus 37, 181
Armillariella mellea (Fr.) Karst. 33, 35, 49, 53, 78, 92, 165, 184
Ascochyta sp. 79, 115, 143
Aspergillus sp. 33, 44
Aulacorthum solani (Kalt.) 171

Banana Bunchy Top Virus 81
Barley Stripe Mosaic Virus 181
Barley Yellow Dwarf Virus 27, 40, 61, 68, 115, 167
Barley Yellow Mosaic Virus 170
Bean Common Mosaic Virus 166
Bean Golden Mosaic Virus 171
Beet Curly Top Virus 68, 181
Beet Mild Yellowing Virus 115
Beet Necrotic Yellow Vein Virus 40
Beet Western Yellows Virus 23
Beet Yellows Virus 23
Belonolaimus longicaudatus Rau 122
Bemisia tabaci (Genn.) 171, 187, 188
Black Currant Reversion Virus 181
Botryodiplodia theobromae Pat. (B. rhodina (Berk. & Curt.) v. Arx) 51, 53, 91, 150
Botrytis cinerea Pers ex Fr. 60, 61, 63–68, 79, 141, 142, 153, 180, 187
B. fabae Sard. 138
Botrytis sp. 38
Bremia lactucae Regel 24, 92, 133, 183, 196

Capitophorus elaeagni, see Myzus elaeagni
Carnation Mottle Virus 192
Carnation Ringspot Virus 81
Carrot Motley Dwarf 68, 170
Carrot Thin Leaf Virus 68
Cassava Common Mosaic Virus 25, 81, 171
Cauliflower Mosaic Virus 115
Cavariella aegopodii 170
Cephalobus sp. 42
Cephalosporium gramineum Nis. & Ika. (Hymenula cerealis Ell. & Ev.) 33, 129, 172, 173
C. gregatum All. & Chamb. (Phialophora gregata (Allington & Chamberlain) W. Gams) 167
Cephalosporium sp. 44
Ceratocystis fimbriata Ell. & Halst. 17, 23, 85, 86, 179, 192
Cercospora apii Fres. 163, 177, 196
C. beticola Sacc. 85, 163, 174, 183, 200
C. coffeicola Berk. & Cooke 17, 23
C. zeae-maydis Tchou & Daniels 129
Cercospora sp. 26, 60
Cercosporidium personatum (Berk. & Curt.) Deighton 163, 164
Cherry Raspleaf Virus 81
Chrysanthemum Virus B 81
Citrus Exocortis Disease 86, 185
Citrus Tristeza Virus 81, 185
Claviceps paspali Stev. & Hall. 94
C. purpurea (Fr.) Tul. 94, 108, 201
Clitocybe tabescens (Scop. ex Fr.) Bres. (*Armillaria tabescens* (Scop. ex Fr.) Emel) 184
Clover Yellow Mosaic Virus 61
Cochliobolis miyabeanus (S. Ito & Kuribayashi) Drechs. ex Dastur (*Drechslera oryzae* (van Breda de Haan) Subram. & Jain) 53, 137, 139, 141, 143

C. sativus (S. Ito & Kuribayashi) Drechs. ex Dastur 107, 108
Colletotrichum coffeanum Noack 17, 86, 91, 148, 149, 163
C. falcatum Went 192
C. gloeosporioides Penz. 17
C. lindemuthianum (Sacc. & Magn.) Briosi & Cac. 40
C. linicola Pethyb. & Lafferty 79
C. musae (Berk. & Curt.) v. Arx 90
C. phomoides (Sacc.) Chester 138, 148, 153
Colletotrichum sp. 79
Coniella diplodiella (Speg.) Petr. & Syd. 62, 63
Corticium sasaki (Shirai) Matsumoto (*Thanatephorus sasakii*) 94
C. salmonicolor Berk. & Br. 17
Corynebacterium insidiosum (McCulloch) Jensen 108
C. michiganense (E. F. Smith) H. L. Jens. 86
C. nebraskense Schuster, Vidaver & Mandel 124
Coryneum carpophilum (Lév.) Jauch (Stigmina carpophila (Lév.) M. B. Ellis) 88
Cotton Leaf Curl Virus 23
Cotton Virescence Mycoplasma 25
Cowpea Aphid-Borne Mosaic Virus 68
Cowpea Mosaic Virus 68, 171
Criconemoides xenoplax Raski 39, 185
Crinipellis perniciosa (Stahel) Singer 175
Cronartium ribicola Fisch. v. Waldh. 87
Cucumber Mosaic Virus 23, 64, 67, 68, 81, 140, 167, 170, 171, 186–189
Cucumber Vein Yellowing Virus 188
Curly Top, see Beet Curly Top
Cuscuta sp. 68, 94
Cylindrocarpon destructans (Zins.) Scholten 41
Cylindrocarpon sp. 38, 44
Cylindrocladium crotolariae (Loos) Bell & Sobers 84, 86
Cytospora leucostoma Sacc., see Leucostoma persoonii

Dacus oleae (Gmel.) 63
Dahlia Mosaic Virus 81
Diaporthe vaccinii Shear 94
Didymella lycopersici Kleb. 85
Diplodia maydis (Berk.) Sacc. 124, 163
D. natalensis, see Botryodiplodia theobromae
D. zeae (Schw.) Lév. 55, 124, 137
Diplodia sp. 176
Diplodina gossypina Cke. 49
Ditylenchus angustus (Butler) Filip. 94
D. dipsaci (Kuehn) Filip. 81, 89, 142, 192
Ditylenchus sp. 37, 38
downy mildews 22, 23, 25, 26, 51, 54, 58, 61, 63, 72, 79, 82, 140, 150, 153, 163, 180, 181, 183
Drechslera dictyoides (Drechs.) Shoemaker 180
D. graminea (Rabenh. ex Schlecht.) Shoemaker 49
D. maydis (Nisikado & Miyoke) Subram. & Jain 59, 124, 172
D. oryzae, see Cochliobolus miyabeanus
D. poae (Baudys) Shoem. 180
D. sativa, see D. sorokiniana
D. sorokiniana (Sacc.) Subram. & Jain 49, 53, 112, 129, 151, 172, 180
D. turcica (Pass.) Subram. & Jain 124, 129, 137, 141
Dysdercus sp. 63

Erwinia amylovora (Burr.) Winslow et al. 65, 70, 83, 88, 90, 116, 151, 182
E. carotovora (L. R. Jones) Bergey et al. 37, 176
E. carotovora (L. R. Jones) Bergey et al. var. atroseptica (Hellmers & Dows.) Dye 139, 142
E. phytophthora (Appel) Margrow, see E. carotovora var. atroseptica
Erwinia sp. 38
Erysiphe cichoracearum DC. ex Mérat 40, 48, 70, 151
E. graminis DC. ex Mérat 70, 139, 141, 169, 172, 198
E. graminis DC. f. sp. *hordei* Em. Marchal 16, 22, 24, 70, 151, 169, 190
E. pisi DC. ex Saint-Amans 52, 53
E. polygoni DC. ex Saint-Amans 151, 173
Exobasidium vexans Massee 17, 163
Exocortis see *Citrus Exocortis* Disease

Fiji disease, see Sugarcane Fiji Disease
Fomes annosus (Fr.) Cke. 107
F. lignosus Klotzsch 187
Fomes sp. 78, 129
Frankliniella sp. 170
Fusarium avenaceum (Corda ex Fr.) Sacc. 41, 43, 45, 51, 108
F. coeruleum (Lib.) Sacc. 63
F. culmorum (W. G. Smith) Sacc. 29, 33, 43, 49, 50, 52, 53, 79, 202, 206
F. equiseti (Corda) Sacc. sensu Gordon 29
F. graminearum Schw. 33, 49, 115
F. moniliforme Sheld. 53–55, 85, 108, 112, 124, 137
F. moniliforme var. *subglutinans*, see *F. sacchari* var. *subglutinans*
F. nivalis, see *Gerlachia nivalis*
F. oxysporum Schlecht. 33, 39, 41, 42–44, 55, 97, 106, 108, 120, 180
F. oxysporum Schlecht. f. sp. *albedinis* (Kilian & Maire) Gordon 84, 116, 141, 153
f. sp. *batatas* (Wollenw.) Snyd. & Hans. 43, 110
f. sp. *coffeae* Garcia 17
f. sp. *conglutinans* (Wollenw.) Snyd. & Hans. 141
f. sp. *corianderi* Nanila & Joshi 120

f. sp. *cubense* (E. F. Smith) Snyd. & Hans. 30, 43, 95, 96, 120
f. sp. *cucumerinum* Owen 43
f. sp. *cyclaminis* Gerlach 43
f. sp. *dianthi* (Prill. & Del.) Snyd. & Hans. 43, 84, 142
f. sp. *elaeidis* Toovey
f. sp. *gladioli* (Massey) Snyd. & Hans. 192
f. sp. *lini* (Bolley) Snyd. & Hans. 43, 140
f. sp. *lycopersici* (Sacc.) Snyd. & Hans. 29, 31, 39, 43, 45, 53, 97, 110, 111, 136, 138, 143, 150, 185, 201
f. sp. *melonis* Snyd. & Hans. 43, 44, 141, 142, 185
f. sp. *nicotianae* (J. Johnson) Snyd. & Hans. 110
f. sp. *phaseoli* Kendr. & Snyd. 43, 100, 271
f. sp. *pisi* (van Hall) Snyd. & Hans. 110
f. sp. *raphani* Kendr. & Snyd. 43
f. sp. *vasinfectum* (Atk.) Snyd. & Hans. 29, 39, 97, 110, 121, 136, 141, 201
F. roseum Lk. 33, 44, 120, 179
F. sacchari (Butler) W. Gams var. *subglutinans* Nirenberg 83
F. solani (Mart.) Sacc. 39, 42, 43, 44, 105, 108, 120, 176
F. solani (Mart.) Sacc. f. sp. *batatas* McClure 33
f. sp. *phaseoli* (Burk.) Snyd. & Hans. 54, 109, 127, 138
f. sp. *pisi* (F. R. Jones) Snyd. & Hans. 33, 127, 143, 173
Fusarium sp. 5, 8, 34, 38–40, 45, 55, 70, 76, 82, 84, 85, 93, 104, 112, 129, 137, 141, 150, 169, 172, 176, 201
Fusicoccum amygdali Delacr. 153

Gaeumannomyces sp. 38
Gaeumannomyces graminis (Sacc.) v. Arx & Olivier 33, 36, 43, 45, 54, 71, 76, 94, 106, 107, 113, 120, 121, 124, 126, 128, 130, 137, 140
Ganoderma sp. 78, 91, 92, 129
Gerlachia nivalis (Ces. & Sacc.) W. Gams & E. Muell. 49, 55, 94, 111, 120, 124, 141, 169, 171, 172
Gibberella fujikuroi (Saw.) Wollenw. 136, 141, 163
G. stilboides Gordon ex Booth 53, 91
G. zeae (Schw.) Petch 48, 55, 124, 141
Globodera pallida (Stone) 99, 117, 176
G. rostochiensis (Woll.) 88, 99, 111, 112, 176, 178, 179
Globodera sp. 112
Gloeosporium fructigenum Berk. 153
G. limetticola Clausen 17
Gloeosporium sp. 85
Gloeotinia temulenta (Prill. & Delacr.) Wilson, Noble & Gray 94
Glomerella cingulata (Stonem) Spaulding & Schrenk 91, 158
G. tucumanensis (Speg.) v. Arx & Mueller 25, 176
Godronia cassandrae Peck 94
Grapevine Fanleaf Virus 81
Groundnut Rosette Virus 166, 167, 170
Guignardia bidwelli (Ell.) Viala & Ravaz 91, 130, 199
Gymnosporangium spp. 87

Helicotylenchus sp. 38, 112, 122
Helminthosporium sp. (see also Cochliobolus, Drechslera) 46, 79
H. sigmoideum Cav. (Nakataea sigmoidea Hara) 141
Hemileia coffeicola Maubl. & Roger 149
H. vastatrix Berk. & Br. 17, 149
Hendersonia toruloidea Nattrass 179
Heterodera avenae Filip. 31, 71, 84, 117
H. glycines Ichinohe 39, 42, 100, 112, 138
H. marioni (Cornu) Marcinowski 122
H. schachtii Schmidt 31, 84, 112, 173
H. solanacearum 41
H. tabacum Lownsbery & Lownsbery 122
H. trifolii Goffart 42
Heterodera sp. 38, 105, 112
Hoja blanca Rice Virus, see Rice Hoja Blanca Virus
Hoplolaimus columbus Sher. 29, 54, 71, 127
H. galeatus (Cobb) Filip. 31
H. tylenchiformis Daday 122
Hylemyia cilicrura Rond. 37

Iris Mild Mosaic 81

Kabatiella zeae Narita & Hiratsuka (*Aureobasidium zeae* (Narita & Hiratsuka) Dingley) 129, 153
Laodelphax striatella Fallen 170
Leptosphaeria coniothyrium (Fckl.) Sacc. 63
L. maculans (Desm.) Ces. & de Not. 58, 173
Leptosphaeria sp. 94
Lettuce Big Vein Disease 40
Lettuce Mosaic Virus 133
Leucostoma cincta (Fr.) Hoehnel 51
L. persooni (Nits.) Hoehnel 51, 53
Leveillula taurica (Lév.) Arn. 55, 67, 69, 70, 176, 193
Lily Symptomless Virus 81
Longidorus elongatus (de Man) Thorne & Swanger 37
L. macrosoma Hooper 37
Longidorus sp. 37, 38

Macrophomina phaseolina (Tassi) Goid. 9, 33, 34, 49, 50, 52, 53, 55–57, 64, 67, 89, 120, 131, 138, 161

Maize Chlorotic Dwarf Virus 27, 68
Maize Dwarf Mosaic Virus 27, 68
Maize Rough Dwarf Virus 68, 170
Meloidogyne arenaria (Neal) Chitwood 116, 117
M. hapla Chitwood 42, 50, 54, 67, 84
M. incognita (Kofoid & White) Chitwood 29, 39, 41, 67, 71, 103, 107, 116, 117, 122, 142, 192, 196, 197
M. javanica (Treub) Chitwood 39, 42, 67, 111, 117, 122
Meloidogyne sp. 38, 39, 95, 105, 112, 122, 141, 150, 185
Microsphaera grossulariae (Wallr.) Lév. 90
Mistletoe, see Topinanthus bangwensis
Monilia cinerea Bonn. 63
M. roreri Cif. 90, 175
Monilochaetes infuscans Halst. ex Harter 192
Mycosphaerella pinodes (Berk. & Blox.) Stone (Didymella pinodes (Berk. & Blox.) Petr. 51, 183
Myrothecium sp. 44
Myzus (Capitophorus) elaeagni D. Guer 167
Myzus persicae (Sulz.) 23

Narcissus Mosaic Virus 81
Nectria cinnabarina (Tode ex Fr.) Fr. 88
Nectria galligena Bres. 63
Nectria sp. 51
Nematospora gossypii Ashby & Nowell 63

Oat Mosaic Virus 40
Oidium sp. (rubber) 16, 17, 58
Olpidium brassicae (Woron.) Dang. 40, 43
Onion Yellow Dwarf Virus 26, 81
Oospora pustulans Owen & Wakefield 178
Ophiobolus graminis, see *Gaeumannomyces graminis*
Orobanche aegyptiaca Pers. 97
Orobanche sp. 68, 95, 116, 117, 177, 178

Paratrichodorus minor (Colbran) 37, 39
Passionfruit Woodiness Virus 81
Pea Early Browning Virus 37
Pea Seed-Borne Mosaic Virus 68
Peach Mosaic Virus 81
Peach Rosette and Decline Virus 90
Peach Yellows Virus 192
Peanut Mottle Virus 181
Pear Red Mottle Virus 192
Pear Vein Yellows Virus 192
Pectobacterium carotovorum var. *atrosepticum*, see *Erwinia carotovora* var. *atroseptica*
Pelargonium Flower Break Virus 81
Penicillium sp. 33
Peronospora brassicae Gaeum. 24, 183
P. destructor (Berk.) Casp. 23–25, 71, 89, 183, 192
P. farinosa (Fr.) Fr. 24, 25, 163, 183, 201
P. parasitica (Pers. ex Fr.) Fr. 141, 163
P. pisi, see P. viciae
P. tabacina Adam 24, 25, 52, 53, 89, 163, 173, 183, 196
P. viciae (Berk.) de Bary 25, 92
Phialophora radicicola Cain 36
Phleospora idahoensis Sprague 94
Phoma betae Frank 169
P. lingam (Tode ex Fr.) Desm. 183
P. trifolii E.M. Johnson & Valleau 176
Phoma sp. 38, 84, 85
Phomopsis sojae Lehman 89
P. sclerotioides van Kesteren 43
Phyllosticta maydis Arny & Nelson 124, 129
Phymatotrichum omnivorum (Shear) Duggar 6, 29, 33, 67, 105, 108, 113, 121, 133, 138, 140
Physalospora urenae 53
Physoderma maydis Miyabe 124
Phytophthora cactorum (Leb. & Cohn) Schroet. 49, 71, 108, 153, 185, 187, 201
P. cinnamomi Rands 6, 43, 49, 54, 120, 185
P. colocasiae Rac. 78
P. citrophthora (Smith & Smith) Leonian 15, 64, 71, 91, 130, 138, 153, 159, 160, 185, 187
P. cryptogaea Pethyb. & Laff. 53, 132
P. infestans (Mont.) de By. 16, 20–24, 27, 51, 59, 63, 64, 89, 141, 143, 148, 153, 154, 156–158, 163, 177, 178, 193
P. megasperma Drechs. var. *sojae* Hildeb. 41, 49
P. palmivora (Butl.) Butl. 17, 90, 91, 199
P. parasitica Dast. (P. *nicotianae* van Breda de Haan var. *parasitica* (Dast.) Waterhouse) 41, 91, 121, 159, 185
P. parasitica Dast. var. *nicotianae* (van Breda de Haan) Tucker (P. *nicotianae* van Breda de Haan var. *nicotianae* (van Breda de Haan) Tucker) 39, 84, 95, 142, 196, 197
P. syringae (Kleb.) Kleb. 71
Phytophthora sp. 34, 38, 40, 45, 82, 84, 120
Plasmodiophora brassicae Wor. 22, 29, 30, 71, 108, 117, 136, 140, 142, 153
Plasmopara halstedii (Farl.) Berl. & de Toni 173
P. viticola (Berk. & Curt.) Berl. & de Toni 90, 133, 141, 150, 166
Pleiochaeta setosa (Kirchner) Hughes 94
Plum Pox Virus 81
Podosphaera leucotricha (Ell. & Ev.) Salm. 5, 20, 22–24, 90, 116, 150
Polymyxa betae Keskin 40, 84
P. graminis Ledingham 40
Polyspora lini Laff. 79, 173
Poria sp. 78
Potato Leafroll Virus 81, 176
Potato Mop Top Virus 39
Potato Virus X 39, 166, 176
Potato Virus Y 176, 187–189

Pratylenchus alleni Ferris 31, 117
P. brachyurus (Godfr.) Filip. & Schuurm. 104, 107
P. crenatus Loof. 31
P. indicus Das. 103, 107
P. neglectus (Rensch.) Filip. & Schuurm. 117
P. penetrans (Cobb) Sher. & Allen 31, 107, 116, 117, 122, 123
P. thornei Sher. & Allen 39, 97
Pratylenchus sp. 38, 84, 117, 173
Protomyces macrosporus Unger 143
Prune Dwarf Virus 81
Prune Necrotic Ringspot Virus 181
Pseudocercosporella herpotrichioides (Fron) Deighton 8, 34, 77, 94, 111, 113, 120, 121, 124, 202
Pseudomonas lachrymans (E.F. Smith & Bryan) Carsner 86, 153, 178
P. mors-prunorum Wormald f. sp. *persicae* Prunier et al. 50, 51
P. phaseolicola (Burkholder) Dows. 153, 174
P. savastanoi (E.F. Smith) F.L. Stevens 63
P. solanacearum (E.F. Smith) E.F. Smith 29, 43, 45, 104, 107, 128, 133, 138, 185, 196, 197
P. syringae van Hall 29, 39, 51, 91, 141
P. tomato (Okabe) Alstatt 174
P. woodsii (E.F. Smith) F.L. Stevens 89
Pseudomonas sp. 38
Pseudoperonospora cubensis (Berk. & Curt.) Rost. 22, 23, 145, 148, 151, 160, 183
P. humuli (Miy. & Tak.) Wils. 90
Pseudopeziza medicaginis (Lib.) Sacc. 176
Pseudoplea trifolii (Rostr.) Petr. (*Leptosphaerulina trifolii* (Rostr.) Petr.) 176
Puccinia asparagi DC. 94
P. caricis (Schum.) Schroet. var. *grossulariata* Arth. 87
P. carthami (Hutzelmann) Cda. 96
P. cerasi (Bérenger) Castagne 87
P. coronata Cda. 24, 87
P. graminis Pers. 67, 70, 87–89, 137
P. graminis avenae Eriks. & Henn. 68
P. graminis tritici Eriks. & Henn. 22, 141, 186
P. hordei Otth 87
P. menthae Pers. 94, 176, 192
P. poae-nemoralis Otth 94
P. recondita Rob. ex Desm. 22, 87, 131, 169, 186, 193
P. sorghi Schw. 87
P. stakmanii Presley 87
P. striiformis Westend 24, 70, 89, 137
Pyrenochaeta lycopersici Schneider & Gerlach 97, 185
P. terrestris (Hans.) Gorenz, J.C. Walker & Larson (Phoma terrestris Hansen) 97
Pyrenophora bromi (Died.) Drechs. 129
Pyricularia oryzae Cavara 52, 53, 136, 137, 139, 142, 143
Pythium aphanidermatum (Edson) Fitzpatrick 105, 169
P. myriotylum Drechs. 104, 176
P. oligandrum Drechs. 36
P. splendens Braun 17
P. ultimum Trow 39, 41, 43, 49, 127
Pythium sp. 32, 34, 38–40, 43–45, 47–49, 84, 97, 106, 120, 121, 128, 137, 138, 143

Radish Mosaic Virus 115
Radopholus similis (Cobb) 95, 150
Raspberry Bushy Dwarf Virus 81
Raspberry Ringspot Virus 37
Rattle viruses 38
Rhizobium japonicum (Kirchner) Buchanan 41, 42
Rhizobium sp. 41
Rhizoctonia crocorum (Pers.) DC. ex Fr. 176
R. lamellifera Small 17
R. leguminicola Gough & Elliot 176
R. solani Kuehn 33, 36, 39, 43, 47–49, 71, 97, 106, 108, 120–122, 127, 128, 131–133, 137, 138, 142, 163,169, 172, 176, 177, 187, 201
Rhizoglyphus echinopus (Fumouze & Robin) 37
Rhizopus stolonifer (Ehrenb. ex Fr.) Lind 179, 187
Rhizopus sp. 44, 85, 159
Rhopalosiphum sp. 170
Rhynchosporium secalis (Oudem.) J.J. Davis 99, 124, 131, 181, 183
Rice Hoja Blanca Virus 68
Rice Tungro Virus 68
Rosellinia bunodes Berk. & Br. 29
R. necatrix (Hartig) Berl. ex Prill 84, 129, 158, 160
R. pepo Pat. 29
Rotylenchulus reniformis Linf. & Oliv. 107, 111
Rotylenchulus sp. 117
Rotylenchus sp. 38

Schizophyllum sp. 50, 91
Scleroderma bovista Fr. 40
Sclerophthora rayssiae Kenn., Kolt. & Wahl var. zeae Payak & Renfro 70
Sclerospora graminicola (Sacc.) Schroet. 23, 89
S. sacchari Miyake 88, 89
S. sorghi (Kulk.) Weston & Uppal 70, 92, 172
Sclerospora sp. 120
Sclerotinia fructicola (Winter) Rehm 26, 130, 141, 179
S. fructigena Aderh. & Ruhl. 91, 179
S. homoeocarpa F.T. Bennett 53
S. laxa Aderh. & Ruhl. 60
S. minor Jagger 151

S. sclerotiorum (Lib.) de Bary 33, 67, 82, 84, 85, 95, 96, 131, 133, 143, 151, 164, 181–183, 200
Sclerotinia sp. 5, 92, 104, 108, 163
Sclerotium cepivorum Berk. 151
S. oryzae Catt. 94, 128, 136, 143, 152
S. rolfsii Sacc. 6, 9, 29, 33, 36, 49, 50, 58, 60, 67, 82, 84, 85, 89, 97, 102–109, 111, 116, 124, 129, 131, 132, 139, 140, 142, 145, 151, 163, 164, 171, 187, 201
Sclerotium spp. 5, 38, 92, 163
Scutellonema bradys (Steiner & LeHew) Andràssy 192
Selenophoma bromigena (Sacc.) Sprague & A. G. Johnson 94
Septoria apii (Briosi & Cav.) Chester 153, 196
S. avenae Frank. 94
S. nodorum (Berk.) Berk. 94, 124, 129, 131, 139, 140, 169, 183, 198
S. tritici Rob. ex Desm. 115, 124, 129, 148, 164, 183, 202
Septoria sp. 60, 76, 94, 181
Sitona sp. 171
Soybean Mosaic Virus 166, 167
Sphacelotheca sorghi (Link) Clint. 161, 162
Sphaerotheca fuliginea (Schlecht. ex Fr.) Poll. 67, 87
S. humuli (DC.) Burr. 90, 130
S. pannosa (Wallr.) Lév. 90, 151, 196
Sphaerulina oryzina Hara 141
Spongospora subterranea (Wallr.) Lagh. 39, 40, 117
Stagonospora curtisii (Berk.) Sacc. 192
Stemphylium botryosum Wallr. 145, 176
Stemphylium sp. 68, 174, 181
Stereum purpureum (Pers. ex Fr.) Fr. 92, 142, 165, 184
Strawberry Crinkle Virus 81
Strawberry Latent Ringspot Virus 37
Streptomyces ipomeae (Person & W. J. Martim) Waks. & Henrici 33
S. scabies (Thaxt.) Waks. & Henrici 29, 33, 43, 113, 121, 139, 140, 142, 143, 146, 147, 161, 201
Striga asiatica (L.) Kuntze 117
Subterranean Clover Red Leaf Virus 171
Sugarbeet Yellows Virus 171
Sugarcane Chlorotic Streak Virus 192
Sugarcane Fiji Virus 81
Sugarcane Mosaic Virus 192
Sugarcane Ratoon Stunting Disease 61, 192
Synchytrium endobioticum (Schilb.) Pers. 49, 83

Take-all, see Gaeumannomyces graminis
Taphrina deformans (Berk.) Tul. 90
Thielaviopsis basicola (Berk. & Br.) Ferraris 29, 33, 41, 49, 97, 138
Thielaviopsis sp. 120
Thomasiniana theobaldi 63
Thrips tabaci Lind. 170
Tilletia caries (DC.) Tul. 161, 162, 172, 173
T. foetida (Wallr.) Liro 172, 177
T. laevis, see T. *foetida*
T. tritici, see T. *aries*
Tobacco Mosaic Virus 86, 136
Tobacco Necrosis Virus 39, 40
Tobacco Rattle Virus 37
Tobacco Ringspot Virus 37
Tobacco Stunt Virus 40
Tomato Leaf Curl Virus 167
Tomato Mosaic Virus 40, 49, 64
Tomato Ringspot Virus 37
Tomato Spotted Wilt Virus 170
Tomato Yellow Leaf Curl Virus 171, 183, 188, 189, 193
Topinanthus bangwensis 17
Trachysphaera fructigena Tabor & Bunting 90
Trametes sp.
Tranzschelia pruni-spinosae (Pers.) Diet. var. *discolor* (Fuck.) Dunegan 55, 192
Tranzschelia sp. 90
Trichoderma harzianum Rifai 36
Trichoderma sp. 38, 44
Trichodorus christiei, see *Paratrichodorus minor*
T. similis Seinhorst 37
Trichodorus sp. 37, 38, 95, 117
Tristeza, see Citrus Tristeza Virus
Tulip Breaking Virus 81
Tungro Virus, see Rice Tungro Virus
Tylenchorhynchus dubius (Buetschli) Filip. 31
Tylenchorhynchus sp. 38, 84, 95, 112, 122
Tylenchulus semipenetrans Cobb 84, 122, 123
Tylenchulus sp. 38
Typhula idahoensis Remsberg 49, 55, 94, 111, 120, 124, 169, 171, 172

Uncinula necator (Schw.) Burr. 90
Urocystis agropyri (Preuss.) Schroet. 94, 161, 162
U. tritici Koern., see U. agropyri
Uromyces phaseoli (Pers.) Wint. 40
U. striatus Schroet. var. *medicaginis* (Pass.) Arth. 176
Ustilago hordei (Pers.) Lagerh. 161, 162
U. kolleri Hille 30, 31
U. maydis (DC.) Cda. 141, 167, 172
U. scitaminea Syd. 25, 89
U. tritici (Pers.) Rostr. 60

Venturia inaequalis (Cke.) Wint. emend. Aderh. 20, 23, 63, 90, 133, 139, 151, 174
Verticillium albo-atrum Reinke & Berth. 29, 31, 34, 37, 39, 43, 67, 70, 71, 83, 86, 105, 107, 110, 112, 116, 117, 119, 120, 137, 138, 142, 165, 166, 178, 179, 185, 202–204

V. dahliae Kleb. 6, 39, 63, 67, 70, 71, 79, 83–85, 92, 94–97, 102–108, 110, 112, 116, 120, 128, 129, 136, 138, 141, 142, 150, 151, 166, 179, 202
Verticillium sp. 5, 24, 38, 45, 50, 61, 112, 129, 130, 136, 138, 141, 142

Watermelon Mosaic Virus 68, 187
Wheat Rosette Dwarf Virus 89
Wheat Soil-Borne Mosaic Virus 40
Wheat Spindle Streak Virus 39, 40
Wheat Streak Mosaic Virus 129

Xanthomonas albilineans (Ashby) Dows. 25
X. ampelina Panagopoulos 184
X. campestris (Palm.) Dows. 64, 151, 153
X. citri (Hasse) Dows. 108, 186
X. coryli P.W. Miller et al. 53
X. malvacearum (E.F. Smith) Dows. 25, 84, 85, 89, 145, 174
X. manihotis (Arth.-Ber.) Starr 100, 141
X. oryzae (Uyeda & Ishiyama) Dows. 71, 139, 141, 143
X. phaseoli (E.F. Smith) Dows. 84
X. rubrilineans (Lee et al.) Starr & Burkh. 89
X. translucens (L.R. Jones, A.G. Johnson & Reddy) Dows. 51
X. vesicatoria (Doidge) Dows. 174
Xiphinema americanum Cobb 37, 122
X. coxi Tarjan 37
X. diversicaudatum (Micoletzky) Thorne 37
X. index Thorne & Allen 37
Xiphinema sp. 37, 38, 129
Xyloporosis 86

Subject Index

Aegilops sp. 70
aeration 9, 17, 58, 59, 62, 64
Agropyron repens 68, 70, 71, 132
Agrostis sp. 201
Alopecurus myosuroides 68
Alopecurus sp.
Amaranthus graecizans 110
A. retroflexus 110
A. spinosus 71
Amaranthus sp. 68
alfalfa, see lucerne
almond 63, 107, 136, 153
alternate hosts 78, 87, 88
Anchusa sp. 87
anemone 37, 192
Anemone coronaria 87
apple 20, 22–24, 36, 49–51, 63, 72, 81, 90, 91, 107, 116, 123, 124, 129, 133, 139, 150, 151, 153, 158, 174, 179, 185, 196
apricot 63, 129, 141, 142, 179
asparagus 94, 117
Aspilia africana 71
Avena sp. 71
avocado 6, 13, 29, 43, 49, 54, 107, 120, 185
azalea 120
Azidarachta indica 122

banana 13, 23, 29, 43, 81, 87, 90, 93, 95, 96, 120, 133, 206
bark ringing 92, 184
barley 8, 16, 22, 24, 36, 40, 49, 51, 61, 68, 70, 74, 83, 87, 89, 94, 101, 106, 107, 112, 113, 115, 119–122, 124, 127, 130, 131, 136, 140, 141, 151, 161, 167, 169, 170, 172, 180, 190
bean 40, 43, 49, 54, 64, 80, 84, 101, 105–107, 109, 110, 120, 121, 138, 143, 151, 153, 163, 169, 171, 174, 177
bees 182
beet (see also sugarbeet) 24, 36, 37, 40, 76, 85, 89, 107, 117, 163, 174
Benincasa cerifera 185
bentgrass (*Agrostis palustris*) 151
Berberis sp. 87, 88
black currant 37
blueberry (*Vaccinium* sp.) 94
bluegrass (*Poa pratensis*) 53, 94, 179, 180
Bouteloua sp. 87
Brassicae 22, 24, 25, 29, 64, 80, 115, 117, 136, 140, 151, 163, 174, 183
Bromus inertis 94
buckwheat 120, 122
burning plant residues (see also flaming) 60, 93–95, 183

cabbage 61, 71, 100, 141, 153
calcium 142
Cardamine pratensis 71
Carex sp. 87
carnation 43, 45, 81, 84, 89, 142, 192
carrot 25, 68, 89, 97, 100, 107, 111, 164, 170, 174, 177, 183
cassava 25, 81, 88, 100, 105, 141, 171
castor bean 117, 122
cauliflower 30, 141
celery 25, 37, 80, 95, 131, 163, 164, 177
cereals 7, 8, 11, 33, 67, 71, 74, 75, 79, 84, 93, 105–107, 113–115, 118, 124, 126, 128, 130, 134, 135, 144, 148, 151, 161, 168, 177, 181, 183, 195, 198, 202, 206
Chenopodium album 110
C. amaranticolor 68
Chenopodium sp. 68, 70
cherry 60, 63, 80, 81, 87
chickpeas 80, 117
chilli, see pepper
chilling injury 50, 66, 158, 202
chrysanthemum 36, 117, 165
Cirsium arvense 70
Cirsium sp. 70
citrus 6, 15, 17, 39, 41, 51, 64, 72, 80, 81, 84, 91, 123, 130, 138, 150, 153, 159, 160, 185, 186, 189, 202
Citrus aurantifolia (lime) 189
C. aurantium, see sour orange
clove 56
clover 33, 42, 55, 61, 80, 115, 121, 168, 176, 180
Coccinia grandis 68
cocoa 16, 17, 23, 29, 80, 85, 87, 90, 91, 114, 116, 175, 199
coffee 7, 15–17, 23, 29, 53, 80, 91, 148, 149, 163, 175, 184
communal pest management 1, 5, 6, 9, 18–27, 78, 83, 190, 191, 195, 202, 204, 205
coriander 120, 143

corn, see maize
cost/benefit 73, 74
cotton 6, 23, 25, 29, 33, 39, 41, 50, 53, 54, 56, 57, 63, 65, 71, 84, 85, 87, 92, 95, 96, 97, 100, 103, 106, 107, 110, 112, 115, 120, 121, 128, 133, 136, 138, 140, 141, 145, 161, 165, 166, 168, 174, 183, 201, 202
cowpea 68, 71, 110, 111
crop age 56, 58–64
crop canopy, see density of crop
crop climate 1, 10, 11, 14–18, 20, 61, 62, 72, 126, 130, 131, 133, 146–148, 150, 158, 163, 176, 200
crop insurance 62, 75
crop nutrition 1, 14, 16, 19, 20, 35, 54, 55, 75, 76, 99, 109, 133–143, 205
crop rotation, see crop sequence
crop sequence 6, 7, 19, 35, 45, 71, 74, 76, 77, 98–112, 136, 143, 144, 171, 200, 202, 204
Crotalaria sp. 116, 117, 122
crucifers 108, 117, 120
cucumber 22, 40, 41, 43, 50, 80, 86, 87, 148, 151, 153, 160, 167, 178, 185, 188
cucurbits 23, 43, 70, 78, 87, 131, 164, 183, 184
cyclamen 43
Cynodon dactylon 71
C. transvalensis 94
Cyperus esculentus 71
C. rotundus 71

Dactylis glomerata 71
dahlia 81
damping-off 36, 58, 66, 161, 163, 169
Datura stramonium 117
date palm 84
Daucus sp. 68
debris management 35
defoliation strain 55, 180
density of crop (see also sowing density) 7, 14, 15, 64, 72, 147, 149, 150, 160
dew 14, 16, 66, 145, 148, 154, 156–160, 162, 165, 166, 182, 186
Digitaris sanguinalis 68, 70, 110
disease-free period 5, 175, 182, 202, 205
dormancy (including dormancy breaking) 18, 23, 65, 90, 169, 174, 184, 202
drainage 54, 83, 96, 125, 143–145, 152

Easter lily 120
Echinocloa sp. 68
eco-fallowing 103
eggplant 24, 64, 80, 97, 107, 110, 117, 122, 163, 185
Eranthis sp. 87
erosion 76, 146, 205

fallow 8, 95, 99, 103–105, 118, 144
fertilization (see also crop nutrition) 18, 26, 31, 64, 77, 92, 99, 102, 103, 117, 118, 133, 135, 150, 158, 190, 193, 194, 202, 207
Festuca arundinacea 94
F. rubra 94
filbert nut 53
flaming 93–95
flax 43, 71, 79, 80, 117, 122, 140
Fleurya estruans 71
flooding 46, 54, 83, 95, 96, 102, 128, 152, 162
flower crops 14, 26, 74, 82, 86, 163, 165, 194
flowering period, manipulation of 18, 148
forage crops 14, 60, 74, 168, 169, 176
frost 12, 13, 46, 50, 51, 62, 65, 66, 124, 149, 151, 158, 168, 173, 202, 206
fruiting period, manipulation of 18, 148

Gaillardia 116
garlic 26, 80, 81
gladiolus 37, 187, 192
gooseberry 87, 90
grapefruit 51, 159
grape vine 8, 37, 62–64, 66, 81, 86, 90, 91, 129, 130, 132, 141, 150, 166, 184, 199
green manure 102, 106, 111, 118, 121, 125, 144
groundkeepers 78, 88, 132
groundnuts 6, 7, 27, 36, 49, 56, 57, 80, 84, 89, 97, 100, 103, 104, 107, 116, 117, 131, 132, 134, 142, 163, 164, 166–168, 170, 176
guttation 151, 165

hail 58, 59, 62, 63, 65, 168, 202
harvesting practices 14, 19, 25, 27, 62, 65, 85, 134, 144, 145, 175–180, 204
harvesting season 5, 25, 65
heat treatment 91, 191, 192
Hellenium sp. 116
herbicides 71, 79, 200–203
Heteropogon contortus 70
Holcus lanatus 71
hop 71, 80, 83, 84, 90, 129, 130, 202–205
Hordeum sp. 70
host-free period 25, 26, 103
host range 34, 37, 45, 66, 67, 70, 81, 87, 180, 195, 196, 200

insolation, see radiation
intercropping 7, 116, 118, 206
iris 81
iron 142
irrigation 1, 2, 8–10, 14–16, 18, 19, 22, 26, 27, 35, 47, 50–52, 55, 59, 64–66, 70, 72, 75–77, 83, 98, 102–104, 111, 115–119, 125, 126, 130–132, 135–137, 145–160, 162, 169, 190, 192–195, 200, 205–207
Isopyrum sp. 87

Juniperus sp. 87
jute 56

kenaf 23, 80

lawn 27, 180
leek 80
Leersia hexandra 68
L. oryzoides 71
legumes 75, 79, 81, 92, 107, 114, 115, 118, 121, 149, 176, 183
lentils 43, 45, 80
lespedeza 122
lettuce 24, 37, 40, 43, 64, 92, 100, 109, 131, 133, 164, 183, 187
lima bean 141
lily 81
Lolium perenne 68, 71, 94
L. rigidum 94
Lolium sp. 201
lotus 95
lucerne 24, 27, 29, 37, 41, 56, 58, 61, 69, 80, 93, 94, 116, 120, 122, 136, 141, 142, 165, 176, 178
lupin 93, 94, 123

maize 16, 23, 27, 46, 48–50, 53, 55, 56, 59, 68, 70, 76, 79, 80, 101, 104, 107, 109, 110, 112, 115, 117, 120, 121, 124, 129, 132, 136, 137, 141, 151–153, 162, 163, 167, 169, 170, 172, 198–200, 206
Malva sylvestris 110
mandarin 185
manganese 143
mango 6, 9, 29, 53, 107, 142, 150
mangold 27
marrow, see squash
Melilotus officinalis 121
melon 23, 43–45, 80, 87, 112, 116, 141, 142, 145, 185, 186
millet 80, 89, 161
mint 14, 43, 93, 94, 176, 192
misting 10, 51, 65, 149, 151
mistletoe 17
molybdenum 143
Momordia charantia 68
M. dioica 68
monoculture 75, 111–113, 118, 206
monsoon 145
mulching 6, 9, 10, 13, 16, 18, 50, 64, 86, 92, 96, 118–125, 158, 160, 186–189
mustard 115, 121, 122
mycorrhiza 35, 40, 41

narcissus 81, 89, 117, 192
nematodes 6, 34, 37–40, 42, 79–82, 89, 96, 101, 103, 105–107, 112, 116, 119, 125, 127, 129, 131, 133, 142, 144, 150, 152, 180, 181, 191
Newbouldia laevis 71
nitrogen 31, 47, 51, 61, 64–66, 76, 77, 113, 114, 121, 130, 134–140, 163, 198, 203, 204
non-hosts 109, 110
nutrition, see crop nutrition

oat 8, 24, 30, 31, 40, 46, 67, 68, 71, 80, 81, 87, 89, 94, 113, 117, 121, 122
oilpalm 9, 17, 92, 116, 141, 153, 184, 201
oil-radish, see *Raphanus oleiferus*
okra 39, 80, 110, 117, 122
olive 8, 63, 83, 84, 107, 116, 117, 129, 134, 136, 150, 179, 198
onion 23, 25, 66, 72, 80, 81, 89, 97–99, 109, 114, 132, 183, 192
opium poppy 56
Ornithogalum sp. 87
Oryza sp. 68
Oxalis stricta 87

Panicum sp. 68
Papaver rhoeas 117
parsley 50
Paspalum dilatatum 94
passion fruit 81
pasture 93
pea 25, 29, 33, 37, 43, 51, 52, 68, 80, 92, 105, 107, 111, 112, 120, 121, 127, 136, 138, 141, 173
peach 23, 29, 36, 37, 49–51, 80, 81, 87, 90, 107, 116, 122, 153, 179, 182, 192
pear 22, 36, 51, 70, 80, 83, 87, 90, 151, 182, 185, 192, 196
pecan 40
pelargonium 81
pepper 55, 56, 69, 70, 80, 163, 186–189, 201
peppermint, see mint
Phaseolus lathyroides 68
phosphorus 130, 140
pigeon pea 115
pineapple 82, 83, 107, 117
Pinus sp. 87
pistachio nut 97
plum 80, 81, 87, 179
Poa pratensis, see bluegrass
Poa sp. 201
poinsettia 120
pomegranate 8
pomelo 187
Portulaca oleracea 110
potato 16, 20–24, 27, 29, 31, 33, 37, 39, 40, 43, 49–51, 56, 57, 59, 63, 64, 75, 76, 80, 81, 83, 86, 87, 89, 92, 94, 97, 99–102, 104–107, 110–112, 117, 119, 120, 122, 127, 128, 131, 139–143, 146, 147, 150, 151, 153, 154, 158, 161–164, 166, 168, 176, 177, 179, 183, 186, 193, 198, 199, 201, 206
potassium 57, 134, 140–142

proximity (to source of inoculum) 6, 7, 19, 22, 26, 27, 180–183, 199
prune, see stone fruits
pruning 14, 26, 62, 65, 85, 86, 91, 162, 166, 169, 174, 184
pumpkin 143

radiation 10–12, 46, 58, 59, 97, 116, 125, 146, 159, 184, 197
radish 43, 80, 115
rain 4, 6, 8–11, 14–17, 52, 74, 76, 77, 135, 137, 144, 145, 147–150, 159, 162, 165, 169, 174, 175, 196, 197, 206
rape 58, 121, 173, 182
Raphanus oleiferus 117, 121
raspberry 36, 37, 63, 80, 81, 90
ratooning 25
red currant 37, 87
regional pest management, see communal pest management
Reseda odorata 117
rhizomania 40, 84
rhododendron 120
rhubarb 37, 89
rice 52, 53, 68, 71, 79, 80, 94–96, 103, 105–107, 114, 115, 136, 137, 139, 142, 143, 151, 152, 168, 173, 193
risk assessment 74, 75
rose 36, 49, 81, 90, 109, 151, 196
rubber 16, 17, 187
rye 8, 80, 113, 121, 123, 151, 172
rye-grass 117, 141

safflower 53, 80, 96, 101, 117
sanitation 1, 5, 7, 18, 19, 26, 27, 60, 77–98, 100, 132, 202, 205, 206
scald 46, 58, 165, 184
seed crops 25, 61, 64, 65, 88, 89, 103
seed transmission of disease 37, 79–81, 172, 174, 182, 191
senescence 55, 57, 58, 60–66, 88, 131, 133, 134, 141, 145, 149, 150, 160, 163–165, 176, 182
sesame 107, 117
setaria 89
sewage sludge 85, 118, 122, 123
shade 10, 15–18, 57, 114, 145, 146
shifting cultivation 7, 105
Sinapis arvensis 71
slopes, see topography
soil amendments 35, 118–125
soil climate 13
soil conditioner 132
soil reaction 29–31, 35
soil suppressiveness 42–45
Solanum integrifolium 185
S. nigrum 70, 110
S. torvum 185
solar soil disinfestation 96–98, 102, 107
Sonchus oleraceus 70
Sophia sp. 65
sorghum 27, 33, 53, 54, 56, 70, 80, 87, 101, 103, 115, 120, 137
Sorghum halepense 68, 70, 132
sour orange 159, 185
sowing density 1, 9, 10, 14, 15, 19, 20, 26, 59, 64, 66, 77, 82, 145, 162–168, 194, 195, 198–200
sowing depth 19, 20, 26, 55, 161
sowing season 1, 5, 10, 11, 13, 18–20, 26, 27, 59, 65, 74, 76, 77, 102, 128, 130, 144, 145, 148, 155, 168–171, 193, 195, 200, 202
soybean 23, 39, 41, 42, 49, 56, 71, 80, 84, 89, 100, 106, 110, 112, 117, 127, 138, 143, 166, 167, 186
spice crops 60, 74, 134
spinach 47–49, 140
splash dispersal 15, 64, 72, 119, 130, 153, 154, 156, 158, 159, 164, 181, 196
steam disinfestation 96
Stellaria media 71
stone fruit 36, 53, 81, 91, 130, 142, 150, 158, 179, 185, 202
strain 18, 45–57, 116, 131, 158, 196, 197
  excess water strain 55, 144
  high temperature strain 18, 47–51, 56, 57
  low temperature strain 13, 18, 47, 48, 50, 66, 191
  nutrient strain 18, 31, 55, 64, 134
  water deficit strain 52–54, 56, 57, 76, 77, 127, 131, 144, 149, 150, 158–160, 163, 165, 169, 191, 196, 202
strawberry 37, 41, 64, 66, 81, 89, 158, 186, 187, 192
stress, see strain
subsistence farming 7, 20, 74, 78, 90, 100, 114, 168
sudan grass 107, 117, 122
sugarbeet (see also beet) 23, 25, 61, 68, 75, 84, 87, 89, 112, 115, 138, 144, 163, 165, 168–171, 173, 183, 198–201
sugarcane 46, 61, 80, 81, 88, 89, 94, 120, 176, 192
sunflower 56, 80, 109, 115, 117, 132
sunscald, see scald
sweet potato 33, 43, 81, 105, 107, 110, 192
symptomless carriers of disease 67, 70, 71, 200, 202

Tagetes minuta 117
T. patula 117
Tagetes sp. 116
taro 78
tea 7, 17, 60, 163, 176
Thalictrum sp. 87
Thlaspi arvense 110

tillage 1, 6, 10, 18, 19, 26, 28, 35, 54, 72, 74–78, 86, 88, 93, 96, 99, 102, 103, 123, 125–133, 145, 146, 161, 190, 191, 195, 198–200, 203
tobacco 24–26, 33, 37, 39–41, 52, 53, 57, 79, 80, 84, 110, 117, 138, 141, 142, 163, 173, 183, 196, 197
tomato 16, 24, 29, 31, 39–41, 43, 45, 49, 50, 53, 54, 64–66, 70, 71, 79, 80, 85, 86, 92, 97, 100, 101, 104, 107, 109–111, 117, 122, 131, 136, 138, 141, 143, 145, 148, 150–155, 158, 159, 163, 167, 168, 170, 171, 174, 176, 183, 185–188, 193, 201
topography 5, 9–15, 18, 25, 28, 47, 51, 66, 143–145, 158, 185, 206
trefoil, see clover
trickle irrigation 13, 156, 158, 159, 160, 162, 206
Tripleurospermum inodorum 71
triticale 51
Triticum sp. 70
tulip 40, 81, 89
turfgrass 151

urena 53

Vaccinium sp., see blueberry
ventilation, see aeration
vetch 80
Vicia villosa 68
virus vectors 5, 9, 20, 24, 26, 27, 55, 67, 72, 88, 114, 115, 118, 155, 166, 167, 169–171, 180–182, 185–189, 196
volunteer plants 26, 27, 78, 88, 125, 132

walnut 53
water bamboo 95
water chestnut 95
watermelon 47–49, 68, 80, 110
water spinach 95
weeds 6, 23, 27, 66–72, 76, 78, 85–87, 96, 98–103, 110, 113, 114, 118, 121, 123, 129, 130, 132, 151, 158, 165, 190, 200, 202–204
wheat 8, 22, 24, 29, 33, 36, 40, 43, 48, 49, 51–55, 67, 70, 71, 74–77, 80, 87, 89, 94, 103, 106, 107, 110, 111, 113, 115, 120, 121, 124, 125, 128–131, 135–137, 139–141, 148, 150, 151, 161, 164, 168, 169, 172, 173, 177, 186, 193, 198–202, 206
wind 10, 11, 14, 16, 46, 58–60, 62, 66, 78, 88, 144, 145, 155, 156, 160, 162, 165, 181, 183, 186, 187, 191, 196
windbreak 11, 92, 145, 185, 186

yam 192

Advanced Series in Agricultural Sciences

Co-ordinating Editor: B. Yaron
Editor: G.W. Thomas, B.R. Sabey, Y. Vaadia, L.D. van Vleck

Volume 1
A.P.A. Vink
Land Use in Advancing Agriculture
1975. 94 figures, 115 tables. X, 394 pages
ISBN 3-540-07091-5

Contents:
Land Use Surveys. - Land Utilization Types. - Land Resources. - Landscape Ecology and Land Conditions. - Land Evaluation. - Development of Land Use in Advancing Agriculture.

"...The author provides a remarkable overview of the utility and character of land-use surveys, the different types of land use, the diversity of natural resources and ecological requirements, and methods of land evaluation. The citation and use of so many references means that the volume forms a valuable introduction to the conditions and literature of all parts of the world. The organisation of chapters and sections, and the clear presentation of figures and tables, makes it an especially attractive book to handle..." *Biological Conservation*

Volume 2
H. Wheeler
Plant Pathogenesis
1975. 19 figures, 5 tables. X, 106 pages
ISBN 3-540-07358-2

Contents:
Concepts and Definitions. - Mechanisms of Pathogenesis. - Responses of Plants to Pathogens. - Disease-Resistance Mechanisms. - Genetics of Pathogenesis. - Nature of the Physiological Syndrome.

"...This concise and very clearly written book presents an excellent account of the process of plant pathogenesis. ...The book is very stimulating in that is presents not merely facts (in a very systematic way), but also the author's interpretation thereof. The author is to be congratulated on his use of examples to illustrate the phenomena described and to make his points clear. ...The excellent integration of biochemical, ultrastructural and genetic information is an important aspect of this book." *Netherlands Journal of Plant Pathology*

Volume 3
R.A. Robinson
Plant Pathosystems
1976. 15 figures, 2 tables. X, 184 pages
ISBN 3-540-07712-X

Contents:
Systems. - Plant Pathosystems. - Vertical Pathosystem Analysis. - Vertical Pathosystem Management. - Horizontal Pathosystem Analysis. - Horizontal Pathosystem Management. - Polyphyletic Pathosystems. - Crop Vulnerability. - Conclusions. - Terminology.

"..because *Plant Pathosystems* is already a very good book and unique in many ways. The text is based on long and varied experience of field plant pathology in Africa and essentially is a series of perceptive analyses of interactions between populations of host plants and populations of pathogens. ...The book is well and clearly written but must be taken slowly because each paragraph on each page contains one or more substantial points which must be understood if the main themes which are propounded are to have their full impact. But the effort called for is very well rewarded because few books published on plant pathology during the past three decades are so stimulating, challenging, and have been written with such worthy objectives. *Plant Pathosystems* will undoubtedly cause quite a stir among plant pathologists and plant breeders, for most of whom it must be regarded as compulsory reading." *Nature*

Volume 4
H.C. Coppel, J.W. Mertins
Biological Insect Pest Suppression
1977. 46 figures, 1 table. XIII, 314 pages
ISBN 3-540-07931-9

Contents:
Glossary. - Historical, Theoretical, and Philosophical Bases of Biological Insect Pest Suppression. - Organisms Used in Classical Biological Insect Pest Suppression. - Manipulation of the Biological Environment for Insect Pest Suppression. - A Fusion of Ideas. - Index.

This cohesive review of many areas of entomology describes the potential and practical aspects of suppressing insect pest populations by a variety of biological methods. These include the introduction and encouragement of natural enemies, host resistance, hormones, pheromones, antifeedants, genetics, and integrated systems.

Springer-Verlag Berlin Heidelberg New York

Volume 5

J.J. Hanan, W.D. Holley, K.L. Goldsberry

Greenhouse Management

1978. 283 figures, 117 tables. XIV, 530 pages

ISBN 3-540-08478-9

Contents:

Introduction. - Light. - Greenhouse Construction. - Temperature. - Water. - Soils and Soil Mixtures. - Nutrition. - Carbon Dioxide and Pollution. - Insect and Disease Control. - Chemical Growth Regulation. - Business Management. - Marketing. - Appendices: Conversion Tables. Symbolism. Definitions.

This is the first comprehensive American treatment of greenhouse production since Post's book on Florist Crop Production and Marketing, published in 1950. The authors develop principles of crop behaviour in greenhouses on basic fundamentals of plant physiology, illustrating the practical application of these principles with numerous examples. For the first time in a greenhouse management text, attention is given in detail to the effects of climate, disease and insect control, pollution and the use of herbicides, personnel management, growth regulation, greenhouse engineering and all phases of marketing. The treatment of each major phase provides a "bench-mark", or reference point, for the state-of-the-art. It is intended as a text upon which serious practitioners of greenhouse production may build for the future.

Volume 6

J.E. Vanderplank

Genetic and Molecular Basis of Plant Pathogenesis

1978. 3 figures, 36 tables. XI, 167 pages

ISBN 3-540-08788-5

Contents:

Variation in the Resistance of the Host and in the Pathogenicity of the Parasite. - The Gene-for-Gene and the Protein-for-Protein Hypotheses. - The Protein-for-Protein Hypothesis: Temperature Effects and Other Matters. - Common Antigenic Surfaces in Host and Pathogen. - Other Large Molecules in Relation to Gene-for-Gene Disease. - Population Genetics of the Pathogen. - Horizontal Resistance to Disease. - Selective Pathotoxins in Host-Pathogen Specificity. - A Molecular Hypothesis of Vertical and Horizontal Resistance. - Biotrophy, Necrothrophy, and the Lineage of Symbiosis.

This book brings together much information on the genetic and molecular background of plant disease. Variation in the disease resistance of the host and in the pathogenicity of the parasite feature largely, both as the expression of mutation, and the substance of disease control by plant breeding. Within the phenotype, proteins are the great stores of mutational change and the governors of mutational effects. Host-parasite interactions are studied largely in terms of the thermodynamics of the host protein-parasite protein polymerization, and a molecular hypothesis of disease resistance is worked out. A special study is made of the population genetics of pathogens, based on analyses of detailed plant-disease surveys.

Volume 7

J.K. Matsushima

Feeding Beef Cattle

1979. 31 figures, 23 tables. IX, 128 pages

ISBN 3-540-09198-X

Contents:

Nutrients. - Classification of Feeds. - Procedures in Ration Formulation. - Processing Feeds for Beef Cattle. - Systems of Feeding. - Feed Additives. - Growth Stimulants.

The author, whose experience spans a period of thirty years, has had a far-reaching impact on the growth and development of the feddlot industry. His up-to-date and comprehensive book offers scientists, beef cattlemen and students all the facts they need to understand beef cattle nutrition and to plan feeding programs suited to individual needs.

Volume 8

R.J. Hanks, G.L. Ashcroft

Applied Soil Physics

Soil Water and Temperature Applications

1980. 55 figures, 19 tables. VI, 159 pages

ISBN 3-540-09457-1

Contents:

Water Quantities. - Water Potentials. - Water Flow in Soil. - Soil-Plant-Atmosphere Relations. - Soil Heat Flow and Temperature. - Appendix. - Example Index. - Subject Index.

This text was developed from a series of lectures delivered by two experts in agriculture and soil science. It deals with all aspects of the physical properties of soils. Special emphasis is placed on the quantities, potentials and flow of water in soils, the relation between soils, plants and atmosphere, and temperature and heatflow in soils.
Both teachers and students of agronomy and related fields will find this book an important and useful tool. It covers the subject concisely yet thoroughly and contains several "problems" to be solved at the end of each chapter.

Springer-Verlag
Berlin
Heidelberg
New York

MIX
Papier aus verantwortungsvollen Quellen
Paper from responsible sources
FSC® C105338

If you have any concerns about our products,
you can contact us on
ProductSafety@springernature.com

In case Publisher is established outside the EU,
the EU authorized representative is:
Springer Nature Customer Service Center GmbH
Europaplatz 3, 69115 Heidelberg, Germany

Printed by Libri Plureos GmbH
in Hamburg, Germany